DISCIPLINE & EXPERIENCE

DISCIPLINE

SCIENCE AND ITS CONCEPTUAL FOUNDATIONS
DAVID L. HULL, EDITOR

& EXPERIENCE

THE MATHEMATICAL WAY IN THE SCIENTIFIC REVOLUTION

PETER DEAR

THE UNIVERSITY OF CHICAGO PRESS

CHICAGO AND LONDON

Peter Dear is associate professor in the Departments of History and of Science and Technology Studies, Cornell University. He has taught the history of science at Imperial College, London, and has held a research fellowship at Gonville and Caius College, Cambridge. He is the author of *Mersenne and the Learning of the Schools* (1988) and editor of *The Literary Structure of Scientific Argument* (1991).

The illustration on the title page is taken from Franciscus Aguilonius, *Opticorum libri sex* (1613), p. 452. Courtesy of Division of Rare and Manuscript Collections, Carl A. Kroch Library, Cornell University.

The University of Chicago Press, Chicago 60637
The University of Chicago Press, Ltd., London
© 1995 by The University of Chicago
All rights reserved. Published 1995
Printed in the United States of America
04 03 02 01 00 99 98 97 96 95   1 2 3 4 5
ISBN: 0-226-13943-3   (cloth)
       0-226-13944-1   (paper)

Library of Congress Cataloging-in-Publication Data

Dear, Peter Robert.
    Discipline & experience : the mathematical way in the scientific revolution / Peter Dear.
       p. cm.—(Science and its conceptual foundations)
    Includes bibliographical references and index.
    1. Science—Europe—History—17th century. 2. Mathematics—Europe—History—17th century. I.Title. II. Series.
Q127.E8D43   1995
501—dc20   95-12996
                                                                                  CIP

⊗The paper used in this publication meets the minimum requirements of the American National Standard for Information Sciences—Permanence of Paper for Printed Library Materials, ANSI Z39.48-1984.

To P.C.M.C.D., again.

Quand nous citons les auteurs, nous citons leurs démonstrations, et non pas leurs noms; nous n'y avons nul égard que dans les matières historiques.

Blaise Pascal, letter to Noël

This story Sir R. *Moray* affirmed to have received from the *Earl* of *Weymes*, Brother in Law to the Lord *Sinclair*, as it was written to him from *Scotland*.

Report in *Philosophical Transactions* (1666)

CONTENTS

	List of Figures	ix
	Acknowledgments	xi
	Note on Citations and Translations	xiii
	Introduction: The Measure of All Things	1
One	Induction in Early-Modern Europe	11
Two	Experience and Jesuit Mathematical Science: The Practical Importance of Methodology	32
Three	Expertise, Novel Claims, and Experimental Events	63
Four	Apostolic Succession, Astronomical Knowledge, and Scientific Traditions	93
Five	The Uses of Experience	124
Six	Art, Nature, Metaphor: The Growth of Physico-Mathematics	151
Seven	Pascal's Void, Natural Philosophers, and Mathematical Experience	180

Eight	Barrow, Newton, and Constructivist Experiment	210
	Conclusion: A Mathematical Natural Philosophy?	245
	Bibliography	251
	Index	281

FIGURES

Figure 1   Table of height and duration of fall for heavy bodies, from Riccioli 82

Figure 2   Grassi's apparatus for detecting the motion of air, from Galileo 88

Figure 3   Culverin and screens, from Cabeo 128

Figure 4   The difficulty of judging distance with one eye, from Aguilonius 148

Figure 5   Pascal's void-in-the-void apparatus 194

Figure 6   Pascal's hydrostatical apparatus 204

ACKNOWLEDGMENTS

The arguments in this book have been forming over the course of more than a decade. As a consequence, it is impossible properly to thank all those who have played a role, usually unwitting, in its production. There are, however, two people with whom I have discussed much of the material over that entire period, who have read much earlier draft material, and who have in addition read versions of the book manuscript itself. Steven Shapin has been a source of wisdom, advice, and encouragement throughout, even when my approach has differed from his own; he has a breadth of historiographical vision and an intellectual seriousness found in very few historians, and I owe him a great debt. Tom Broman has been an enthusiastic and challenging thorn in my side since our earliest graduate school days, and provides an intellectual kinship for which I am always grateful: it is endlessly reassuring to know that there is someone else who perceives and values the same historical problems as oneself.

Parts of this book in earlier forms, whether as talks or as material in articles, have been heard or read by many people who have commented in writing or orally. There have been seminars, workshops, and conferences at All Soul's College, Oxford; Harvard University; Hobart and William Smith College; Indiana University; Princeton University; the Sorbonne; the University of California, San Diego; the University of Chicago; the University of Minnesota; the Van Leer Foundation in Jerusalem; and Virginia Polytechnic Institute. Material has also been presented at a number of annual meetings of the History of Science Society. A few paragraphs have appeared, in a slightly different version, in Peter Dear (ed.), *The Literary Structure of Scientific Argument*

(University of Pennsylvania Press, 1991) and in the journals *Isis* and *Studies in History and Philosophy of Science*. I received helpful suggestions and valuable questions at all those gatherings. I have received specific pieces of help (references or information) from Rivka Feldhay, Daniel Foukes, Steven Harris, James Lattis, Albert Van Helden, and others whom, alas, I cannot now recall. Moti Feingold read the entire manuscript late in the process and made enormously helpful suggestions that I benefited from in final revisions. Margaret Rogers and Laura Linke, the librarians of the History of Science Collections at Cornell, are continually helpful and interested in the materials that are of concern to me, and deserve, out of all the library staff at numerous libraries from whose services I have benefited, special thanks.

For support of the work represented here I am also very grateful to the National Science Foundation, grant DIR-8821169, and to the National Endowment for the Humanities Fellowship for University Teachers, FA-31605.

Finally, of all things, the attention and involvement, as well as excellent advice, of Susan Abrams, the natural sciences editor of the University of Chicago Press, are beyond measure.

Ithaca, New York

NOTE ON CITATIONS AND TRANSLATIONS

The footnotes in this book give abbreviated titles (and, on first appearance in a chapter, publication dates) for referenced items. Full citations appear in the bibliography. All translations throughout are my own unless otherwise noted; with a few exceptions, the original untranslated passage is given in the relevant footnote only when it comes from a source not readily available in a modern edition.

INTRODUCTION: THE MEASURE OF ALL THINGS

Modern science is an enterprise that developed into something approximating its current institutional form in nineteenth-century Europe. It was emulated in European settlements in the Americas and pursued in colonial outposts around the world, to become in the course of the twentieth century a universal endeavor.[1] The characteristic features of science that carry the traces of this development are found most obviously in its social structure. The procedures of training and professional certification, together with the loci of its practice (both localized, as with laboratories, and dispersed, as in its disciplinary networks), came into being as part of a nineteenth-century process of professionalization that took hold first in France and then, especially, in Germany, transforming by 1900 the ways in which natural knowledge was made in Western societies. But this institutional growth was allowed by, and lent itself to, a particular kind of knowledge that stressed the extensive measuring and accounting of the world rather than its intensive apprehension.[2] The cultural and intellectual prerequisites for the nineteenth-century explosion of organized science were the operational ideal, which made the world into something to be mastered, and a quantitative epistemology, which held that such an ideal exhausted everything accessible to human knowing.[3]

1. This is too large a matter for adequate documentation here, but for detailed entrées into science in various colonial regions, see the pioneering work by Lewis Pyenson: *Cultural Imperialism and Exact Sciences* (1985); *Empire of Reason* (1989); *Civilizing Mission* (1993).
2. Cannon, "Humboldtian Science" (1978), is a classic discussion of these matters.
3. On the latter, see Frängsmyr, Heilbron, and Rider, *The Quantifying Spirit* (1990).

It has long been argued that these elements of modern science are rooted in the so-called Scientific Revolution of the sixteenth and seventeenth centuries, Francis Bacon standing for operational or utilitarian knowledge, Isaac Newton for mathematization. There was an alternative cognitive ideal, however, with which the new departures of the seventeenth century had intimate relations; this alternative has received little attention. To those who saw a "scientific revolution," there was a new beginning that sloughed off the past. To those who denied such a "revolution," there was continuity with such things as medieval quantifying tendencies in natural philosophy, or with scholastic talk of logical techniques for acquiring knowledge from experience of appearances.[4] That there could have been genuine novelty in seventeenth-century developments, but a novelty that possessed wholesale continuity with what went before, has been a rather trivial proposition that has fallen between the cracks of explicit discussion. The important thing is not to recognize the proposition, however, but to explicate it in detail. That means paying attention to contemporary conceptions of the roots of natural knowledge and to the ways in which those conceptions related to the making of concrete pieces of knowledge.

In 1660, John Wilkins described the business of the fledgling Royal Society as the promotion of "Physico-Mathematicall-Experimentall Learning."[5] This book may be seen as an attempt to understand Wilkins's term as a summary of themes that, during the course of the seventeenth century, had developed into a powerful formulation of ways to address and to know nature. One of the lessons of recent historiography has been that some Fellows of the early Royal Society, most prominently Robert Boyle, were indifferent at best to the mathematical sciences;[6] Wilkins's remark, however, warns us against any overbroad interpretation of this point. In acknowledging the relevance of mathematics to his own philosophical enterprise, Wilkins pointed to the dominance in other forums of an approach to natural knowledge that would ultimately subsume much of the Society's own endeavors. When the Royal Society gave itself up, at the end of the century, to the self-labeled "mathematical" natural philosophy of Isaac Newton, it reconnected with the enterprise that Wilkins had earlier identified

---

4. See the recent discussion of these matters in Barker and Ariew, *Revolution and Continuity* (1991), esp. "Introduction"; note also the classic study by Crombie, *Robert Grosseteste and the Origins of Experimental Science* (1953), which has a frequent concern with mathematical sciences.
5. B. Shapiro, *John Wilkins* (1969), p. 192.
6. See especially Shapin, *A Social History of Truth* (1994), chap. 7.

as its intended calling. Boylean experimental philosophy was not the high road to modern experimentalism; it was a detour.

The elements of Wilkins's expression—"physics," "mathematics," and "experiment"—were central parts of a new ideology of natural inquiry that took its life from the scholastic philosophy of early-modern European colleges and universities, both Protestant and Catholic; none of the three was a newcomer to academic discourse or scholarly practice. The first two continued to conform to scholastic-Aristotelian formal definitions, in apparent continuity with medieval precedents. The third, "experiment," also relied, explicitly as well as implicitly, on scholastic-Aristotelian definitions. What had changed were the characterizations that many philosophers, especially practitioners of the classical mathematical sciences (such as astronomy, mechanics, and optics), had begun to give of their mutual relationships.

Aristotelian physics (also called "natural philosophy") was the qualitative science of the natural world that explained *why* things happen in terms of the essential natures of bodies; it became increasingly denigrated in the seventeenth century on the grounds that it was usually capable only of yielding probable accounts. The mathematical sciences, by contrast, were allegedly capable of certain demonstration of quantitative relations, and were on that head held to be superior to merely probable physics. This new evaluation was justified by arguments that relied on Aristotelian commonplaces—as was the earlier, and opposite, evaluation that held physics superior to mathematics on account of its superior subject matter. The view that physics was more important than the mathematical sciences depended on the Aristotelian observation that it concerned the natures of things rather than merely their quantitative characteristics, and was therefore more noble. The inverted view, by contrast, depended on the Aristotelian position that the highest form of knowledge, *scientia* (*epistēmē* in Greek), demanded certain demonstration, at the provision of which the mathematical sciences were uncontroversially acknowledged to be supreme.[7] The third element of Wilkins's triptych, however, was his use of the word "experimentall," an English form derived from the Latin *experimentum* or *experientia*. Those Latin terms were implicated in the linguistic practices of the sciences in ways that were often ambiguous.

It used to be thought that experiments were unproblematic, indeed commonsensical, ways to learn about nature. Recently, however, historians have begun to notice that, even within those cultural traditions

---

7. For details of these matters, see chapter 2, below.

that produced the modern institution known as science, experimental behavior has not always been central or, indeed, desirable. Throughout the seventeenth century, the touchstone for definitions of experience in the literate philosophical discourse of Western Christendom remained the writings of Aristotle. An "experience" in the Aristotelian sense was a statement of *how things happen* in nature, rather than a statement of *how something had happened* on a particular occasion: the physical world was a concatenation of established but sometimes wayward rules, not a logically integrated puzzle. But the experimental performance, the kind of experience upheld as the norm in modern scientific practice, is unlike its Aristotelian counterpart; it is usually sanctioned by reports of historically specific events. While events sometimes found their place in premodern natural philosophy, they did not serve the same function.[8] Steven Shapin and Simon Schaffer's *Leviathan and the Air-Pump* is a historiographical landmark in problematizing and explaining the specific form of experimental activity found in the Royal Society of Restoration England: the new English experimental philosophy conformed to the model of the reported event experiment, and the sources of its legitimacy, rooted as they were in local settings, thus allow us to appreciate what the making of experimental knowledge can imply.[9]

There is, however, much more to say about the available sources of meaning attaching to the various kinds of philosophical experience found in the Europe of that period. Kinds of experiential activities in the making of natural knowledge, some analogous to those of Robert Boyle and the early Royal Society, some differing in significant ways, may be found in diverse settings across the Continent and as far afield as the European settlements in the New World.[10] But all were beholden, whether explicitly or implicitly, to the familiarity of Aristotelian teachings for their legitimation. Purely local explanations for each, comparable to Shapin and Schaffer's for England, would leave a constellation of inexplicable coincidences stretching from London to Rome, from Paris to Warsaw, and beyond.[11] The present study therefore adopts a different, complementary approach.

This book establishes linkages through an examination of common-

---

8. See chapter 2, below.
9. Shapin and Schaffer, *Leviathan and the Air-Pump* (1985); also Shapin, *A Social History of Truth*.
10. Studies in Baroncini, *Forme di esperienza e rivoluzione scientifica* (1992), investigate further the usage and meaning of experiential terminology in this period.
11. A similar point is made by Schuster and Watchirs, "Natural Philosophy, Experiment and Discourse in the 18th Century" (1990), esp. pp. 38–39. They liken the relation-

alities of linguistic philosophical practice. The social and institutional basis for much of the book's argument is rooted in educational institutions and their curricular and pedagogical structures, and emphasis is placed on what those settings shared rather than on their local idiosyncrasies. Thus one can see how people who were a part of such institutions (or had learned their ways of discussing from them) talked about how to make natural knowledge or justified their philosophical work by reference to what they took to be proper procedure. Through such means, common characterizations can be given that allow the re-creation of a nonlocal philosophical culture. The approach requires the suspension of the presupposition that meaning is constructed only in local situations of immediate use; it assumes that forms of discourse have their own, albeit limited, agency.[12]

It should be stressed that in speaking of "discourse" (a term that I do not wish to labor), I mean the broad notion familiar from the work of Michel Foucault.[13] This book, in other words, is about a kind of *practice*, the watchword of science studies these days, rather than "just" about language.[14] Although my evidence and arguments will chiefly deal with how people talked about what they did, I take it that what they did can only be characterized and understood through their forms of speech about it. One cannot describe a set of experimental practices unless one first determines that they *are* experimental practices. In order to decide whether Galileo performed any "experiments" relating to bodies rolling down inclined planes, for example, it is necessary to consider not only one's own definition of the term, but also how Galileo typically formulated his experience of falling bodies in language. An intelligent fly on Galileo's wall would not be able to say whether what transpired was an "experiment" in any epistemologically or sociologically well-defined sense, any more than an observer

---

ship between the broader picture and its local instantiations to that between a language and its dialects, which at least provides an empirically solid analogy, if not an actual understanding of the matter. Cf. Ophir and Shapin, "The Place of Knowledge" (1991), pp. 15–16. A local approach (certainly a valuable corrective to the usual general accounts) is taken by Porter and Teich (eds.), *The Scientific Revolution in National Context* (1992).

12. This point should scarcely need to be made, but some developments in science studies, exemplified in their most extreme form by ethnomethodological approaches, have sometimes seemed to call it into question: see, however, Lynch, *Scientific Practice and Ordinary Action* (1993), pp. 28–30, for clarification.

13. See, e.g., Foucault, "The Discourse on Language" (1972), with its integration of "discourse" with disciplinary structures.

14. On "practice": Pickering, *Science as Practice and Culture* (1992); Golinski, "The Theory of Practice and the Practice of Theory" (1990).

from Mars could be expected to characterize the activity of (what we would call) a game of chess in the same way as the players being observed.[15]

The "experiment" as a hallmark of modern experimental science, then, is constituted linguistically as a historical account of a specific event that acts as a warrant for the truth of a universal knowledge-claim. "Experiments" in this sense only became a part of a coordinated knowledge-enterprise during the course of the seventeenth century. Understanding how the change came about, and discovering its philosophical meaning, amounts to investigating the cognitive and disciplinary categories that constrained and allowed it. Doing this then draws attention to another crucial difference between scholastic "experience" and modern "experiments" as warrants for statements about nature: the former could only be observational perceptions of nature's ordinary course, whereas the latter by design subverted nature.

Chapter 1 addresses the most fundamental problem of all: how were individual experienced events related to universal knowledge claims about the world? This, a version of the famous "problem of induction," was not generally seen as a problem at all in the seventeenth century; various ways of connecting the singular to the universal (or leaving them disconnected) were almost naturalistically allowed, in keeping with Aristotle's nonanalytical approach. Above all, throughout the century the universal experience reigned virtually unchallenged as the irreducible touchstone of empirical adequacy. The question that this raises is the following: how could reports of singular events become integrated into philosophical practice in the study of nature, if they thus had no philosophical standing?

Chapter 2 begins to address the problem through an investigation of the ways in which Jesuit mathematical scientists—specifically, astronomers and opticians—in the early part of the century attempted to justify these disciplines against criticism of their scientific status. The criticisms were based on Aristotelian definitions of science as demonstrative knowledge of real objects, and the defense involved an unquestioning acceptance of the propriety of those definitions. The importance of the arguments lies in the enormous role of Jesuit scholar-

---

15. This general point may be related to the ideas, appealing to the later philosophy of Wittgenstein, of Peter Winch, *The Idea of a Social Science* (1958). Winch's discussion of the philosophy of action has recently been articulated into a more formal terminological distinction between "behavior" and "action," the former referring solely to spatiotemporally defined physical change, the latter to a meaning-laden interpretation of such change that imputes agency, by Collins, *Artificial Experts* (1990), chap. 3; also idem, "The Structure of Knowledge" (1993).

ship in the mixed mathematical sciences throughout this period; even non-Catholics, as well as Catholics who had not been trained in the Jesuits' extensive educational system, studied Jesuit philosophical writings as the most accomplished and up-to-date available. That, no doubt, was their purpose.[16]

In chapter 3, the appearance of historical reports of specific, usually contrived, experiences in Jesuit writings is shown to be rooted in the considerations introduced in the previous chapter. These reports imputed to the writer a competence to speak, confirming a general claim to expertise in the relevant matters; they did not provide the raw material of philosophical assertions. Used as tokens of empirical faith, they generally figured in situations of controversy and conflict: by adducing a specific case, the writer threw down a challenge to those who would dissent from the associated universal empirical claims (which constituted the real philosophical content). Disagreeing with a specific, experienced factual claim was morally much harder than questioning a piece of reasoning.

The use of the techniques of the mathematical sciences to handle empirical novelty—the antithesis of common experience—is explored in chapter 4. Galileo's controversy over sunspots with the Jesuit Scheiner (one of the figures examined in chapter 2) throws additional light on Galileo's arguments about Jovian moons in the *Sidereus nuncius*, and shows how his use of well-understood procedures places his discoveries within an existing, long-standing tradition of astronomical practice despite his avowed contempt for the authority of the past. The chapter goes on to relate such matters to general notions of tradition in the sciences of the sixteenth and early seventeenth centuries, associated with humanist attempts at "renovation."

Chapters 5 and 6 go farther afield, to show the ubiquity of the handling of experience in scientific argument hitherto examined with special reference to Jesuit mathematicians, and to expose more of its characteristics. Chapter 5 suggests, in addition, the problem of identifying whose knowledge counted as "evident"; that is, what it took to render the knowledge of a subset of society "common" for the purposes of making certified natural knowledge. The special empirical knowledge of mathematical astronomers usually counted as "common" in this sense, but so did that of sailors, or children. Chapter 6 extends the analysis of experimentation and its growth in this period by focusing

---

16. Tommaso Campanella cannily recommended natural philosophy as a stalking horse designed to reconvert the Protestants without their being aware of what was happening: see Jacob, *Henry Stubbe* (1983), p. 86.

on the aspect of *contrivance*, rather than simply historical specificity, that was first introduced in chapter 2. This involves consideration of the Aristotelian distinction between art and nature, which integrates with that between mathematics and natural philosophy that so exercised mathematical scientists in particular. The appearance and rapid spread of the new label "physico-mathematics" (what one might with propriety call a "term of art") bears witness to the increasing ambitions of mathematicians as the century progressed to absorb the cognitive territory of the natural philosophers.

Blaise Pascal's famous work on the so-called "Torricellian experiment" then serves, in chapter 7, as an exemplification and further confirmation of the complex of issues surrounding mathematical sciences, scientific argument, and the nature of scientific experience in the seventeenth century. His contrivances, culminating in his brother-in-law's renowned ascent of the Puy-de-Dôme to see the effect of altitude on the height of mercury in a Torricellian tube, were integrated into the formal argumentative structure usual in the mixed mathematical sciences. What seem like event experiments, specific trials undertaken to provide the grounds for universal knowledge-claims, turn out to exemplify the mathematician's procedure of adducing the lessons of contrived experience.

Finally, chapter 8 confronts the locus classicus of early-modern experimentation, Restoration England. The historiographical importance of England in anglophone history of science, especially in the wake of *Leviathan and the Air-Pump*, has tended to magnify the apparent significance of the "experimental philosophy" promoted by Robert Boyle. This chapter shows how the event-centered empiricism of the early Fellows of the Royal Society stood somewhat at odds with the dominant modes of experience in contemporary philosophy of nature. Events had never been able, by their nature, to suffice as guarantors of universal propositions (Spinoza criticized Boyle on just these grounds); Isaac Newton provided the English experimenters with a way of representing their activity as philosophically meaningful.[17] Newton's work drew directly from the tradition of mathematical sciences examined previously in this book, and it is shown further to be premised on a constructivist conception of mathematical objects themselves. Geometrical figures, according to a dominant line of argument in the seventeenth century, were things *to be drawn* rather than preexisting in a Platonic realm; for Newton and others before him, experimental con-

---

17. On Spinoza's complaints, see A. R. Hall and M. B. Hall, "Philosophy and Natural Philosophy" (1964).

trivance made experiential data in just the same foundational way. An event, furthermore, could be a reliable indicator of certain kinds of natural regularity insofar as its actuality in some particular case determined its attendant conditions of possibility—just as (to use an example touted by Isaac Barrow, Newton's mathematical predecessor at Cambridge) inspection of a single triangle sufficed to draw conclusions true of all triangles. The Royal Society's experimental philosophy had not been conceived as a mathematical enterprise, but with Newton's apotheosis it became associated with a legitimatory methodology that had been provided by the mathematicians.

This book, then, presents a view of physical science in the seventeenth century that is rooted in a milieu of academic scholarly endeavor: it does not need repeating here that most of the major figures associated with the Scientific Revolution were either themselves for a good part of their careers university professors (as, for example, Galileo and Newton) or had been trained to think philosophically in such academic settings.[18] Other dimensions of the large changes in views of knowledge and its purposes in this period are also of crucial importance, including most immediately the increasing social status of craft and artisanal knowledge long epitomized by Francis Bacon.[19] Developments in the life sciences, meanwhile, seem to have followed a somewhat analogous, but often distinct, academic track, rooted as they were in medical rather than mathematical traditions of thought and practice.[20] Nonetheless, the story of how the classical mathematical sciences and their seventeenth-century practitioners became the physico-mathematical vanguard of a new natural philosophy, one that stressed contrived, often witnessed events as the experimental justification of a science of appearances, contributes much to our understanding of the spiritual core of Western scientific development. A mathematical philosophy that had ambitions to the measurement of all things became a science that attempted to grasp everything.[21]

18. Gascoigne, "A Reappraisal of the Role of the Universities" (1990).

19. For important perspectives on this issue, see, e.g., Rossi, *Philosophy, Technology, and the Arts* (1970); Eamon, *Science and the Secrets of Nature* (1994).

20. See especially Wear, "William Harvey and the 'Way of the Anatomists'" (1983); Baroncini, *Forme di esperienza e rivoluzione scientifica,* chaps. 1 (on Achillini) and 5 (on Harvey); see also Schmitt, "William Harvey and Renaissance Aristotelianism" (1984), on Harvey's Aristotelian methodological talk.

21. My argument can be seen as being in partial agreement with Cunningham, "Getting the Game Right" (1988), and Cunningham and Williams, "De-centring the 'Big Picture'" (1993), which argue that the culturally specific "origins" of modern science lie broadly in the early nineteenth century. I want to consider some of the earlier developments that made the nineteenth-century story possible.

One INDUCTION IN EARLY-
MODERN EUROPE

I. Warranting Universals

Europeans at the beginning of the seventeenth century lived in a world of precarious intelligibility. To understand meant to grasp regularities, to know what to expect and how things went; yet most regularities contained no guarantee of their own reliability. The authority of ancient texts had become increasingly ineffective in the face of an expanded world. Religion still held out the promise of infallible certainty, but the importance invested in that promise fueled the battles, spiritual and military, of the great age of conflict between Catholic and Protestant. For the philosophically well-educated, small in number though they were, mathematics appeared as one of the few refuges of eternal verity untainted by the possibility of dissent, while all around them the natural world displayed a variety and impenetrability that mocked attempts at taming it. If awareness of the tenuousness of knowledge had not been so acute, the premium placed on certainty would not have been so high.[1]

Knowledge had to be all-encompassing. A knowledge of past events was not true knowledge; a knowledge of the current state of affairs was itself mere history. The question "Why?" in the sense of Aristotle's "Why thus and not otherwise?"—expecting the answer "because it cannot be otherwise"—haunted would-be knowers, heirs to the Western philosophical tradition. "Experience" was understood as a field

1. On the compromised authority of ancient texts, see Grafton, *New Worlds, Ancient Texts* (1992); Pagden, *European Encounters with the New World* (1993), esp. chap. 2. Popkin, *The History of Scepticism* (1979), details the "sceptical crisis" of this period.

11

from which knowledge was constructed, rather than a resource for acquiring knowledge, because "experience" was in itself incapable of explaining the necessity of those things to which it afforded witness. By the end of the seventeenth century, however, a new kind of experience had become available to European philosophers: the experiment.

Although it has sometimes been claimed, perhaps most prominently by Alexandre Koyré, that the Scientific Revolution of the seventeenth century was above all a matter of metaphysics rather than of empirical advance, few scholars nowadays would contest the proposition that experimentation should be accorded a central place in the new European culture of natural knowledge that arose in that period.[2] At the most fundamental level, however, Koyré was clearly right: his crucial point was that it is impossible for nature to speak for herself. Even with novel deployments of apparatus and technique to bring about hitherto unknown behaviors, no knowledge can be created unless those new human practices and new natural appearances are rendered conceptually in an appropriate way. Indeed, even to identify a technical practice as new rather than as an unimportant variant upon an old practice, or to identify the resultant appearances as new kinds of natural phenomena rather than variants of previously known ones—or pathological instances—requires particular conceptual and cognitive expectations on the part of the knower. In that sense, the Scientific Revolution was indeed a matter of a cognitive shift rather than the simple acquisition of new information that demanded new theoretical frameworks to accommodate it.

In order to understand the new practices of experimentation that became established in the seventeenth century we must learn to ascribe meanings in correct seventeenth-century ways to what appear to us as experimental actions. Only then will we have a firm grasp of those events which constitute the historical episode in question. First of all, it is necessary to recognize that there is nothing self-evident about experimental procedures in the study of nature. There is no one independently given class of practices that naturally corresponds to the label "scientific experimentation"; there are many different practices, with their associated epistemological characterizations, that relate to experience and its place in the creation of natural knowledge. In the seventeenth century old practices changed and new ones appeared. Those changing practices represent shifts in the meaning of

2. See above all Koyré, *Metaphysics and Measurement* (1968), esp. chap. 2. The implicit parallels with certain aspects of Thomas Kuhn's ideas in Kuhn, *The Structure of Scientific Revolutions* (2d ed., 1970), are made clearer in Kuhn, "Alexandre Koyré and the History of Science" (1970).

experience itself—shifts in what people saw when they looked at events in the natural world.

How we apprehend the world through experience depends on the ways in which we conceptually formulate experience. How we see things is strongly conditioned by the mental categories that we bring to our perceptions. So, to use a well-known example of Norwood Russell Hanson's, at sunrise Kepler sees the earth rolling around in the direction of the sun, where Tycho sees the sun moving around the earth.[3] But the "formulation" of experience goes beyond the simple matter of mental categories. It involves the way in which we relate knowledge about the nature of the world as a whole to our own moment-by-moment sensory awareness and our memories of previous sensory awarenesses. It also, and more fundamentally, involves the way in which we relate that knowledge to what other people tell us about *their* sensory experience. The central question is this: how can a *universal* knowledge-claim about the natural world be justified on the basis of *singular* items of individual experience?

The historical meaning of this question rests on the following claim: it is not until the seventeenth century that singular, contrived events become generally used as foundational elements in making natural knowledge; modern experimental science appears in the seventeenth century. Before then, Western natural philosophy used singular, historically reported experiences mostly as illustrations of general knowledge-claims, or as occasions to investigate some issue, but not as arguments to justify universal propositions about nature.[4] At the beginning of the seventeenth century, a scientific "experience" was not an "experiment" in the sense of a historically reported experiential event.[5] Instead, it was a statement about the world that, although

---

3. Hanson, *Patterns of Discovery* (1958), p. 5.

4. The modern English terms "experiment" and "experience" invoke a distinction (not terminologically represented—or in the same way—in many other European languages) that begs the questions at issue here. In translating seventeenth-century Latin usage, therefore, I have endeavored to give literal renderings, "experience" for *experientia* and "experiment" for *experimentum*, even while mindful of the dangers of misleading the reader. When a Latin author uses the word *experimentum* in a way that clearly means "experience" in the usual Aristotelian sense, I have tried to indicate this sense in the text. Nothing, however, should be taken for granted regarding the precise terms used by seventeenth-century writers; the reader should try to suspend a sense of familiarity whenever a word like "experiment" appears in a quotation.

5. Sometimes a systematic difference may be discerned between, for example, the uses of *experimentum* and *experientia*; more often, none is evident, and each seems simply to mean "experience" of some kind. The use of *experimentum* in reference to "experience" in the Aristotelian sense is found, for example, in the Jesuit astronomer Riccioli: "Viguit iam inde à viginti & amplius saeculis in Academiis Physicorum, praesertim Peripatet-

known to be true thanks to the senses, did not rest on a historically specifiable instance—it was a statement such as "Heavy bodies fall" or "The sun rises in the east." Singular, unusual events were of course noticed and reported, but they were not, by definition, revealing of how nature behaves "always or for the most part," as Aristotle said; instead, they might be classified as "monsters" or even "miracles."[6]

By the end of the seventeenth century, by contrast, it had become routine, especially in English natural philosophy, to support a knowledge-claim by detailing a historical episode. Thus Isaac Newton attempted to discredit Descartes's belief in a pervasive etherial medium by presenting a historically reported account of an elaborate test he had conducted using a pendulum. He included details of the measurements he had made and the precautions he had taken to ensure their reliability:

> I suspended a round deal box by a thread 11 feet long, on a steel hook, by means of a ring of the same metal, so as to make a pendulum of the aforesaid length. The hook had a sharp hollow edge on its upper part, so that the upper arc of the ring pressing on the edge

---

icorum cum Aristotele I. de caelo cap. 88. illud ab experimentorum inductione collectum principium & axioma, per se satis notum sensui: Gravia naturali motu descendentia per medium levius, eò velociùs ac velociùs continuè moveri, quò propiùs accedunt ad terminum, ad quem tendunt." Riccioli, *Almagestum novum* (1651), part 2, p. 381 col. II. On the other hand, a work of 1648 by Iohannes Chrysostomus Magnenus on atomism allows the following use of *experientia*, recalling (below, pp. 44–45) Aguilonius's methodological remarks: "Experientias accurate factas tanquam principia per se nota admittere" (quoted in Meinel, "Early Seventeenth-Century Atomism" [1988], p. 80). Scheiner's practice in his book *Oculus* (1619), as seen in chapter 2 below, whereby a practical distinction seems to be made between the two terms, does not seem to have been usual, although it would be possible to read it into the passage from Riccioli just quoted. Jung, *Logica Hamburgensis* (1638/1681), under the heading "De experientiâ" (Book IV, chap. 4; pp. 207–209), distinguishes various applications of the word *experientia*, including *experientia singularis*, as in "Hoc ignis comburit, Hic sal aqua liquescit" (p. 207), and *experientia universalis*. The latter can be either *externa*, as in "Omnis cera calore liquatur," or *interna*, as in "Omne phantasma Centauri ex phantasmatibus hominis et equi est conflatum" (p. 208). Experience can be either accidental (the examples here are universal, as "water dissolves salt") or deliberately acquired, when it is known as *observatio* (the Greek equivalent that Jung uses here is found in Ptolemy; see chapter 2, section II, below). The simple terms *experientia* and *experimentum* had no unequivocal distinct reference.

6. Lorraine Daston has recently stressed the emergence of a concentration on deviant, singular "facts" in the seventeenth century as a novel (and quite anti-Aristotelian) development: Daston, "The Factual Sensibility" (1988); idem, "Marvelous Facts and Miraculous Evidence" (1991); idem, "Baconian Facts, Academic Civility, and the Prehistory of Objectivity" (1991). The admission of singulars regarded as "typical" rather than "deviant" into a central place in philosophical discourse, however, represents an even more fundamental development, as will be seen below.

might move the more freely.... I accurately noted the place.... I marked three other places.... All things happened as is above described.[7]

That is an experimental report—an example of the characteristically seventeenth-century genre of the *event experiment*.

II. Hume's Analysis

The issue of the "influence" of medieval scholastic thought on the Scientific Revolution has ceased to be as pressing as it once was. Firmly rooted in untenable historiographical assumptions concerning "scientific method," and directed towards particular apologetic ends whereby the prize went to the originators of that method, it nowadays lacks cogency. However, the conduct of the debate has certainly yielded some valuable discoveries and useful arguments.[8] That those usually seen as standing in the forefront of the development of new approaches to nature and knowledge in the seventeenth century were acquainted with, and intellectually shaped by, the "learning of the schools" is surely an unquestionable proposition.[9] But to formulate its meaning in terms of either an essential "continuity," however qualified, or a "discontinuity" is unnecessary and unilluminating. What is called for is an understanding of the intellectual culture within which new ideas and manners of formulating knowledge *made sense* in the seventeenth century in ways that they had not before. If we can identify specific changes of this sort, and then locate their contemporaneous meanings, we will also be able to ask questions about the means by which they came about. It is the purpose of this book to investigate one such change.

The "problem of induction" as it has come down to us in various forms since Hume is not a timeless philosophical difficulty. Within the Western tradition, it appeared only after its constitution in the practice of seventeenth-century natural philosophy; Hume codified it, and did not even see it as a *problem*. Before the seventeenth century the modern problem of induction did not exist. Hume stabilized the very concept that he came to be seen as having made into a problem: modern induction was created by its claimed inadequacies. Words carry a deceptive

---

7. Newton, *Principia* (1687/1972), vol. 1, pp. 461–463, trans. Cajori, *Principles* (1934), pp. 325–326.

8. See the valuable review by Eastwood, "On the Continuity of Western Science" (1992); Barker and Ariew, *Revolution and Continuity* (1991), "Introduction," pp. 1–19.

9. See, for example, Gilson, *Études sur le rôle de la pensée médiévale* (1951); Wallace, *Galileo and His Sources* (1984); Dear, *Mersenne and the Learning of the Schools* (1988).

sense of their own continuity of reference, however. Thus those words, in various languages, identified in the Middle Ages and Renaissance with the Latin *inductio* corresponded to linguistic and other cognitive practices different from their modern counterparts.[10]

This is more than a quirk in the history of philosophical terminology. Instead, it serves to introduce a crucial, and perhaps determining, feature of the development of modern physical science. One way of framing the issue is negative: Why was induction in its post-Humean guise never formulated prominently until the eighteenth century? Philosophers had tangled with thornier problems for millennia; why had this one escaped their attention?[11] The other, more fruitful way is positive: What had happened during the course of the seventeenth century to make this kind of induction problematic? Why was it now important to formulate these issues and to address them? The answer, broadly, is that experimental science had been invented, and had gained a considerable degree of cultural authority. The practice of experimental science was predicated on a different set of practical assumptions regarding the relationship between singular experience and universal generalizations from those prevalent before the seventeenth century. The question of why Hume, and others after him, chose to create and sustain a "problem of induction" down to the present day is far beyond the scope of this study.[12] But it remains significant that Hume developed particular ideas on the relation between singular experiences and universals that removed it from a world of necessary causal connections.

The practices of experimental science permitted Humean induction with its philosophical characteristics because they directly implicated the central issue of universals and their status. An understanding of how this came about thus provides at the same time an understanding of experimental science itself as a cultural product of a particular time

---

10. Milton, "Induction before Hume" (1987); Pérez-Ramos, *Francis Bacon's Idea of Science* (1988), part IV, esp. pp. 216–224. Hacking, *The Emergence of Probability* (1975), chap. 19, presents an alternative account to that offered by Milton on Hume and the reasons for the previous nonemergence of the "problem of induction."

11. This is not to say that questions to do with the justification of a universal proposition by a number of specified events had never been discussed in Western philosophy (that is what so-called rhetorical induction was about—see section IV, below); it is to recognize that such questions had never become elevated to the status of major conceptual difficulties in the logic of demonstration.

12. See the useful presentation of arguments on the reasons for Hume's novelty in Milton, "Induction before Hume," pp. 63–73. The problem of induction does not appear to have acquired its canonical status much before the twentieth century in any case; Hume does not exactly regard it as a *problem*.

and place, seventeenth-century Europe. "Induction" before Hume designated various kinds of cognitive practice; the word did not label a single, problematic kind of philosophical inference. Yet a brief examination of Hume's own formulation of the issue serves to illuminate some of the salient features of pre-Humean conceptions of inference and practice in natural philosophy.

Hume's objection to induction in the form of simple inference from the past behavior of some entity to its future behavior has traditionally been interpreted as reflecting a so-called "regularity" view of causation. According to this view, there are no genuine, immanent causal relations in the world, but only our perceptions of regular concurrences of events that lead us to impute a "causal" relationship among them. Thus past observation of regularities can give no guarantee that those regularities will continue to hold in the future, since there is no reason to suppose that they are rooted in ontologically real causes. However, a number of scholars, including most recently and forcefully Galen Strawson, have denied that Hume held such a view.[13] With respect to induction at least, the texts seem to support the argument that Hume regarded it as problematic not because of its lack of ontological foundation but because of a more straightforward epistemological reason. This reason is particularly interesting for its integration into general themes found earlier in seventeenth-century philosophy.

In the *Enquiry Concerning Human Understanding* (first edition 1748), Hume observes that past experience of objects tells us "that those particular objects, at that particular time, were endowed with ... powers and forces" of particular kinds. That experience can in no way guarantee that the objects in question, much less other objects deemed to be similar, will continue in the future to possess those powers: "the secret nature" of the bodies, "and consequently all their effects and influence, may change, without any change in their sensible qualities."[14] It is our unavoidable ignorance, not any indeterminacy in the physical world, that compromises inferences from the past to the future, from particular entities at particular times to other entities at other times. Much as Newton asked Richard Bentley in 1692 not to ascribe the notion of action at a distance to him, Hume wrote in a letter of 1754: "But allow me to tell you, that I never asserted so absurd a Proposition as *that anything might arise without a Cause.*"[15] On Hume's view, only the brute

13. Strawson, *The Secret Connexion* (1989), esp. pp. 182–183.
14. Hume, *Enquiries Concerning Human Understanding* (1975), p. 37; cf. Strawson, *The Secret Connexion*, p. 182.
15. Newton to Bentley in Cohen, *Isaac Newton's Papers and Letters* (1958), p. 298; Hume quoted in Strawson, *The Secret Connexion*, p. 5.

empirical fact of the uniformity of nature underwrites our notions of necessary connections between objects or events.[16] But he did not deny that there are "constant and universal principles" known empirically.[17]

The original Humean problem of induction, then, relied on the proposition that certain causal properties of things are, as scholastic-Aristotelians would have put it, *occult*—that is, hidden from direct observation.[18] Only the manifest behavior of things allows us to infer the existence of the occult causal properties that produce it. Hume then made the further argument that, precisely because they are occult, there could be no guarantee that such properties might not imperceptibly change, thereby changing in turn subsequent behavior. So inductive inference from observed behavior to observed behavior is unreliable, because not all the relevant factors can *be* observed.

The ingredients and concerns making up Hume's position were by no means novel, as the ease of translation into scholastic terminology indicates. Roughly contemporaneously with Hume's writings on this issue, a Jesuit philosopher, Louis Bertrand Castel, drew on long-familiar Aristotelian conceptualizations to criticize the practice of using a single experiment as the basis for a universal claim in natural philosophy. He argued, with specific reference to Newton's work on colors, that a singular fact will not necessarily represent the ordinary course of nature, perhaps being instead monstrous: "A unique fact is a monstrous fact."[19] Aristotelian "monsters," unlike Baconian ones, were not illuminating rarities that served to reveal nature's workings, but were instead nature's mistakes, in which regular processes had become spoiled through adventitious causes. The "mechanical philosophy" that had developed in the seventeenth century had encouraged the opening up of occult causes to the mediated gaze of experiment,[20]

---

16. Cf. Hume, *Enquiries Concerning Human Understanding*, p. 82.

17. E.g., ibid., p. 83.

18. Ironically, during the seventeenth century "occult" properties had come increasingly to be regarded as objects of investigation, and in that sense not "occult" at all. Hutchison, "What Happened to Occult Qualities in the Scientific Revolution?" (1982); Millen, "The Manifestation of Occult Qualities" (1985).

19. See Schier, *Louis Bertrand Castel* (1941), pp. 103–108. In an unpublished essay Castel says that his philosophy "ne veut pourtant que des faits, mais naturels journaliers, constants, mille fois répétés, faits habituels plutôt qu'actuels, faits d'humanité plutôt qu'un homme. Un fait unique est un fait monstrueux" (quoted in ibid., p. 107); Castel made similar remarks in his *L'optique des couleurs* (Paris, 1740), quoted in ibid., p. 108. See also Hankins, "The Ocular Harpsichord of Louis-Bertrand Castel" (1994), esp. pp. 148–150.

20. Bacon, *Novum organum* 1, aph. 50, in *The Works of Francis Bacon* (1860–64), vol. 1, p. 258; trans. Spedding in ibid., vol. 8, p. 83: "The sense decides touching the experiment only, and the experiment touching the point in nature and the thing itself."

but Hume clearly doubted the possibility of such a project. Like Castel and the scholastics in whose footsteps Castel trod, Hume thought that there was no way to control for unobserved changes in occult causes.

Unlike Hume, however, scholastic-Aristotelians throughout the preceding century and a half had not treated the issue as one of particular philosophical significance for their project. Instead, they had treated it as a minor practical matter. It was discussed in an unusually forthright manner by Franciscus Aguilonius, an Antwerp-based Jesuit mathematician and pedagogue, in a 1613 optical textbook:

> If the senses always cheated the mind's acuteness, there could be no science, which [is what] the Academics[21] strive to assert; and if they never failed, the most certain experience would be had by a single act. Now, the matter is, however, in the middle [of these extremes]: for although the senses are sometimes deceived, usually however they do not err. Hence it is, that whenever an experience is certain the first time, it is confirmed by the repetition of many acts agreeing with it.[22]

The fundamental difficulty from this perspective was not one of capriciously changing occult principles underlying observable behavior, but the characterization of an experienced behavior as typical of the "ordinary course of nature." Castel had highlighted monstrous instances, whereas Aguilonius highlights the fallibility of the senses; but both seem to regard "the repetition of many acts," all exhibiting the same behavior, as the appropriate solution. For Hume, that was not good enough: an inference was either certain or uncertain. For an Aristotelian, by contrast, the student of nature could not be expected to establish its regularities with a degree of reliability that exceeded the degree of regularity inherent in nature itself. Castel had drawn attention to monstrous deviations because of the uncertainty as to whether any given phenomenon actually exemplified the ordinary course of nature.

Where Francis Bacon regarded deviations from the ordinary course

---

21. A reference to the ancient philosophical school of "academic scepticism": see Stough, *Greek Scepticism* (1969); Popkin, *The History of Scepticism;* Schmitt, "The Rediscovery of Ancient Skepticism" (1983); also the review by Popkin, "Theories of Knowledge" (1988).

22. Aguilonius, *Opticorum libri sex* (1613), p. 215: "Si perpetuò mentis aciem sensus eluderent, nulla scientia dari posset, quod Academici persuadere contendunt; & si numquam fallerent, unico actu certissima haberetur experientia. Nunc autem medio modo res se habet: nam tametsi nonnumquam labantur sensus, plerumque tamen non errant. Hinc sit, ut tum primùm certa sit experientia, cùm plurium actuum sibi consentientium repetitione firmatur."

of nature as providing privileged insights into its workings,[23] to the scholastic Aristotelian they were "monsters." If sufficiently spectacular, they might be taken as portents or omens, literally supernatural occurrences due to God's intervention.[24] But whether portents or not, monsters were by definition contrary to nature, and hence not illuminating of the natural order. Although typically applied to monstrous births, the term was used quite generally. Thus the prominent Jesuit astronomer and mathematician Christopher Clavius, when arguing against sceptics doubtful of the reality of astronomers' mathematical devices, extended its connotations in a revealing way, by concluding: "From all these things, therefore, I judge it to be established that eccentrics and epicycles are not so monstrous and absurd as they are feigned [to be] by adversaries, and that they are not introduced by astronomers without great cause."[25] Epicycles and eccentrics were not contrary to nature, monstrous, but seemed to be an integral part of it. More intriguingly, when, in a letter of 1611, Clavius confirmed Galileo's recent telescopic discoveries and predicted that other such novelties would be found in the planets, he described them using the word "monstrosity" (*monstruosita*).[26] The new features of the heavens were contrary to what would be *expected* on the received view of their nature. The fact that they were constant features previously unobserved, and not occasional anomalies such as new stars, meant that they were not truly monsters; they were instead indications that the ordinary course of nature might be different from what had hitherto been supposed.

The sense of "monster" exploited by Clavius illustrates well the scholastic understanding of the use of experience in sciences that dealt with the physical world. Experience taught how nature usually behaved; it did not consist of knowledge of discrete events, because such

23. Bacon, *Novum organum*, II, aph. 29.
24. Céard, *La nature et les prodiges* (1977); Daston and Park, "Unnatural Conceptions" (1981). Cf. the critical review of this and related literature by Hanafi, "Matter, Machines, and Metaphor" (1991), chap. 1, and esp. ibid., chap. 2 on monsters in natural magic literature and their relation to Aristotelian accounts of nature.
25. Clavius, "In sphaeram Ioannis de Sacro Bosco commentarius," in Clavius, *Opera mathematica* (1611–12), vol. 3, p. 304: "Ex his ergo omnibus constare arbitror, Eccentricos, & Epicyclos non esse adeo monstrosos, & absurdos, ut ab adversarijs finguntur, eosque ab Astronomis non sine magna causa inductos esse." On the context of Clavius's remark, see N. Jardine, "The Forging of Modern Realism" (1979); this provides a more accurate reading than that in Duhem, *To Save the Phenomena* (1969), pp. 92–96. Duhem's account (first published in 1908) is largely followed by Ralph M. Blake, "Theory of Hypothesis among Renaissance Astronomers," chap. 2 of Madden, *Theories of Scientific Method* (1960), pp. 32–35.
26. Quoted in Naux, "Le père Christophore Clavius" (1983), p. 193, apparently from a letter of 29 January 1611 (the citation seems to be in error).

events might be anomalous, "monstrous." Thus, in 1626, the Jesuit Orazio Grassi replied to Galileo's assertions about cometary paths in his *Assayer* (1623) with the observation that a comet is a "monster" (*monstrum*, which can also mean "wonder" or "portent"), being "the rarest sight in the world." "It is no wonder, therefore, if motion is assigned to it beyond those stable and perpetual [motions] which I might consider to be in the eternal stars. For he does not know the power of nature, as Seneca says, who does not believe to be sometimes allowed to it anything that it has not done often."[27] Nature can deviate from its normal paths, and the generalizations of natural philosophy cannot then capture it.

The reason why a minor practical difficulty for a scholastic Aristotelian became a major philosophical concern for Hume is, as remarked earlier, that a new way of coming to understand nature had emerged: experimental science. Although, famously, Aristotle's philosophy was rooted in experience—exemplified in the scholastic motto *nihil in intellectu quod non prius in sensu*—the framing and construing of that experience differed radically from the emergent new experience of the seventeenth century.

III. Experiences and Experiments

One common way of distinguishing an "experiment" from simple "experience" is to define the former as involving a specific question about nature which the experimental outcome is designed to answer; by contrast, the latter merely supplies raw information about phenomena that has not been deliberately solicited to interrogate a theory or interpretation. One premodern, scholastic use of "experience" in natural philosophy, for example, tended to take the form of selective presentation of instances. These illustrated the conclusions of philosophizing that had itself been conducted on the experiential basis of common knowledge. It was not a matter, therefore, of employing deliberately acquired experience to test philosophical propositions. From this perspective, it is true to say that "experiment" became a characteristic feature of the study of nature only in the seventeenth century.[28]

27. Orazio Grassi [Lothario Sarsi], *Ratio ponderum librae et simbellae* (1626), in Galileo, *Opere*, vol. 6, pp. 373–500, on pp. 404–405.

28. Two particularly clear discussions of this are Schmitt, "Experience and Experiment" (1969), and Rossi, "The Aristotelians and the Moderns" (1982), each of which examines Zabarella as a representative Aristotelian on the threshold of the seventeenth century. See, however, Baroncini, *Forme di esperianza e rivoluzione scientifica* (1992), chap. 2, which calls into question some of Schmitt's specific conclusions regarding Zabarella but does not contest the general point at issue here. M. R. Reif, "Natural Philosophy"

Such an analysis of the nature of "experiment," however, tends to obscure understanding of its historical emergence. The dictionary definition of an experiment as a test of a theory fails to capture the meaning and diversity of the new practices of the seventeenth century: Robert Hooke's and Isaac Newton's terms *instantia crucis* and *experimentum crucis* were certainly intended to pick out an aspect of Francis Bacon's teaching suitable to the notion of "experiment" as a test of alternative propositions, but Boyle's "experimental histories," also indebted to Bacon, took the form of collections of facts intended to underpin subsequent elaboration.[29] Again, the "experiments" of the Accademia del Cimento were frequently designed to test hypotheses or decide between alternatives, but the empirical work of the Accademia's Florentine forebear, Galileo, seems to have been directed towards establishing premises for formal scientific demonstrations—a quite different function that involved use of a broadly Aristotelian conception of scientific knowledge.[30] As a novel feature of the Scientific Revolution, therefore, "experiment" is difficult to pin down.

In the academic world inherited by seventeenth-century Europe, an "experience" was a universal statement of how things are, or how they behave. It did not refer to immediate perception because, as Aristotle said in the *Posterior Analytics*, "One necessarily perceives an individual and at a place and at a time, and it is impossible to perceive what is universal and holds in every case." Therefore, "since demonstrations are universal, and it is not possible to perceive these, it is evident that it is not possible to understand through perception."[31] Instead, perception produced scientific experience via memory: "from perception there comes memory ... and from memory (when it occurs often in connection with the same thing), experience; for memories that are many in number form a single experience."[32] But one did not need to have acquired such experiences personally in order to use them in argumentation, provided that they were commonly accepted, either through daily familiarity or through the statements of a weighty authority.

(1962), pp. 288–289, remarks on the use of "experience" as uncritical ordinary experience rather than contrived experiment in scholastic philosophy.

29. Hooke, *Micrographia* (1665/1962), p. 54; the term apparently derives from Bacon's "instantia crucis."

30. See Waller, *Essayes of Natural Experiments* (1684/1964), a translation of the Accademia's *Saggi* into English under the auspices of the Royal Society. Galileo is examined in further detail on this point in chapter 5, below.

31. Aristotle, *Posterior Analytics* I.31, trans. Barnes.

32. Ibid., II.19. See also Schramm, "Aristotelianism" (1963), esp. pp. 104–105.

For Aristotle, the nature of experience depended on its embeddedness in the community; the world was construed through communal eyes. Experience provided phenomena, and phenomena were, literally, *data*, "givens"; they were statements about how things behave in the world, and they were to be taken into account when discussing topics concerning nature. The sources of phenomena were diverse, including common opinion and the assertions of philosophers as well as personal sense perception. Given these statements (critically evaluated, to be sure), a system of syllogistic reasoning might yield, in principle, a theoretical description and explanation of them. True scientific knowledge should be demonstratively certain, which required that its premises themselves be certain. Natural philosophy, however, as later scholastic writers often admitted, tended to fall short of this ideal because of the merely "probable" status of many of its experiential principles; it then represented "dialectical" rather than demonstrative reasoning. "Probable" here, as Ian Hacking has argued, means "worthy of approbation" rather than simply "likely." The probability of a statement was therefore intimately bound up with matters of authority: the more authoritative the source, the more probable the statement. Experience and social accreditation were never sharply distinguished.[33]

This approach towards experience and the phenomena it claimed became embedded in scholastic pedagogical procedure. There the emphasis was on disputation and the proper form of an argument; the truth of the argument's premises was typically subjected to much less scrutiny.[34] The establishment of a premise often consisted of quoting an authority for it, such as Aristotle himself; for medieval and Renaissance scholastics, as for Aristotle, attaching the name of an authority to a statement of experiential fact rendered it probable (although not necessarily *likely*), and hence suitable for use in argument. R. W. Southern has pointed out the inadequacy of the sense-experience of individ-

---

33. See Owen, "*Tithenai ta phainomena*" (1986); also idem, "Aristotle" (1970) esp. pp. 252–254; Gaukroger, *Explanatory Structures* (1978), pp. 91–92; cf. Baroncini, *Forme di esperienza e rivoluzione scientifica*, p. 176. On premodern concepts of probability, see Hacking, *The Emergence of Probability*, chaps. 1–5. On scholastic "probability" in natural philosophy, see Wallace, "The Certitude of Science" (1986). Moss, *Novelties in the Heavens* (1993), is structured around a sharp analytical distinction between the three contemporary categories of dialectical, demonstrative, and rhetorical argument.

34. On scholasticism as a pedagogical "method," and the role of the disputation, see Makdisi, "The Scholastic Method in Medieval Education" (1974), esp. pp. 642–648. See also Marenbon, *Later Medieval Philosophy* (1987), pp. 12–14, 19–20, 27–34, on *quaestiones* and disputations; Murdoch, "The Analytical Character of Late Medieval Learning" (1982).

uals within the framework of medieval philosophical discourse: Robert Grosseteste rebuked those who would "form their own opinions from their experiments without a foundation of doctrine."[35]

Authoritative texts by such as Aristotle or Galen thus made up the accepted framework of scholastic natural philosophical discourse and interpretation. This framework was supported by the position of the commentary as the standard scholastic philosophical genre. If one wished, for example, to write about the nature of the heavens, an appropriate procedure was to compose a commentary on *De caelo*. Because the text determined the character and function of statements of universal experience, the authority of such statements derived from the place they held in the text, and the place the text held in natural philosophical inquiry, rather than from a blind faith in the assertions of Aristotle.[36] The way in which an authoritative text could inform the structure of natural philosophical discourse even without a commentary structure appears in an example from Albertus Magnus: Albertus, discussing trees, says that his procedure will be to "consider first the whole diversity of the parts."

> First, however, we shall only cite these differences and afterward assign the causes of all the differences. If we did not follow Aristotle, however, but others, we would surely proceed otherwise. We say, therefore, with Aristotle, that certain plants called trees have gum—as the pine tree, resin and almond gum, myrrh, frankincense, and gum arabic.[37]

In effect, Aristotle was an authority because he was the author of texts used habitually as loci for the discussion of particular subjects. Within that framework, "experience" had a particular meaning and role.[38]

---

35. See R. W. Southern, commentary on medieval science, in Crombie, *Scientific Change* (1963), pp. 301–306, on p. 305; Baroncini, *Forme di esperienza e rivoluzione scientifica*, p. 92 n. 76; and for a comparison of Grosseteste's concept of *experimentum* to "experience" in the Aristotelian sense see Eastwood, "Medieval Empiricism" (1968), esp. p. 321. See also McEvoy, *The Philosophy of Robert Grosseteste* (1983), esp. pp. 206–211.

36. Grant, "Aristotelianism and the Longevity of the Medieval World View" (1978), attributes the survival of the Aristotelian world-picture well into the seventeenth century to the status of the commentary as the prime vehicle of natural philosophical discussion; the form tended to atomize topical discussions and preserve the overall structure from scrutiny.

37. Albertus Magnus, *On Plants* I.2, trans. in Grant, *A Source Book in Medieval Science* (1974), p. 692 (the text Albertus used is no longer believed to be by Aristotle).

38. On the continuing dominance of commentaries on Aristotle during the Renaissance see Schmitt, *A Critical Survey and Bibliography* (1971); idem, "Toward a Reassessment of Renaissance Aristotelianism" (1973); and idem, *Aristotle in the Renaissance* (1983). For a short discussion of medieval commentary see Gilbert, *Renaissance Concepts of*

For the scholastic philosopher, the grounding in experience of the statements that acted as premises in his arguments was guaranteed by their universality—"heavy bodies fall" is a statement to which all could assent, through common experience embodied in authoritative texts. Jean Buridan, in the course of a well-known fourteenth-century discussion of the motion of the earth, remarked that "an arrow projected from a bow directly upward falls again in the same spot of the earth from which it was projected. This would not be so if the earth were moved with such velocity." The "experience" here is a general statement about how things habitually behave; furthermore, like a purely rational argument, Buridan introduces it by ascribing it to Aristotle: it is an "appearance that Aristotle notes." Buridan does not regard the point as conclusive, and he marshals counterarguments to the interpretation of the experience as proof of the earth's stability, but he does not question the truth of the experience itself.[39]

In characteristically seventeenth-century philosophical discourse, by contrast, experience increasingly took the form of statements describing specific events. These are exemplified most famously by the research reports found in countless contributions to the *Philosophical Transactions* by the early Fellows of the Royal Society, but they did not appear there de novo.[40] Natural philosophers and, especially, mathematical scientists increasingly used reports of singular events, explicitly or implicitly located in a specific time and place, as a way of constructing scientifically meaningful experiential statements. But as evidential weight came to be attached to that very singularity, the kind of common assent by which the truth of scientific experiences had formerly been established could no longer be anticipated—most starkly in cases of contrived experiences using special apparatus. The new scientific experience of the seventeenth century established its legitimacy by rendering credible its historical reports of events, often citing witnesses. The singular experience could not be *evident*, but it could provide *evidence*.

This, then, is the context to which Hume's criticism of induction pointed. It is what must be understood if the emergence of modern scientific practices, and the world that modern science inhabits, are to become comprehensible.

Method (1961), pp. 27–31. For a description of various genres of commentary, see Lohr, "Renaissance Latin Aristotle Commentaries: Authors A–B" (1974), on pp. 228–233.

39. Jean Buridan, *Questions on the Four Books on the Heavens and the World of Aristotle,* in Clagett, *The Science of Mechanics in the Middle Ages* (1961), p. 596.

40. On the practice of the early Royal Society, see chapter 8, below.

IV. *Inductio* and Certainty

The Latin word *inductio* was coined by Cicero to translate the Greek *epagôgê*, first adopted into technical philosophical usage by Aristotle.[41] Aristotle used the word in a number of ways, all nonetheless united by the common function of designating the generation of universal statements capable of serving as premises in a demonstrative syllogism. Thus, for example, *epagôgê* could refer to a rhetorical induction where a general claim is made plausible by the presentation of a few purportedly typical examples. It also referred to an induction by complete enumeration of instances. But most revealingly, perhaps, it could cover the recognition of an essential, necessary truth regarding some class of things when the mind grasps the universal in the particular by inspection: thus, for example, one might soon realize, after having encountered a few triangles, that not only do the internal angles of each add up to two right angles, but that this must be true of all triangles whatever, by the very nature of a triangle. Aristotle's belief in the reality of universals as entities existing above and beyond their individual instances played a crucial part in establishing this very influential sense of "induction." It is a sense that shaped many of the problems with which we shall subsequently be concerned.[42]

Two major strands of Aristotelian induction were represented in European scholastic traditions by the late sixteenth century. One was the rhetorical, found especially in humanist textbooks and commentaries on such works as Aristotle's *Topics*; it encompassed the first two senses just mentioned. The other strand comprised something akin to the third sense; the great Paduan logician and philosopher Jacopo Zabarella, near the end of the century, called it "demonstrative induction."[43] The essentially mysterious character of this procedure, whereby the mind comes to grasp a universal, relates closely to Zabarella's most famous logico-scientific procedure, the "demonstrative regress." Also

---

41. See, e.g., Milton, "Induction before Hume," pp. 51–53; McKirahan, "Aristotelian Epagoge" (1983).

42. Milton, "Induction before Hume"; Pérez-Ramos, *Francis Bacon's Idea of Science*, esp. pp. 208–211; Wallace, *Galileo's Logic of Discovery and Proof* (1992), pp. 165–170. Francis Bacon's use of the term differed in some important ways: he understood "induction" in its legal sense, deriving from its legal application by Cicero and Quintilian. A legal "induction" involved the review of a case in order to determine the precedents that should properly characterize it, and thus to discover the legal principles that would enable its resolution: Martin, *Francis Bacon* (1992), pp. 167–168.

43. Wallace, *Galileo's Logic of Discovery and Proof*, pp. 166–167; Milton, "Induction before Hume," p. 71, stressing that the thesis of the existence of necessary relations among singulars must be accepted for Zabarella's view to make sense. Cf. Jung, *Logica Hamburgensis*, p. 219, mentioning this term with reference to Zabarella.

known to modern scholars as *"regressus* theory," this was a logical technique of analysis and subsequent synthesis of propositions designed to generate true scientific knowledge, which for an Aristotelian had to be certain knowledge. By no means wholly original with Zabarella but closely associated with his name throughout Europe in the later sixteenth and seventeenth centuries, the technique had developed from a commentary tradition that focused on Aristotle's *Posterior Analytics,* and in particular on Aristotle's distinction between two forms of demonstration: *apodeixis tou dioti* and *apodeixis tou hoti,* usually latinized as *demonstratio propter quid* and *demonstratio quia.*[44]

Demonstratio propter quid was true scientific demonstration, that is, deductive syllogistic demonstration of an effect, or phenomenon, from an immediate cause.[45] (It is crucial to note that the effect or phenomenon is not itself something the existence of which needs to be established through any special confirmatory procedure.) Correspondingly, Zabarella—whether or not being faithful to Aristotle's intentions—treated *demonstratio quia* as a kind of deductive move from an effect back to its proper cause. Once identified, the cause could then be used as the premise of a *demonstratio propter quid.* The demonstrative regress was the union of these two moves, which Zabarella called "resolution" and "composition," that would result in scientific demonstration of the original effect. The relevance to "demonstrative induction" appears in the inferential step required to link the two halves of the process. For logical reasons, *demonstratio quia* could serve only to discover concomitants of the original phenomenon or effect. It could not say anything about any causal relationship that might be involved. But the demonstrative regress required that a newly found concomitant be established as a cause if a *demonstratio propter quid* were to be based upon it.[46]

A simple example (not itself Zabarellan) may serve to illustrate the problem. Take the effect, or phenomenon, of cold weather in winter. A constant concomitant of that situation is the evening visibility in the northern hemisphere of the constellation Orion. Nonetheless, it might

---

44. I base my account on N. Jardine, "Epistemology of the Sciences" (1988); see also idem, "Galileo's Road to Truth" (1976), esp. pp. 280–303. I have romanized the Greek. For other discussions, see Poppi, *La dottrina della scienza in Giacomo Zabarella* (1972); Risse, *Die Logik der Neuzeit,* vol. 1 (1964), pp. 278–290; idem, "Zabarellas Methodenlehre" (1983); L. Jardine, *Francis Bacon* (1974), pp. 54–58.

45. There were, of course, various restrictions on what should count as proper principles in such a demonstration: for a clear account see Wallace, *Galileo and His Sources,* pp. 111–116. For the immediate medieval background to these ideas, see Serene, "Demonstrative Science" (1982).

46. See also McMullin, "The Conception of Science in Galileo's Work" (1978), pp. 213–217.

seem implausible to reverse the analysis that identified this constant concomitant so as to say that coldness in winter occurs *because* of the visibility of Orion. However, it might seem a good deal less implausible if a different constant concomitant were taken as the cause of winter coldness, namely, the fact that the sun in winter is much lower in the sky. The difficulty lies in codifying the procedure whereby causal status is assigned or denied in any given case.

In fact, neither Zabarella nor anyone else succeeded in producing such a codification. Instead, Zabarella described the attribution of causal status as the outcome of a mysterious process usually known as *consideratio* or *negotiatio*. This was a form of contemplation aimed at creating conviction in the mind, and it relied on certain metaphysical assumptions—themselves strictly non-Aristotelian—about the mind's innate ability to grasp universals.[47] The similarity to "demonstrative induction" is therefore direct: in both cases the mind grasps by contemplation the existence of a universal, whether a necessary property of triangles or a necessary causal antecedent of a class of phenomena. *Negotiatio*, or *contemplatio*, could not be reduced to a formal logical technique, but it encompassed in this period a central meaning of "induction." It was also sometimes known in the sixteenth century as *meditatio*, a use of the term that illuminates Descartes's practice.[48]

Induction in this sense thus allowed the establishment of universal truths practically by inspection and took for granted the proposition that universals are real rather than being nominalistic categories. It remained an available conception throughout the seventeenth century.[49] The Port Royalists Arnauld and Nicole, in their Cartesian-Pascalian treatise *La logique ou l'art de penser* of 1662, acknowledged it as the only form of induction that could produce "perfect science"

---

47. See N. Jardine, "Epistemology," esp. pp. 686–693. Cf. Wallace, *Galileo and His Sources*, pp. 125–126. A cause known 'materially' as a result of *demonstratio quia* needed to be turned into a cause known 'formally' if a *demonstratio propter quid* were to be produced (ibid., p. 125). See also Edwards, "Randall on the Development of Scientific Method" (1967); also diagram in Pérez-Ramos, *Francis Bacon's Idea of Science*, p. 236.

48. See Dear, "Mersenne's Suggestion" (1995).

49. For example, Jung, *Logica Hamburgensis*, p. 168, discusses *inductio primaria* and *inductio secundaria*. The former refers to incomplete enumeration (akin to rhetorical induction); the latter refers to the formation of a universal general conclusion from universal special assumptions. It should be noted that even in the former case we are dealing with the legitimacy of inducing a statement such as "all wood floats" from other universal statements such as "ash floats," "oak floats," "elm floats," and so forth, rather than inducing a statement such as "all swans are white" on the basis of statements, concerning individuals, that "swan 1 is white," "swan 2 is white," "swan 3 is white," and so forth. In other words, these considerations remain quite distinct from the modern Humean problem of induction.

(that is, certain knowledge), "the consideration of singular things serving only as the occasion for our mind to give attention to its natural ideas, according to which it judges the truth of things in general."

> For it is true, for example, that I would perhaps never think of considering the nature of a triangle if I had not seen a triangle that gave me the occasion to think about it. But nevertheless it is not the particular examination of all triangles that makes me conclude generally and certainly of all of them that the area that they contain is equal to that of a rectangle of base equal to their circumference [*de toute leur base*] and half their height (for this examination would be impossible), but the sole consideration of what is contained in the idea of 'triangle' that I find in my mind.[50]

The echo by Cartesian innatists of an example of "induction" sanctioned by Aristotle forces the modern mind to recognize a surprising unfamiliarity in early-modern conceptualizations of the nature of universals. This unfamiliarity is all the more striking for the further counterecho found in the writings of the empiricist mathematician Isaac Barrow in the same decade. Barrow's *Mathematical Lectures* of 1664 argue that mathematical, and particularly geometrical, concepts are rooted in practical human experience of space and motion. Barrow's position was later to be repeated in the preface to the *Principia* of Isaac Newton.[51] But the *Lectures* also discuss the formation of general mathematical concepts through "induction." Barrow, who frequently cites Aristotle, says that although particular objects (Socrates, Bucephalus) are perceived through the senses, general truths (such as the principles of mathematical demonstrations) are perceived by contemplation.[52] This, he admits, seems to contradict Aristotle's opinion that all universal propositions, including demonstrative first principles, derive solely from induction: "Nihil est in intellectu quod non fuit prius in sensu."[53]

> From which it appears that according to *Aristotle*, the Principles of all Science depend wholly upon the Testimony of the Senses and particular Experiments: Which, if granted to be true, then scarce any human Reason can be supposed to come up to strict Certainty, or even arise above a probable Conjecture. But where any Proposition is found agreeable to constant Experience, especially where it seems not to be conversant about the Accidents of Things, but pertains to their princi-

---

50. Arnauld and Nicole, *La logique ou l'art de penser* (1662/1965–67), vol. 1, pp. 277–278.
51. See chapter 8 for details.
52. Barrow, *The Mathematical Works* (1860/1973), vol. 1, "Lectiones mathematicae," p. 81.
53. Ibid., p. 82.

pal Properties and intimate Constitution, it will at least be most safe and prudent to yield a ready Assent to it.[54]

However, induction from sense experience is not the source of the first principles of mathematics. Although sense provides the occasion for such conceptions, they are formed by reason: no one has ever seen an exact right line or a perfect circle.[55]

Where the Port Royalists found innate ideas, Barrow found contentless reason working on material suggested by the senses. The outcome, however, remained the same: the mind could make universally true propositions without risking an encounter with a contrary instance. Barrow, like Aristotle, could allow the mind to recognize universals even when the instantiations of those universals relied upon the practical experience of geometrical construction rather than upon a mental recognition of innate ideas.[56] Such a view of geometry was commonplace in the seventeenth century, appearing even in the work of Descartes, the arch-innatist.[57]

V. The Mathematical Model

Aristotle's use of "induction," with its vague reference, left little room for discrete experiences—experiments—to act as foundations for scientific argument. The "problem of induction" did not exist for Aristotle and his followers (witting or unwitting) because they did not regard universal propositions as being derived from specifiable singular propositions regarding individual instances. That would have been akin to rhetorical induction, and thus not scientific.[58] By definition, the first principles of an Aristotelian or Euclidean axiomatic deductive system are not themselves deductively justifiable, but they must be accepted as true in order to show that the conclusions deduced from them are not the product of purely arbitrary suppositions.[59] From a practical viewpoint, precisely how this was done mattered little. In the absence of formal rules for demonstrating first principles, therefore, appeal to universal "experience," whether inner and intuitive or outer and sen-

---

54. Ibid.; trans. from Barrow, *The Usefulness of Mathematical Learning*, p. 73.
55. Barrow, *The Mathematical Works*, "Lectiones mathematicae," p. 83.
56. See chapter 8, below.
57. The word "induction" routinely bore a slew of meanings in the seventeenth century—Descartes used it to mean what looks like a species of deduction: see Clarke, *Descartes' Philosophy of Science* (1982), p. 70.
58. Cf. Galen, *On Medical Experience*, in Galen, *Three Treatises* (1985), which discusses the issue of the relation of singulars to universal knowledge-claims in medicine, comparing the positions of the Dogmatists and Empiricists.
59. E.g., Aristotle, *Posterior Analytics* I.2.

sory, usually served. Mathematics was an especially strategic focus for such epistemological concerns in the seventeenth century because of the privileged reputation for certainty that attached to mathematical demonstration. That reputation rested not only on the clarity of the deductive steps proving a geometrical theorem, but also on the perceived self-evidence of the fundamental principles, or premises, from which the demonstrations were constructed.[60] It was in accordance with such expectations that the mathematical sciences pertaining to the natural world, disciplines such as astronomy, optics, and mechanics, were construed and developed. Their most notable practitioners were the Jesuits.

60. Studies include De Angelis, *Il metodo geometrico* (1964); Schüling, *Die Geschichte der axiomatischen Methode* (1969); Risse, *Die Logik der Neuzeit, 2. Band,* esp. chap. 8; idem, *Die Logik der Neuzeit, 1. Band.* N. Jardine, "Epistemology of the Sciences," pp. 693–697, discusses the debate in this period on the source of the certainty routinely attributed to mathematical demonstration. See also Mancosu, "Aristotelean Logic and Euclidean Mathematics" (1992); and chapter 2, n. 14, below.

Two

EXPERIENCE AND JESUIT MATHEMATICAL SCIENCE: THE PRACTICAL IMPORTANCE OF METHODOLOGY

I. Jesuit Mathematical Science and Empirical Principles

Many local contexts of knowledge-making bear witness to the gradual process by which appeal to discrete experiences became culturally dominant in European philosophy of nature. One of the most revealing, however, surrounds the practice of the classical mathematical sciences by members of what was long regarded as a bastion of reaction—the Jesuit order. The shifts in the concept of experience among Jesuit mathematicians impinge directly on the implications of moving from a scholastic to a characteristically early-modern natural philosophical paradigm. They also help to explain how mathematical models of scientific practice became so closely implicated in the new ideology of natural knowledge that had emerged by the end of the seventeenth century. Jesuit mathematicians began to validate the use of singular experiences made using contrived apparatus because the existing practices of their disciplines, besides needing to be defended methodologically against Jesuit critics, restricted the availability of their results. This and the following chapter will track some of the outcomes of this predicament.

Jesuit colleges were among the most important and prestigious of all educational institutions in early-modern Europe. Established throughout Catholic territories from the middle of the sixteenth century onwards as part of the Jesuits' Counter-Reformatory mission, they provided a high level of academic training focused on rational theology, missionary skills, and general cultural excellence aimed at the intimidation of Protestants.[1] By the early seventeenth century the mathematical disciplines had come to hold a comparatively prominent place in the courses of study offered by the Jesuits at the larger colleges and in the ideal curriculum enshrined in the 1599 *Ratio studiorum*.[2] The exact pattern varied from college to college, and over time, but the only real constraint appears to have been an insufficient number of competent teachers to go around.[3] Throughout the seventeenth century, Jesuit mathematicians carried out research in their subjects, produced major treatises and wrote textbooks that were widely read.[4] Christopher Clavius, professor of mathematics at the Collegio Romano from 1565 until his death in 1612, was the prime mover in establishing mathematics in the curriculum; his work powerfully shaped the style and attitudes manifest in subsequent Jesuit mathematical writing.[5]

Clavius's textbooks formed the standard introduction to the mathe-

---

1. See fundamental works by Dainville, *La naissance de l'humanisme moderne* (1940) and *La géographie des humanistes* (1940), as well as the collected articles in Dainville, *L'éducation des Jésuites* (1978); Codina Mir, *Aux sources de la pédagogie des Jésuites* (1968). Snyders, *La pédagogie en France* (1965), provides a useful account of the ethos of these colleges, as does Ducreux, *Le collège des Jésuites de Tulle* (1981). Chartier, Julia, and Compère, *L'éducation en France* (1976) contains useful material on French Jesuit colleges and their structure, as does Brockliss, *French Higher Education* (1987), esp. material in chap. 1.

2. For a general account of Jesuit mathematical education, see, apart from material in Dainville, *L'éducation des Jésuites*, Heilbron, *Electricity in the 17th and 18th Centuries* (1979), pp. 101–114. Cosentino, "Le matematiche nella 'Ratio studiorum'" (1970); idem, "L'insegnamento delle matematiche" (1971). For comprehensive listings and biographical entries of mathematicians in the colleges see Fischer, "Jesuiten Mathematiker in der deutschen Assistenz bis 1773" (1978); idem, "Jesuiten Mathematiker in der französischen und italienischen Assistenz bis 1762, bzw. 1773" (1983); idem, "Die Jesuitenmathematiker des nordostdeutschen Kulturgebietes" (1984). See also Koutná-Karg, "Experientia fuit, mathematicum paucos discipulos habere" (1991). Krayer, *Mathematik im Studienplan der Jesuiten* (1991), pt. I, chap. 2, concludes (p. 41) with the observation that, at least at the institutional level, Clavius's promotion of mathematics in the curriculum had little impact, contrary especially to the impression given by Dainville. The fact remains, however, that mathematics was taught in the Jesuit colleges, using the textbooks of Clavius, and that those Jesuit mathematicians who emerged from such training were imbued, as their writings attest, with the views and values expressed by Clavius. These mathematicians were, furthermore, widely known, and their works influential, in mathematical-scientific circles throughout Europe.

3. Heilbron, *Electricity in the 17th and 18th Centuries*, pp. 102–103.

4. Harris, "Transposing the Merton Thesis" (1989), stresses the sheer quantity of Jesuit work in the sciences in the early-modern period, as well as its utilitarian bias.

5. On Clavius's work and career see above all Baldini, "Christoph Clavius" (1983); also Lattis, "Christoph Clavius and the *Sphere* of Sacrobosco" (1989), chap. 1; Knobloch, "Sur la vie et l'œuvre de Christophore Clavius" (1988); idem, "Christoph Clavius—Ein

matical sciences for pupils of the Jesuits such as Descartes.⁶ Although he praised the widespread applicability of mathematics to all areas of learning, Clavius restricted his own writings to the more practical aspects of the subject: his work on the Gregorian calendar was very much in keeping with the tenor of his texts on geometry, astronomy, arithmetic, and algebra—the last in its sixteenth-century guise as a department of arithmetic providing analytical computational techniques.⁷ Jesuit mathematical teaching, following Clavius, included considerable emphasis on practical matters such as spherical astronomy, geography, surveying, and the concomitant use of mathematical instruments.

Clavius played a seminal role in the formation of a Jesuit tradition of work in the mathematical disciplines running throughout the seventeenth century. His promotion of their intellectual status was of paramount importance because it provided a basis for a treatment of aspects of the natural world that would stand on an equal methodological footing with Aristotelian natural philosophy (physics). The strict disciplinary structure of the Jesuit colleges instantiated a conceptual structure that placed mathematics in a clearly defined position: it was *not* natural philosophy.⁸ The arts curriculum of the medieval universities had derived from the late-antique classification of the trivium,

---

Astronom zwischen Antike und Kopernikus" (1990). Lattis, *Between Copernicus and Galileo* (1994) appeared too late for consideration here.

6. A document written by Clavius for his superiors in about 1579 or 1580, detailing a projected mathematics course with subjects and authors, is reproduced in Baldini, *Legem impone subactis* (1992), pp. 172–175, and also in idem, "La nova del 1604" (1981), on pp. 89–95; see also a similar listing by Clavius of mathematical texts transcribed in Lattis, "Christoph Clavius and the *Sphere* of Sacrobosco," appendix IV.

7. His various works are collected in Clavius, *Opera mathematica* (1611–12). For praise of mathematics' general utility, see, e.g., Clavius, "Geometrica practica," in ibid., vol. 2, "Praefatio," p. 3. On Clavius's calendrical work, see Baldini, "Christoph Clavius"; on his algebra, Knobloch, "Sur la vie et l'œuvre de Christophore Clavius"; Naux, "Le père Christophore Clavius (1537–1612), sa vie et son œuvre" (1983), on pp. 336–338 (although this article is not very reliable or well documented). See also Homann, "Christophorus Clavius and the Renaissance of Euclidean Geometry" (1983). A full listing of Clavius's works may be found in Sommervogel, *Bibliothèque de la Compagnie de Jésus* (1890–1932/1960), vol. 2, cols. 1212–1224. Lattis, "Christoph Clavius and the *Sphere* of Sacrobosco," appendix II, provides a listing of editions of Clavius's commentary on *De sphaera*, while Knobloch, "Christoph Clavius: Ein Namen- und Schriftenverzeichnis" (1990), pp. 136–139, lists the various editions of the works that went to make up the *Opera mathematica*.

8. See, for the official statement in the Jesuits' 1599 *Ratio studiorum* of the disciplinary and conceptual distinction between natural philosophy and mathematics, Salmone, *Ratio studiorum* (1979), p. 66. See also, on these issues of the relationship between mathematics and natural philosophy among the Jesuits, Baldini, *Legem impone subactis*, chap.

comprising the headings grammar, logic, and rhetoric, and the quadrivium, consisting of arithmetic, geometry, astronomy, and music.[9] The Jesuits in effect elevated the quadrivium from its former propaedeutic place as an arts subject to the second or third year of their advanced three-year philosophy course, where it was usually taught alongside either physics or metaphysics (after a year's training in logic.)[10] The precise relationship of mathematics to philosophy had been a matter of contention, however, both before and after the formal incorporation of mathematical studies into the official Jesuit curriculum in the later sixteenth century. It was Clavius who had championed the philosophical standing of mathematics against its detractors when a general college curriculum was being debated in Rome during the 1580s.

The opposition he faced in attempting to ensure a respectable place for mathematics in Jesuit pedagogy emerges vividly, if not disinterestedly, in a policy document that he wrote in the 1580s aimed at taking to task disrespectful teachers of philosophy.[11] There are those, he claims, who tell their pupils such scurrilous things as that "mathematical sciences are not sciences, do not have demonstrations, abstract from being and the good etc." Teachers of mathematics must be accorded proper respect and status, whereas these unfortunate doctrines promulgated by philosophers "are a great hindrance to pupils and of no service to them; especially since teachers can hardly teach them without bringing these sciences into ridicule." Clavius therefore proceeds to make suggestions calculated to promote the image of mathematics, such as ensuring that mathematics teachers attend the regular formal disputations, even participating in them, just as do the philosophers. He also asserted the necessity of mathematics in the study of natural philosophy, mentioning as an example the relevance of mathematical astronomy to cosmology. Furthermore, he claimed that there was "an infinity of examples in Aristotle, Plato and their most illustrious interpreters which can in no way be understood without some knowledge

---

1, "*Legem impone subactis.* Teologia, filosofia e scienze matematiche nella didattica e nella dottrina della Compagnia di Gesù (1550–1630)."

9. Essays on the quadrivial disciplines in the earlier Middle Ages may be found in Wagner, *The Seven Liberal Arts* (1983). See also Gagné, "Du quadrivium aux scientiae mediae" (1969).

10. See, in addition to references in n. 2, Rochemonteix, *Un collège de Jésuites* (1889), esp. vol. 4, pp. 27, 32.

11. Clavius, "Modus quo disciplinae mathematicae in scholis Societatis possent promoveri," in *Monumenta Paedagogica Societatis Jesu* (1901), pp. 471–474.

of the mathematical sciences."[12] This way of increasing the esteem of philosophers in the colleges for their mathematical colleagues evidently had some appeal: in 1615 Clavius's follower Josephus Blancanus (Giuseppe Biancani) published a 283-page work that proceeds through Aristotle's works in turn picking out passages that admit of mathematical elucidation.[13]

The stakes were considerable: Clavius was not inventing adversaries. Prominent Jesuit natural philosophers denied to mathematics the status of *scientia*, true scientific knowledge (the highest cognitive ideal for methodological Aristotelians). In the sixteenth century a number of Italian philosophers, beginning with Alessandro Piccolomini and including the Jesuit Pereira, had maintained that pure mathematics (geometry and arithmetic) was not a true science in Aristotle's sense because it did not demonstrate its conclusions through causes. They had been followed by the authors of the important Coïmbra commentaries on Aristotle, which were explicitly designed for use in Jesuit colleges. As a rule, teachers of mathematics in Jesuit colleges did not double as teachers of philosophy; if not admitted as an integral part of philosophy, and treated instead as a mere set of calculatory techniques, mathematics, and its practitioners, would suffer accordingly.[14]

An Aristotelian science employed causal demonstrations the ideal unit of which was a syllogism having a middle term that expressed the operative cause (whether efficient, material, formal, or final). This cause should be both necessary and sufficient to account for the effect or property attributed in the conclusion. Any discipline that did not demonstrate its conclusions through causes was, therefore, not scientific; if a geometrical demonstration was just an exposition of logical relations between propositions, it represented an inferior grade of

---

12. Ibid., pp. 471–472. Nicholas Jardine traces an apparent increase towards the end of the sixteenth century in the perceived relevance of mathematical astronomy for natural philosophy in N. Jardine, *The Birth of History and Philosophy of Science* (1984), p. 246.

13. An exercise suggested by Clavius himself in "Modus," p. 473. Blancanus, *Aristotelis loca mathematica* (1615), is usually bound following another work by Blancanus, *De mathematicarum natura dissertatio*; the works bear identical publication details.

14. See Galluzzi, "Il 'Platonismo' del tardo Cinquecento" (1973); see also Wallace, *Galileo and His Sources* (1984), p. 136, for further references to work by G. C. Giacobbe, to which may be added Carugo, "Giuseppe Moleto" (1983). Giacobbe, *Alle radici della rivoluzione scientifica rinascimentale* (1981) presents texts and analysis focused on Catena, with a dominant theme of the replacement, starting in the sixteenth century, of a "dialectical logic" by the "mathematical logic" relating to singular phenomena that comes to characterize (in Giacobbe's view) modern science. See also De Pace, *Le matematiche e il mondo* (1993), esp. chap. 1 on Piccolomini and Pereira. N. Jardine, "Epistemology of the Sciences" (1988), esp. pp. 693–697, provides a valuable guide to these matters.

knowledge.¹⁵ As a consequence, mathematics could not be a proper part of philosophy. Mathematical objects (numbers or geometrical figures) were not real things; they existed only in the intellect. Therefore—so the objection ran—mathematical objects could not have essences: that is, mathematical definitions were not *essential* definitions of real objects from which the characteristic properties of those objects could be deduced.¹⁶

Clavius's approach to maintaining mathematics as a part of philosophy amounted to little more than an argument from authority. He adopted the conventional classification of subject matters found in Aristotle and stemming from Plato, a classification frequently used in the Latin West since the early Middle Ages.

> Because the mathematical disciplines discuss things that are considered apart from any sensible matter—although they are themselves immersed in matter—it is evident that they hold a place intermediate between metaphysics and natural science, if we consider their subject, as is rightly shown by Proclus. For the subject of metaphysics is separated from all matter, both in the thing and in reason; the subject of physics is in truth conjoined to sensible matter, both in the thing and in reason; whence, since the subject of the mathematical disciplines is considered free from all matter—although it [i.e., matter] is found in the thing itself—clearly it is established intermediate between the other two.¹⁷

This classification therefore placed mathematics as an integral part of philosophy, no less than physics or metaphysics. Thus, following the same tack, Blancanus could reply to denials of mathematics' philosophical status by saying that "among Aristotle and all the peripatetics

---

15. Those who denied scientificity to mathematics usually attributed the widely acknowledged certainty of mathematical demonstration to the nature of its subject matter, not to its methodological structure.

16. See references in n. 14.

17. Clavius, "In disciplinas mathematicas prolegomena," in *Opera mathematica*, vol. 1, p. 5: "Quoniam disciplinae Mathematicae de rebus agunt, quae absque ulla materia sensibili considerantur, quamvis reipsa materiae sint immersae; perspicuum est, eas medium inter Metaphysicum, & naturalem scientiam obtinere locum, si subiectum earum consideremus, ut recte à Proclo probatur, Metaphysices etenim subiectum ab omni est materia seiunctum, & re, & ratione: Physices vero subiectum & re, & ratione materiae sensibili est coniunctum: Unde cum subiectum Mathematicarum disciplinarum extra omnem materiam consideretur, quamvis re ipsa in ea reperiatur, liquido constat hoc medium esse inter alia duo." The career of this classification is examined in Weisheipl, "Classification of the Sciences" (1965); idem, "The Nature, Scope and Classification of the Sciences" (1978). On Clavius's arguments, see also Crombie, "Mathematics and Platonism in the Sixteenth-Century Italian Universities" (1977); see also Carugo and Crombie, "The Jesuits and Galileo's Ideas of Science and of Nature" (1983).

nothing occurs more frequently than that there are three parts of philosophy: physics, mathematics, and metaphysics."[18] Indeed, Clavius used the same scheme, on Ptolemy's authority, as a way of suggesting, not just the equality, but the preeminence of mathematics: "For he says that natural philosophy and metaphysics, if we consider their mode of demonstrating, are rather to be called conjectures than sciences, on account of the multitude and discrepancy of opinions."[19]

In his policy document of the 1580s Clavius had tried a similar ploy: mathematical knowledge was not only indispensable for the philosopher, but might be seen as the highest of all intellectual pursuits: "Since therefore the mathematical disciplines in fact require, delight in, and honor truth—so that they not only admit nothing that is false, but indeed also nothing that arises only with probability, and finally, they admit nothing that they do not confirm and strengthen by the most certain demonstrations—there can be no doubt that they must be conceded the first place among all the other sciences."[20] He regarded astronomy as the noblest of all, since it fulfilled Aristotle's criteria of excellence better than any other: it used the most certain demonstrations (those of geometry), while dealing with the most noble subject, the heavens.[21] Clavius was careful to choose criteria of assessment that would elevate his own field at the expense of natural philosophy.

Clavius handled the crucial objection that the mathematical disciplines were not scientific in a similar fashion, by sidestepping it. He chose to appeal to the authority of Aristotle rather than present any positive arguments of his own, relying on the explicit inclusion of the

---

18. Blancanus, *De mathematicarum natura dissertatio*, p.27: "apud Arist. & omnes peripateticos nihil frequentius occurrat, quam tres esse philosophiae partes, Physicam, Mathematicam, & Metaphysicam."

19. Clavius, "In sphaeram Ioannis de Sacro Bosco commentarius," in *Opera mathematica*, vol. 3, p. 4: "Ait enim philolophiam [sic] naturalem & Metaphysicam, si modum demonstrandi illarum spectemus, appellandas potius esse coniecturas, quàm scientias, propter multitudinem, & discrepantiam opinionum." Cf. Ptolemy, *Almagest* I.1; see the English translation by Toomer, *Ptolemy's Almagest* (1984), p. 36. De Pace, *Le matematiche e il mondo*, chap. 4, sect. 5, discusses the use of Ptolemy's arguments in favor of mathematics by Jacopo Mazzoni (and also Galileo); see also Drake, "Ptolemy, Galileo, and Scientific Method" (1978).

20. Clavius, "In disciplinas mathematicas prolegomena," in *Opera mathematica*, vol. 1, p. 5: "Cum igitur disciplinae Mathematicae veritatem adeo expetant, adament, excolantque, ut non solum nihil, quod sit falsum, verum etiam nihil, quod tantum probabile existat, nihil denique admittant, quod certissimis demonstrationibus non confirment, corroborentque, dubium esse non potest, quin eis primus locus inter alias scientias omnes sit concedendus."

21. Clavius, "In sphaeram Ioannis de Sacro Bosco commentarius," in *Opera mathematica*, vol. 3, p. 3.

mathematical disciplines within the domain of Aristotle's general model of an ideal science. According to Aristotle, sciences should be founded on their own unique, proper principles, which provided the major premises for deductive, syllogistic demonstration. Subject matters were therefore strictly segregated to their appropriate sciences, a logical necessity expressed in the methodological rule of homogeneity. Homogeneity required that the principles of a science concern the same genus as its objects, so as to ensure the possibility of a deductive link between them. But disciplines such as astronomy and music clearly violated this rule because they drew on the results of pure mathematics (divided into arithmetic and geometry) so as to apply them to something else, namely celestial motions and sounds. Accordingly, Aristotle made a special accommodation for such subjects by classifying them as sciences *subordinate* to higher disciplines.[22] They were later represented in the quadrivium by astronomy and music (these two standing for a host of others, such as geography and mechanics), and came to be known variously as "subordinate," "middle," or "mixed" sciences.[23] Aristotle's was in some ways an ad hoc solution to the classificatory problem, and it provoked later scholastic discussions, particularly among the Jesuits, on whether demonstrations in a subject such as optics yielded true scientific knowledge if the presupposed theorems of geometry were not proved at the same time.[24] The approach served Clavius's purpose perfectly well, however, because the very attempt to fit the applied disciplines into a general model for a science made clear Aristotle's acceptance of the scientific status of all the mathematical disciplines.[25]

22. See, for valuable discussions, McKirahan, "Aristotle's Subordinate Sciences" (1978); Lennox, "Aristotle, Galileo, and 'Mixed Sciences'" (1986); see also McKirahan, *Principles and Proofs* (1992). Two central sources are Aristotle, *Posterior Analytics* I.7; Aristotle, *Metaphysics* XIII.3 (esp. 1078a14–17).

23. Jesuit discussions include Clavius, "In disciplinas mathematicas prolegomena," pp. 3–4; Blancanus, *De mathematicarum natura dissertatio*, pp. 29–31. See for a useful discussion Laird, "The *Scientiae mediae*" (1983), esp. chap. 8 on Zabarella. Not all "subordinate sciences" were mathematical.

24. See Wallace, *Galileo and His Sources*, p. 134.

25. I refer to the "mixed" mathematical disciplines as "applied" in a somewhat loose, but not altogether misleading, sense: Daston, *Classical Probability in the Enlightenment* (1988), esp. pp. 53–56, maintains that in the eighteenth century mixed mathematics had a character that dwarfed in importance and stature so-called pure mathematics, and that it was not a simple matter of "applying" prepackaged pure mathematics to concrete objects. However that may be, the Aristotelian conceptualization still dominant in the seventeenth century emphasized the subordination of mixed to pure mathematics, and seems clearly different from Daston's portrayal of probability as a mixed mathematical discipline in the following century.

Clavius could therefore present mathematics as genuinely scientific without engaging the tricky question of causes. He said of the mathematical disciplines that "they alone preserve the way and procedure of a science. For they always proceed from particular foreknown principles to the conclusions to be demonstrated, which is the proper duty and office of a doctrine or discipline, as Aristotle, *Posterior Analytics* I, also testifies."[26] He thus kept the issue on a purely methodological plane. In 1615, however, Blancanus tackled the question of causes directly, in a text entitled *De mathematicarum natura dissertatio*. Employing the standard Aristotelian classification of causes into material, efficient, formal, and final, he argued that demonstrations in geometry utilized formal and material causes, since geometry specified both the essences of geometrical objects and their "matter," quantity.[27] Blancanus even made the argument that geometrical optics could provide final causes in the study of the physiology of the eye, in that it explained why the eye needs to be more or less spherical.[28] Blancanus's lengthy apology for mathematics seems to have become well known in the seventeenth century; it was, for example, drawn upon heavily in a similar discussion by another Jesuit, Hugo Sempilius (Hugh Sempill), as part of his own treatise on the mathematical disciplines.[29]

The extent to which Blancanus himself simply followed a path al-

26. Clavius, "In disciplinas mathematicas prolegomena," p. 3: ". . . solae modum rationemque scientiae retineant. Procedunt enim, semper ex praecognitis quibusdam principijs ad conclusiones demonstrandas, quod proprium est munus, atque officium doctrinae sive disciplinae, ut & Aristoteles I.posteriorum testatur." Clavius had commenced this passage by showing that the etymological derivation of "mathematics" linked it to the meanings of "discipline" or "doctrine." He refers to the question of causes, but only in passing, in "Modus," p. 473.

27. Blancanus, *De mathematicarum natura dissertatio*, pp. 7–10. A convenient summary of his discussion may be found in Wallace, *Galileo and His Sources*, pp. 142–143.

28. Blancanus, *De mathematicarum natura dissertatio*, p. 30; pp. 29–31 are on the "middle sciences" in general, presented with the observation that they of course give causal demonstrations, and citing Aristotle to support this characterization. On the shape of the eye, cf. Cabeo, *In quatuor libros Meteorologicorum Aristotelis commentaria* (1646), Lib. III, p. 186 col. 2.

29. Sempilius, *De mathematicis disciplinis libri duodecim* (1635). See Lib. I, pp. 1–20, on the nobility of the mathematical sciences; Lib. II covers another standard topic familiar from Clavius, "De utilitate scientiarum Mathematicarum," pp. 21–53. For more extended accounts of Blancanus's arguments, see Galluzzi, "Il 'Platonismo' del tardo Cinquecento," esp. pp. 56–65; Giacobbe, "Epigone nel Seicento della 'Quaestio de certitudine mathematicarum': Giuseppe Biancani" (1976); Wallace, *Galileo and His Sources*, pp. 141–144. For additional material on Blancanus, see Baldini, *Legem impone subactis*, passim and chap. 6, a new version of idem, "*Additamenta Galilaeana*: I" (1984); and Sommervogel, *Bibliothèque de la Compagnie de Jésus*, s.v. "Biancani." Isaac Barrow mentions Blancanus's

ready trodden by Jesuit mathematicians, however, appears clearly in the introductory material to a master of arts dissertation defended by a student of the Jesuit astronomer and optician Christopher Scheiner at Ingolstadt in 1614. The opening section is headed: "De praestantia, necessitate et utilitate mathematicae," and consists largely of lengthy quotation from Possevino on the value of mathematics for understanding Plato and Aristotle and its use in a number of practical arts. The flavor is exactly that found in Clavius's and Blancanus's writings on the subject; Clavius's own "De utilitate Astronomiae" from his commentary on *De sphaera* is dutifully acknowledged.[30] There follows material on the scientific status of mathematics and the proper objects of its various branches: "Mathematics demonstrates its conclusions scientifically, by axioms, definitions, postulates, and suppositions; whence it is clear that it is truly called a science."[31] A pamphlet of mathematical propositions intended for an academic festival at the French Jesuit college of Pont-à-Mousson in 1622 stressed the same familiar points about the "nature of mathematics," affirming the mathematical disciplines as sciences.[32]

Such discussions by Blancanus and other Jesuit mathematicians of the period served to reinforce Clavius's Aristotelian description of the logical structure of the mathematical disciplines, and they highlight the importance that the Jesuit mathematical tradition stemming from Clavius attached to maintaining the scientific status of mathematical knowledge. It is important to stress that the issue went beyond the mere making of a few apologetic remarks at the beginning of a treatise before proceeding to the real content. The usual, and most effective, approach was to carry on as if the mathematical discipline in question were obviously and unproblematically a science. The Euclidean theorem form provided a structure already conformable to the ideal of scientific demonstration because it had been Aristotle's own model for

---

discussion in his *Mathematical Lectures* (although not approvingly on the specific point at issue): Barrow, *The Mathematical Works* (1860/1973), p. 84.

30. Scheiner, *Disquisitiones mathematicae* (1614), pp. 6–11. This text was evidently written by Scheiner for the student to defend—a common practice in this period. Scheiner sent a copy to Galileo in 1615: Galileo, *Opere*, 12:137–138. Its ninety pages of text include such items as nine diagrams of the positions of Jupiter's satellites at various times in 1612, 1613, and 1614 (p. 79). For the attribution (certain from internal evidence), see Sommervogel, *Bibliothèque de la Compagnie de Jésus*, vol. 7, p. 737 col. 1.

31. Scheiner, *Disquisitiones mathematicae*, pp. 12–15, quote on p. 14: "Mathematica conclusiones suas demonstrat scientificè, per Axiomata, per definitiones, postulata & suppositiones. Unde patet ipsam verè dictam scientiam esse."

32. *Selectae propositiones* (1622), p. 1.

a science.³³ Its mere employment therefore went a long way towards bestowing upon its subject matter the mantle of "science." The difficulties lay in persuading the subject matter to fit the formal structure.

The certainty and necessity of the conclusions in an Aristotelian scientific demonstration were rooted in the premises: it was easy to construct a formally valid syllogism, but hard to invent premises with the right properties. Clavius outlined the problem, for which Aristotle's *Posterior Analytics* was the locus classicus, in the "Prolegomena" to his edition of Euclid:

> While every doctrine, and every discipline, is produced from preexisting knowledge, as Aristotle says, and demonstrates its conclusions from particular assumed and conceded principles, yet no science, according to the opinion of Aristotle and other philosophers, demonstrates its own principles; the mathematical disciplines certainly will have their principles, from which, posited and conceded, they confirm their problems and theorems.³⁴

But if a science could not confirm its own principles, how were those principles, on which the certainty of the science depended, to be established?

The ideal was to have principles, or premises, that were *evident* and therefore immediately conceded by all. In the case of geometry, Euclid's "common opinions"—what Aristotle called "axioms"—represent the concept precisely: statements such as "the whole is greater than its proper part," or "if equals are subtracted from equals, the remainders are equal."³⁵ From a practical standpoint, empirical principles concerning the natural world could be made evident in a similar way, just as (for Aristotle) geometrical axioms themselves ultimately

---

33. Lloyd, *Magic, Reason and Experience* (1979), chap. 2, discusses these matters at length. Giacobbe, *Alle radici della rivoluzione scientifica rinascimentale*, pp. 22–23, notes Aristotle's incorporation of mathematical argument into his scientific methodology while characterizing it as ineffective because fundamentally illustrative. The same could, of course, be said of Aristotle's presentation as a whole in the *Posterior Analytics;* the text is not intended as an instrument of research.

34. Clavius, "In disciplinas mathematicas prolegomena," p. 9: "Cum omnis doctrina, omnisque disciplina ex praeexistente gignatur cognitione, ut auctor est Aristoteles, atque ex assumptis, & concessis quibusdam principijs suas demonstret conclusiones; Nulla autem scientia ex eiusdem Aristotelis, aliorumque Philosophorum sententia sua principia demonstret; habebunt utique & Mathematicae disciplinae sua principia, ex quibus positis, & concessis sua problemata ac theoremata confirment." This passage also serves to place the mixed mathematical sciences on a par with all other sciences, despite their assumption of results from other disciplines, by stressing that *all* principles in *all* sciences are ultimately just conceded.

35. See Lloyd, *Magic, Reason and Experience*, esp. p. 111.

derived from the senses: they would be evident if everyone agreed on their truth and judged argument to be unnecessary in the establishment of that agreement. Experiential statements, therefore, could not play a role in scientific discourse unless they were universal; if they were not, they could never be evident. Formal universality did not in itself establish an experience as "evident," of course; the experience had to express, and derive from, the perennial lessons of the senses.[36]

The classical mixed mathematical sciences appealed to just such empirical principles. The basic empirical premise of the mathematical science of optics, for example, held that light rays (or visual rays) travel in straight lines in homogeneous media. This counted as evident because everyone knew, from common experience, that you can't see around corners.[37] The foundational catoptrical works of Euclid and Ptolemy represent attempts at developing a complete science on such a basis: they present the principle that asserts equality of the angles of incidence and reflection so as to make it appear a necessary corollary of everyday visual experience.[38] The geometrically elaborated axioms of Archimedes' mechanical works were also chosen for their immediate acceptability by all reasonable people: Postulate 1 of *On the Equilibrium of Planes* states that "equal weights at equal distances [from the fulcrum] are in equilibrium, and equal weights at unequal distances are not in equilibrium but incline towards the weight which is at the greater distance."[39]

The one glaring scientific deficiency of such principles was that they fell short of the strict Aristotelian ideal by lacking any obvious *necessity*. This, however, seems to have been ignored by the Jesuit mathematical apologists, no doubt because it could not in its own terms be remedied. Necessity could accrue to empirical statements in physics, on

---

36. A celebrated example is Aristotle's observation that bees appear to reproduce parthenogenetically: Aristotle held his opinion to be open to correction by subsequent experience. Aristotle, *History of Animals* IX.42.

37. The object was to employ principles that were *per se nota*, that is, that were known without recourse to anything else. On the rectilinear propagation of light in antiquity see Lindberg, *Theories of Vision* (1976), pp. 12, 220 n. 79. For an interesting discussion of this principle that shows how much is taken for granted in seeing it as "self-evident," see Toulmin, *The Philosophy of Science* (1953), pp. 17–30.

38. Schramm, "Steps Towards the Idea of Function" (1965), pp. 71–73; note also, however, the difference between Euclid and Ptolemy—Ptolemy is more conscientious about establishing his foundational suppositions. See also Omar, *Ibn al-Haytham's "Optics"* (1977), chap. 1, esp. pp. 17–36.

39. Archimedes, "On the Equilibrium of Planes," Book I, in Heath, *The Works of Archimedes* (1953), p. 189. Archimedes' use of "postulate" is slightly odd in relation to the usual uses of that term as discussed by Proclus; on this point see Dijksterhuis, *Archimedes* (1987), chap. 9, esp. pp. 296–298.

the other hand, owing to the alleged possibility of grasping essences: propositions such as "man is a rational animal," once established, were necessarily true by definition. Actually establishing them was another matter, one that fueled the sixteenth-century methodological discussions concerning the demonstrative regress; the mathematicians kept clear of such niceties.[40]

Mathematical scientists in the early seventeenth century, then, looked to these Aristotelian and other classical sources as their disciplinary models. Jesuits, and those touched by the Jesuit approach, were particularly prone to see these issues through Aristotelian methodological spectacles. Galileo's empirical work on falling bodies, for example, was aimed at finding principles for his "new science of motion" that would imitate the Archimedean model of a mathematical science by commanding immediate assent: it was not aimed at uncovering recondite facts accessible only through elaborate experimental procedures. Galileo saw the construction of a mathematical science of nature in terms of classical mixed mathematics as defined by Aristotle.[41] His attempts to render evident the principles of the "new science of motion" illustrate very neatly the scholastic-Aristotelian concept of experience reflected in the traditional mathematical sciences. To be fully adequate, empirical premises needed to command assent because they were evident, not because particular events were adduced in their support. This was not experimental science in its modern acceptation.

"Experience" in scholastic natural philosophy and mathematics, therefore, in keeping with these criteria for scientific knowledge, typically took the form of universal statements because singular statements, statements of particular events, are not evident and indubitable, but rely on fallible historical reports. The Aristotelian model of a science adopted by the Jesuits took scientific knowledge to be fundamentally public: scientific demonstration invoked necessary connections between terms formulated in principles that commanded universal assent. Singular experiences were not public, but known only to a privileged few; consequently, they were not suitable elements of scientific discussion. Franciscus Aguilonius expressed clearly the Aristotelian perspective of the Jesuits in his 1613 treatise on optics:

> For a single [sensory] act does not greatly aid in the establishment of sciences and the settlement of common notions, since error can exist which lies hidden for a single act. But if [the act] is repeated time and

---

40. On the demonstrative regress, see chapter 1, section IV, above.
41. This last point is made strongly by Machamer, "Galileo and the Causes" (1978). See chapter 5, below, for a more extended examination of Galileo on these questions.

again, it strengthens the judgment of truth until finally [that judgment] passes into common assent; whence afterwards [the resulting common notions] are put together, through reasoning, as the first principles of a science.[42]

Aguilonius's remarks clearly appeal to Aristotle's definition of "experience" in the *Posterior Analytics:* "from perception there comes memory ... and from memory (when it occurs often in connection with the same thing), experience; for memories that are many in number form a single experience."[43] For Aguilonius, then, multiple repetition is essential to creating a proper scientific experience. It guarantees avoidance of accidental deception by the senses or the unfortunate choice of an atypical instance, and ensures a reliable statement of how nature behaves "always or for the most part," as Aristotle put it.[44] The result is experience adequate for the establishment of the empirical "common notions" that form the basis of a science.

The conception that Aguilonius here presents can, perhaps, best be explicated by a contrast with the modern hypothetico-deductive view of scientific procedure. Some version of the latter, whether confirmationist or falsificationist, would place experience, at least as regards its formal justificatory role, at the *end* of a logical structure of deduction from an initial hypothesis: the hypothesis yields conclusions regarding observable behavior in the world, and experiment or observation then steps in to confirm or falsify these predictions—and hence, in a logically mediated way, to confirm or falsify the original hypothesis itself.[45] A methodological Aristotelian, however, approached these issues in a quite different fashion. Since the point of Aristotelian scientific demonstration was to derive conclusions deductively from premises that were already accepted as certain—as with those of Euclidean geometry—there was no question of testing the conclusions against experience. The proper role for experience was to ground the assertions contained in the original premises, as Aguilonius assumed. Once they had been established, so too, from an empirical standpoint, had the conclusions potentially deducible from them.

---

42. Aguilonius, *Opticorum libri sex* (1613), pp. 215–216: "Non enim ad scientiarum primordia, communiumque notionum constitutionem, unicus actus magnopere iuvat; siquidem error huic subesse potest, qui lateat, at saepè ac saepiùs repetitus iudicium veritatis corroborat, quousque tandem in communem assensum transeat. unde [sic] posteà velut ex primis principiis scientiae per ratiocinationem colliguntur."

43. Aristotle, *Posterior Analytics* II.19, trans. Jonathan Barnes in Aristotle, *The Complete Works* (1984); see chapter 1, section III, above.

44. Aristotle, *Metaphysics* VI.2.

45. Classic expositions of variants of this picture are Nagel, *The Structure of Science* (1961); Popper, *The Logic of Scientific Discovery* (1959).

Jesuit philosophers required that a science fulfill a scholastic-Aristotelian conception of experience.[46] The insistence of the Jesuit mathematical school upon the scientific status of their own disciplines therefore obliged them to use only evident and manifestly universal empirical propositions. However, elements of the mathematical sciences themselves as they had been practiced since antiquity demanded that special accommodations be made in the strict Aristotelian framework.

II. Astronomy, Optics, and Expertise

The peculiarities of observational data in astronomy did not intrude themselves into considerations of the scientific role of experience so long as astronomy was regarded as a specialized mathematical discipline concerned with models for the computation of tables.[47] However, by emphasizing that mathematics was a part of philosophy, and by going so far as to promote astronomy itself to the position of a preeminent science, Clavius drew direct attention to its methodological form.

Mathematical, positional astronomy provided potential anomalies for an orthodox scholastic-Aristotelian view of the place of experience in a science. To accord with the formal model of a science, the empirical premises used in astronomical demonstrations would have to take the form of universal statements about how things are or behave in the heavens. In practice, however, astronomy employed data that consisted of discrete observations made at particular times and places with the aid of instruments. Thus astronomical phenomena were clearly not all known through common experience; they were not all, in that sense, *evident*. Precession of the equinoxes, for example, was a phenomenon that could only be constructed from discrete data collected over long periods of time.

Clavius, in keeping with his tendency to avoid head-on confrontations with methodological questions, left the matter alone; his use of

---

46. The ways in which Jesuit philosophers discharged their adherence to Thomism is discussed in Feldhay, "Knowledge and Salvation in Jesuit Culture" (1987).

47. Robert S. Westman has drawn attention to the importance of the disciplinary division between astronomy and natural philosophy in the sixteenth century. See especially Westman, "The Melanchthon Circle" (1975); idem, "The Astronomer's Role in the Sixteenth Century" (1980). See also the important essay by N. Jardine, "The Status of Astronomy," chap. 7 of his *The Birth of History and Philosophy of Science*. Paul Oskar Kristeller made some extremely prescient remarks on this matter in an essay originally published in 1950, "Humanism and Scholasticism in the Italian Renaissance" (1961), on pp. 118–119.

the word "experience" when discussing astronomy, however, while retaining the usual scholastic sense, hints at the complications in its function created by the role of singular experiences in astronomical practice. Regarding questions to do with the rotation of the heavens and precession, for example, Clavius concludes: "Wherefore faith is to be had in the experiences of astronomers, until something else is brought forward to the contrary by which it be demonstrated that what is propounded by astronomers concerning the motion of the stars from the west towards the east above the poles of the zodiac is not true."[48] The "experiences of astronomers" refers to their general accumulated experience in this matter rather than to a body of discrete observations, but the acknowledged possibility that "something else" could be "brought forward to the contrary" admits the practical dependence of astronomical doctrine on such observations. As with the problem of causality in mathematical demonstrations, a more explicit analysis of the role of experience in astronomy was left to Blancanus, in his *Sphaera mundi* of 1620.[49]

Sphaera mundi is an introduction to the elements of astronomy and cosmography that covers much the same ground as Clavius's commentary on *De sphaera*. Blancanus justified this apparent duplication in part by pointing out that the final edition of Clavius's work still failed to take into account the new telescopic discoveries.[50] His regard for the importance of telescopic observations indicates how the special features of astronomical practice could intrude into questions of scientific methodology: telescopic observations were produced by the use of special instrumentation not readily available to all. These "experiences" could certainly be expressed as universal statements—"Venus displays phases"; "the moon's surface appears rough"—but they lacked "evidentness" precisely because only telescopic observers were privy to them. There was thus a sense in which, for everyone else, they

48. Clavius, "In sphaeram Ioannis de Sacro Bosco commentarius," in *Opera mathematica*, vol. 3, p. 33: "Quare experientiis Astronomorum fides habenda est, donec in contrarium aliud quid afferatur, quo demonstretur, vera non esse, quae de motu stellarum ab occasu in ortum super polos Zodiaci traduntur ab Astronomis."
49. Blancanus, *Sphaera mundi* (1620).
50. Ibid., "Praefatio," 2d p.; cf. Clavius, "In sphaeram Ioannis de Sacro Bosco commentarius," in *Opera mathematica*, vol. 3, p. 75. For discussion of the meaning and reception of Clavius's remarks, see Lattis, "Christoph Clavius and the *Sphere* of Sacrobosco," pp. 282–285, 307–317. Clavius's remarks quickly became a standard point of reference: apart from the material in Lattis see, e.g., Scheiner, *Disquisitiones mathematicae*, pp. 50–51, quoting Clavius on the new demands that Galileo's discoveries have put on astronomers to save the phenomena.

stood *outside* the ordinary course of nature: this may be what Clavius had in mind when he called them "monsters."[51]

The key terms in the medieval and early-modern astronomical lexicon were *phaenomenon* and *observatio*, each corresponding to a Greek prototype found classically in Ptolemy's *Almagest*.[52] Ptolemy's usage was fairly straightforward: a phenomenon was any kind of appearance in the heavens, whether the path of a planet or an eclipse of the moon, and an observation was an act whereby a phenomenon became known through the senses.[53] Neither Ptolemy nor his successors had any methodological difficulties with this terminology, because they did not concern themselves with defending the scientific status of astronomical experience along Aristotelian lines. Since Blancanus did, however, his methodological orientation resulted in the presentation of a refined version of the two central Ptolemaic terms.

Following elementary material on the celestial sphere, *Sphaera mundi* gets under way with a section called "Sphaerae materialis et mundanae simul explicatio."[54] The section's opening chapter is headed "Suppositiones." This was a technical term that designated those things that, although not in themselves obvious (as were, in their different ways, definitions and axioms), needed to be accepted prior to the construction of a science. In the case of a subordinate science like astronomy, the results of the pure mathematical sciences on which it relied counted as *suppositiones* because, although they were demonstrable, they were not proved within the subordinate science itself. The Latin term *suppositio* was the equivalent of the Greek *hypothesis*, which was in turn Aristotle's word for the category found in Euclid as postulates.[55] Blancanus explains the suppositions used in astronomy:

---

51. See chapter 1, section II, above.
52. Pedersen, "Astronomy" (1978), is a general survey of medieval astronomy; the secondary literature in general tends to concentrate on either planetary models and tables or on instruments rather than on the interactions between instruments, data, and the process of modeling.
53. See, e.g., his usage in *Almagest* I.3, 4 (Toomer, *Ptolemy's Almagest*, pp. 38, 40). Cf. Joachim Jung's use of the same Greek term to designate "observation" in extra-astronomical contexts: above, chapter 1, n. 5.
54. Blancanus, *Sphaera mundi*, pp. 15 ff.
55. See on this Wallace, *Galileo and His Sources*, pp. 112–113, and, more fully, idem, *Galileo's Logic of Discovery and Proof* (1992), pp. 139–150. On postulates see Lloyd, *Reason, Magic and Experience*, chap. 2, and above, chapter 1. Blancanus seems to use *suppositio* somewhat less strictly than Paulus Vallius (Valla) in the material examined by Wallace. Baliani, *De motu naturali* (1646), p. 9, labels those principles underpinning his science of motion that are known through experience "suppositions," thereby distinguishing them from other "petitions" which, he says, serve construction, but are easy to do and under-

Besides those things supposed by astronomy that it has received from outside, both from geometry and from arithmetic (as has been said in the apparatus at the beginning), it also supposes other principles and, so to speak, foundations intrinsic and proper to it, which are indeed of two kinds, for astronomers call some "Phenomena," or "Appearances," because they appear and are manifest to all, even the vulgar, such as: the rising and setting of the stars, moon, and sun; that all stars move from the east to the west; that the sun moves lower in winter and higher in summer; that the sun does not always ascend from the same place on the horizon; and many other things of that kind which we suppose as very well known to all.[56]

Blancanus, then, unlike Ptolemy, defines "phenomena" not simply as appearances in the heavens, but as appearances that are known generally—they are evident, a part of common experience. That is what one should expect of the empirical suppositions of a true science; that is how they are justified. Blancanus then designates his second kind of internal principle in astronomy using the other central Ptolemaic term, "τηρήσεις, that is, 'Observations'." These are "particular concepts, provided by experiments, which do not become known to all as appearances do, but only to those who apply themselves zealously to the science of the stars with diligent work and with instruments skillfully designed for the purpose."[57] This privileged knowledge includes such things as the apparent diameters of the sun and moon, and the retrogradations and speeds of the planets.[58] Blancanus's terminological distinction thus expresses a distinction in cognitive status: phenomena are evident, while observations are recondite, because while phenomena are more or less given in ordinary experience, only expert astronomers, using their special instruments, are privy to observations.

Blancanus's account of "observations" derives from Ptolemy's de-

---

stand and therefore require no explicit consideration. Cf. on suppositions Aristotle, *Posterior Analytics* I.10.

56. Blancanus, *Sphaera mundi*, p. 15: "Praeter illa, quae extrinsecus accepta tam ex Geometria, quam ex Arithmetica, ut initio apparatus dictum est, supponit Astronomia; adhuc alia intrinseca, & sibi propria Principia, ac veluti fundamenta supponit, quae quidem duplicis sunt generis, alia enim appellant Astronomi Phaenomena, seu Apparentias, eò quod omnibus etiam vulgò appareant, ac manifesta sint, uti sunt; stellas, Lunam, & solem oriri, ac occidere: omnia sydera moveri ab Oriente in Occidentem: solem hyeme humilius incedere, aestate vero altius: non semper solem ex eodem horizontis loco ascendere: & alia id genus complura supponimus ceu cunctis notissima."

57. Ibid., pp. 15–16: "sunt autem cognitiones quaedam ab experimentis comparatae, quae non omnibus, uti apparentiae, innotescunt, sed ijs tantummodo, qui diligenti opera, atque instrumentis ad id artificiosè elaboratis, in stellarum scientiam naviter incumbunt."

58. Ibid., p. 16.

scription of his procedures in the *Almagest*.⁵⁹ Ptolemy used "observation" to refer to the gathering of discrete items of astronomical data rather than to the knowledge manufactured from them. Nonetheless, Ptolemy's account of the production, or construction, of a piece of astronomical knowledge from various sets of data accords exactly with Blancanus's description of the nature of an "observation": Ptolemy described how he had confirmed Hipparchus's belief that the stars within the zodiac maintain their positions relative to those outside it, thus showing precession to be common to all the stars. Hipparchus had formed his conclusion by comparing his positional data with those of earlier astronomers; Ptolemy then compared his own data with Hipparchus's in a similar fashion.⁶⁰

The difference between "phenomena" and "observations" according to Blancanus is not one between universal and discrete experiences: both are experiences expressed as universal statements. An "observation," however, is a universal experience that has been explicitly constructed, using appropriate computational techniques, from discrete experiences, themselves the product of deliberate instrumental manipulation. "Phenomena," by contrast, are universal experiences that do not need to be explicitly constructed, and which therefore fit most closely the Aristotelian definition of an experience. Blancanus's treatment of "observations" as a distinct class of astronomical suppositions addressed the requirements of a genuine science: the privileged experiences of expert practitioners, they created severe difficulties for any attempt to fit astronomy to the Aristotelian model of a science, and therefore required special handling. For anyone not so wedded to an Aristotelian approach, astronomy could appear in a quite different light, as Tycho Brahe exemplifies.

For Tycho, astronomy was not an Aristotelian science. Instead, it was akin to the secret art of alchemy. Like alchemy, its knowledge was private, acquired through personal experience and endeavor. Astronomy/astrology on the one hand and alchemy on the other formed a common pursuit; the astrological linkages between planetary influences and the generation of metals and other minerals on the earth led Tycho to refer to alchemy as "terrestrial astronomy."⁶¹ In 1598, prior to his move to Prague, he wrote regarding projected new astronomical

---

59. See Sabra, "The Astronomical Origin of Ibn al-Haytham's Concept of Experiment" (1971). The passage is Ptolemy, *Almagest* VII.1 (Toomer, *Ptolemy's Almagest*, pp. 321–322).

60. Sabra, "The Astronomical Origin of Ibn al-Haytham's Concept of Experiment," p. 134. On Ptolemy's use of observational data see also Lloyd, *Magic, Reason and Experience*, pp. 183–200.

61. See, e.g., Tycho Brahe, *Opera Omnia* (1972), V, p. 118; VI, p. 145; VII, p. 238.

instruments: "I shall hardly publish anything about these and similar matters that I have invented recently . . . nor about those that I shall invent in future. . . . But to distinguished and princely persons . . . shall I be willing to reveal and explain these matters when convinced of their gracious benevolence, but even then only on condition that they will not give them away."[62] Tycho's perception of his work as private was tempered only by his need for patronage. That astronomical practice lent itself so readily to such a view of knowledge indicates starkly the problems faced by Blancanus.

Jesuit work in others of the mixed mathematical sciences shows a similar concern with establishing and elucidating their methodological structure. Geometrical optics was at least as well established a discipline as astronomy, with a tradition that incorporated the texts of Euclid, Ptolemy, and Alhazen (Ibn al-Haytham). Together with Witelo's contributions in the thirteenth century, Alhazen's remained the major work on the subject at least until Kepler's *Ad Vitellionem paralipomena* of 1604.[63] In the second decade of the seventeenth century there appeared two optical works by Jesuits: the *Opticorum libri sex* of Franciscus Aguilonius in 1613 and Christopher Scheiner's *Oculus* in 1619.[64] Both involve the deliberate creation of instrumentally contrived experiences, something that had been a part of optical science since antiquity and was particularly prominent in Alhazen. Whereas in astronomical treatises the relationship between positional data and derived universal knowledge-statements could be characterized, as it was by Ptolemy, in terms of the comparison of data, optical practice placed a greater emphasis on the particular experiences themselves. Ptolemy's measurement of refraction bears an apparent similarity to astronomical measurements, but it is not typical of his usual approach to optics; furthermore, the data given appear to have been generated with the aid of a computational paradigm of second differences rather than by the bald measuring technique described by Ptolemy.[65] A. I. Sabra and S. B. Omar have each argued that it was Alhazen, not Ptolemy, who first incorporated the deliberate construction of empirical fact into op-

62. Ibid., V, p. 101, trans. adapted from Raeder, Stromgren, and Stromgren, *Tycho Brahe's Description of his Instruments* (1964), p. 101. See Christianson, "Tycho Brahe's German Treatise on the Comet of 1577" (1979); and especially Webster, *From Paracelsus to Newton* (1982), pp. 29–30; Hannaway, "Laboratory Design and the Aims of Science" (1986).

63. Lindberg, *Theories of Vision*, provides a thorough survey.

64. Aguilonius, *Opticorum libri sex*; Scheiner, *Oculus* (1619). Ziggelaar, *François de Aguilòn* (1983), part II, chaps. 1 and 2, examines the *Opticorum libri*. Further references on Scheiner may be found in Baldini, *Legem impone subactis*, p. 115 n. 107.

65. Smith, "Ptolemy's Search for a Law of Refraction" (1982).

tics and explicit optical methodology. As with astronomical data, Alhazen's optical experiences were constructed using special apparatus and were often quantitative.[66]

Alhazen did not employ the standard Arabic translation of the peripatetic (and Galenic) word *empeiria*, "experience." Instead, he used an Arabic astronomical term, one corresponding to Ptolemy's expression for a test or proof involving the comparison of distinct sets of data, akin to the example of Hipparchus and precession. This term, in its various forms, was rendered into Latin as *experimentum* or *experimentatio, experimentare,* and *experimentator.*[67] Alhazen's use of a word distinguished in meaning from the usual "experience" appears to have been deliberate. Sabra maintains that Alhazen's astronomical term retains the technical specificity of Ptolemy's concept of "test by comparison," and that he therefore intended just that notion.[68] Omar disagrees with Sabra, arguing that Alhazen meant "experiment" in the sense of a means of generating, not merely testing, hypotheses; in his support, Omar cites the way in which Alhazen's treatise develops its ideas in tandem with its described "experiments."[69] This interpretation seems doubtful, however: the presentation of material in a systematic treatise need bear little relation to any methodological "logic of discovery," and in any case such a procedure by Alhazen would be egregiously out of keeping with premodern ideals of science.

In fact, Alhazen's contrived experiences (called *experimenta* in the Latin translation) seldom correspond even to the astronomical usage to which Sabra etymologically traces them. Alhazen usually employs such an experience in order to show something to be the case, not to test a hypothesis or to derive sets of data for comparison. When discussing refraction between air and water (to choose a typical example), Alhazen provides, not a report of an attempted test of his claims, but instructions to the reader on how to experience the behavior he describes: "let him take a straight-sided vessel, such as a copper urn, or an earthenware jar, or something similar. . . ." Alhazen concludes by saying, "In this way, therefore, the passage of light through a body of water will be experienced."[70] The sense clearly indicates that the

---

66. Sabra, "The Astronomical Origin of Ibn al-Haytham's Concept of Experiment"; Omar, *Ibn al-Haytham's "Optics"*; see also Sabra, *The Optics of Ibn al-Haytham* (1989).
67. Sabra, "Astronomical Origin," esp. p. 133, referring to Risnerus, *Opticae thesaurus* (1572).
68. Sabra, "The Astronomical Origin of Ibn al-Haytham's Concept of Experiment," pp. 134–135.
69. Omar, *Ibn al-Haytham's "Optics,"* chap. 3, esp. p. 68.
70. Risnerus, *Opticae thesaurus*, pp. 233, 235: "Cum ergo experimentator voluerit experiri transitum luminis in aqua per hoc instrumentum: accipiet vas rectarum orarum, ut cadum cupreum, aut ollam figulinam, aut consimile . . ."; "Hac ergo via experi-

translation "experienced" rather than "tested" or "tried" is appropriate here: Alhazen has given a recipe allowing his reader to see what Alhazen himself already knows, not a protocol to provide the reader with an opportunity of checking his claims. The same point holds throughout. In introducing some theorems in catoptrics (to take another example at random), Alhazen remarks, "What we have said will be evident in spherical mirrors polished on the outside"—there is no sense of contingency about his assertions; he simply provides information about how things always happen.[71]

Although Alhazen used a special technical term, apparently drawn from astronomy, to designate an active production or construction of phenomena, he did not worry about the problems of casting optics as an Aristotelian science.[72] Methodologically, he saw himself within a properly optical tradition stemming from Euclid, who is his main source of references on such matters.[73] For Alhazen, contrived experiences were techniques; they did not constitute a cognitive category. Both Aguilonius and Scheiner, by contrast, wanted to establish optics as a science fulfilling Aristotelian canons: we have already quoted Aguilonius on the place of repeated trials in establishing empirical principles. However, Aguilonius gave no sign of seeing problems in the use of specially constructed experiences in scientific demonstration, despite his acknowledgment that, in his optical work, "I consider things intentionally altered by me," in addition to those provided by occasion.[74] Scheiner, the Ingolstadt astronomer and optician who became embroiled in a dispute with Galileo over sunspots,[75] was much more concerned with the implications of such an admission.

Scheiner begins his preface to the *Oculus* by categorizing optics in standard Aristotelian fashion, taking for granted its subordination to geometry.

---

mentabitur transitus lucis per corpus aquae." This was, of course, the version known to early seventeenth-century optical writers such as Kepler or the Jesuits.

71. Ibid., p. 126: "In speculis sphaericis extrà politis patebit quod diximus."

72. Smith, "Alhazen's Debt to Ptolemy's Optics" (1990), argues that Alhazen's characteristic mixing of mathematical, physical, and physiological concerns is no radical departure from the approach adopted by Ptolemy's *Optics*.

73. See, e.g., material in Risnerus, *Opticae thesaurus*, pp. 30–32. Rashed, "Optique géometrique et doctrine optique chez Ibn al Haytham" (1970), notes that Alhazen uses traditional optical terminology even while departing from Euclidean-Ptolemaic conceptions of the nature of optics.

74. Aguilonius, *Opticorum libri sex*, preface "Lectoris," 9th p.: "Consulto igitur consilio à me mutata res." Aguilonius often seems to follow Alhazen's view of optics, as in his stress on vision as an aspect of cognition—compare, e.g., ibid., preface "Lectoris," 1st p., and Book III *passim*, with Alhazen, book II, chap. 1, in Risnerus, *Opticae thesaurus*.

75. See chapter 4, below.

> Optics, truly and properly called a science,[76] has much distinct from, and much in common with, physics. Common are the object and the things foreknown. For both, as much physicists as opticians, are concerned with visible things and the organ of sight; however, in different ways. For geometry, as the Philosopher declares, 1.2. Phys.t.20. [*Physics* II.2], considers the physical line, but not insofar as it is [the physical line] of the physicist: optics, however, indeed considers the mathematical line, but not insofar as it is physical. They both [optics and physics] investigate the truth of the same thing, therefore, but in different ways.[77]

Both physics and optics, Scheiner continues, deal with those things which enter the senses:

> of which some, which come about naturally and are evident to everyone, and require only the attention of the sedulous investigator, are called "Phenomena" or appearances: others, which either do not occur or do not become evident without the industry of special empirics, are called "Experiences."[78]

Scheiner's dichotomy for optics matches Blancanus's for astronomy, with the slight difference that what Blancanus calls "observations" Scheiner calls "experiences." Scheiner's "experiences" need to be constructed by "special empirics," just as Blancanus's "observations" required the specialized work of astronomers. Since they would be couched in the form of universal statements (if only as conditionals), these optical "experiences" resemble Aristotle's general concept of "experience" sufficiently to allow Scheiner to invoke Aristotle's authority. Thus his description of the contents of Book I, Part II of the *Oculus* reads: "In the second part we bring forward experiences produced for the purpose, so that from them we might establish truth and rebut

---

76. Here a marginal reference to "Arist.I.Post.Text.30 & passim", as well as references to Pereira (an ironic choice), Villalpandus, and Blancanus's *De mathematicarum natura dissertatio*.

77. Scheiner, *Oculus*, "Praefatio," 1st p.: "Optice vera [reading as *Optica vere*] & proprie dicta scientia, multa seiuncta, multa cum Physicâ communia habet. Communia sunt obiectum & praecognita. Utrique enim tam Physici quam Optici circa visibilia, & organum visus versantur; modo tamen diverso. Geometria enim, teste Philosopho, 1.2. Phys.t.20. de Physica linea considerat, sed non quatenus est Physici: Perspectiva autem mathematicam quidem lineam, sed non quatenus Physica est. Veritatem ergo eiusdem rei ambo, sed viis diversis investigant."

78. Ibid.: "quorum alia quae ita contingunt ut naturâ fiant omnibusque obvia sint, solamque seduli speculatoris animadversionem requirant, Phaenomena, sive apparitiones: alia quae absque peculiari Empyrici industriâ aut non fiunt aut non patescunt; Experientiae vocantur."

errors. For one true experience, the Philosopher [i.e., Aristotle] declares, is worth more than a thousand deceitful subtleties of underhand reasons."[79]

Not only were Scheiner's "experiences" such that without the "industry of special empirics" they would fail to become evident; without it they might not occur at all. Indeed, Scheiner planned in his treatise to "bring forward experiences produced for the purpose" to establish his arguments, a point he elaborates upon in part II of book I, entitled "Experientiae variae" (regarding the operation of the eye).

> This other part of the first book is occupied with putting forward and explaining various recondite and well-tried experiments, most diligently investigated with singular industry, tenacious labor, and much exertion, and faithfully brought into the light from the hidden treasury of nature, so that from these as it were foreknown, undoubted [starting points], it may be permitted finally to reach into the true throne of the visual power without obscurities.[80]

Like Aguilonius, Scheiner is concerned with establishing "undoubted" principles as premises for scientific demonstration. In that connection, he speaks of putting forward recondite *experimenta*, not *experientiae*, and the distinction is a functional one embodied in the structure of the presentation. Thus the subsequent text commences with chapter I, "Experientia Prima. Pupillae variatio."[81] This chapter contains three sections, each headed "Experimentum" and consisting of a set of instructions whereby an aspect of the variation in size of the pupil can be evinced. Most of the other chapters in this part of *Oculus* are also labeled as *experientiae*; they deal with a number of matters apart from pupil size, such as the effect of different degrees of illumination on the apparent size of objects. Only the first chapter is broken down into sections explicitly labeled *experimentum*, but the other chapters that bear the heading *experientia* follow the same form, being composed of so-called *experimenta* similar in nature to those of Alhazen (Scheiner is in general covering material quite similar to Alhazen's). They are

---

79. Ibid., p. 1: "Parte secunda experientias pro re nata adferimus: ut ex illis veritatem stabiliamus, refellamus errores. Una enim vera experientia, Philosopho teste, plus valet, quam mille rationum subdolarum fallaces argutiae."

80. Ibid., p. 29: "Occupatur haec libri primi pars altera, in proponendis atque explicandis variis, reconditis, probatisque experimentis, singulari industria, labore pertinaci, sudore plurimo diligentissimè investigatis, atque ex abditis naturae thesauris in lucem fideliter protractis, ut ex ijs tanquam praecognitis indubitatis in verum Visiuae [sic] potentiae thronum absque ambagibus devenire tandem liceat."

81. Ibid., pp. 29–32.

presented in the form of instructions or in the form of a geometrical constructional problem (typically commencing "Sit . . . ").[82]

Scheiner's terminology is, therefore, quite clear: an "experience" is sensory knowledge about an aspect of the world which must be deliberately brought into being, or constructed. It is expressed as a universal statement, thereby corresponding to Aristotle's definition and so fit for use in scientific demonstration. An "experiment," by contrast, is a particular procedure whereby the experience may be instantiated. Thus the experiments are the *means* by which an experience is constructed; one can come into possession of the experience that the pupil contracts when exposed to bright light by performing a number of different experiments that contribute to its formation.[83] Blancanus had similarly described his universal "observations" as "particular concepts, provided by experiments," the latter here being instrumentally manufactured pieces of astronomical data. The contrast with Scheiner's other category, "phenomena," again as with Blancanus's, is that the individual instances constituting phenomena are not codified or made explicit, because they are presented routinely by nature.[84]

At the beginning of book I, part I, Scheiner had considered "the necessity of anatomical inspection concerning the eye." That necessity arose, he explained, from the requirement for firm premises on which to build optical science. As Clavius had said, every science demonstrates its conclusions from "particular assumed and conceded principles," but no science "demonstrates its own principles."[85] Scheiner needed, therefore, to justify his principles through experience, and that justification would, strictly speaking, be *outside* the science itself. He writes, "For our purpose it is not so much phenomena that are foreknown as singular experiments [*experimenta*] derived from study. They

82. Material running up to ibid., p. 52.

83. This terminological distinction, apparently quite consistently applied in the headings under consideration here, is not otherwise fully consistent even for Scheiner himself, however; to find the word *experimentum* where one might expect *experientia* is not uncommon in this work. A striking example appears in Scheiner's *Disquisitiones mathematicae* of 1614, where an item headed "De luce Solis, Experientia notabilis" (p. 72) appears, followed by another, "De igne Experimentum" (p. 74). These two items describe almost identical procedures, each given in universalized terms, involving the spreading out of light passed through holes; nonetheless, one is called *Experientia*, the other *Experimentum*.

84. There is another contrast with Aguilonius here, who, in dealing with this same question, adduces the way in which the pupil is known to dilate habitually in the gloom so as to admit more light and vice versa. He considers it necessary only to refer to common experience, not to formulate discrete, artificial procedures. Aguilonius, *Opticorum libri sex*, pp. 19–20.

85. See above, text to n. 34.

are of two kinds; one from inspection of the eye, the other selected from consideration of appearances diffused out from observable things into the eye. We will gather something from both."[86] What Scheiner calls *experimenta* are not necessarily artificial contrivances; the word refers equally to simple anatomical inspection. It should not, therefore, be confused too readily with the modern English "experiment," since *experimentum* in Scheiner's usage means, as we have seen, those items of empirical knowledge that go to make up a universal generalization—an authenticated experience that can act as a scientific principle. The two parts of Scheiner's book I detail such a procedure for experiences to do both with the eye and with reflection and refraction, making them available as principles for use in properly scientific demonstrations. In book II, the newly established principles are turned to account: propositions are demonstrated on the foundations of the experiences previously constructed. For example, demonstrations concerning the refraction of light at the surface of the cornea involve such forms as these: "Demonstratio II. Per Experientiam citati Capitis 8"; "Demonstratio III. Sumitur ex Experientia 4.c.4.p.2.1.1"; "Demonstratio IV. Obvia est ex Experientia 8. Capite II allata."[87] The experiential roots of the science of optics remain constantly on view.

Thus Scheiner distinguished constructed "experiences," which required the work of "special empirics," from the evident "phenomena" that ought to have been sufficient in an ideal Aristotelian universe but were not in the real one, just as Blancanus distinguished between constructed "observations" and evident "phenomena." Furthermore, Scheiner's "experiences" were constructed from "experiments" just as Blancanus's "observations" were made from "experiments" in a way that echoed the conceptualization (and corresponding attributed Latin terminology) of Alhazen.

At the end of the century, Étienne Chauvin's *Lexicon philosophicum* (1692 and 1713), under the heading "Experientia," displayed a similar distinction as one that had evidently by then become a commonplace. Experience, he says, holds a place among physical principles second

---

86. Scheiner, *Oculus*, p. 1: "Praecognita ad institutum nostrum non tam sunt Phaenomena, quam experimenta singulari studio hausta, eaque duplicis generis; altera ex oculi inspectione; altera e specierum a rebus aspectabilibus in oculum diffusarum consideratione desumpta. De ambobus nonnulla delibabimus." Things "foreknown" (*praecognita*), like *suppositiones*, need to be accepted by a science at the outset, prior to the production of scientific demonstrations; however, unlike *suppositiones* as a general category, this term implies consideration of the grounds on which they are accepted. See Wallace, *Galileo's Logic of Discovery and Proof*, pp. 149–150.

87. Scheiner, *Oculus*, p. 79.

only to reason, "for reason without experience is like a ship tossing about without a helmsman." There are three kinds: the first is the experience that is acquired unintentionally in the course of life; the second is the kind gained from deliberate examination of something but without any expectation of what might happen; and the third is experience acquired deliberately so as to determine the truth or falsity of a conjectured cause (*ratio*). The second and third are the most useful philosophically, especially the third, which determines straightaway the truth or falsity of an opinion.[88] He proceeds to elaborate on what characterizes a properly philosophical experience: it should be based on *experiments* (*experimenta*), of varying kinds and of considerable number, since "those err the most who refer the whole basis of their philosophizing to a few experiments over which perhaps they have reflected studiously," chemists being Chauvin's prime example.[89] Furthermore, such experiments should encompass mechanical artifice as well as natural history. Thus an experience is made from numerous experiments, as a kind of hybrid of the new and the old—it is experiments that stand for the Aristotelian "many memories of the same thing," as Scheiner and Blancanus had exemplified near the century's beginning.[90]

III. Recipes and *Problemata*

Jesuit mathematicians, then, formulated and tried to codify the methodological problems involved in making the mixed mathematical disciplines into genuine Aristotelian sciences. The central problem—that of establishing the principles from which scientific demonstrations could proceed—admitted a welcome degree of flexibility. The principles of a true science had to be evident and acceptable as true, but the means by which that acceptability should be established fell outside the formal procedures discussed in the *Posterior Analytics*. There were some restrictions, in that principles should be primitive and indemonstrable (with the special provisions for subordinate sciences and their borrowed suppositions providing a partial exception); otherwise, if a purported empirical principle could be made evident, however that might in practice be accomplished, then it was fit for use in scien-

---

88. Chauvin, *Lexicon Philosophicum* (1713/1967), p. 229 col. 2: "est enim ratio sine experientiâ velut navis sine rectore fluctuans."

89. Ibid., p. 230 col. 1: "ut illi maximè fallantur, qui ad pauca experimenta, in quibus forsan studiosè versati sint, omnem philosophandi rationem revocant." Francis Bacon had criticized such people as William Gilbert for doing just this, in Gilbert's case building a philosophy from magnetism: Bacon, *Novum organum* 1, aph. 54.

90. And cf. chapter 1, section III, above.

tific demonstration. The very imprecision of Aristotle's talk of "induction" indicates the wealth of possibilities open to those who sought new ways of grounding a science of nature.

The classifications used by Blancanus and Scheiner allowed that certain kinds of supposition were only truly evident to specialized investigators operating with tools designed for the purpose. In the purely pedagogical context such suppositions might be treated as *petitiones*, premises voluntarily granted on the word of the teacher, but such a recourse was inadequate to establish scientific status.[91] This was empirical knowledge that depended on the diligence and expertise of observers and experimenters, with their specialized instruments and skills; it was not a straightforward matter for others to reproduce for themselves the experiences claimed by astronomers or opticians. How could "experiences" be established as common property if most people lacked direct access to them?

A partial solution to the difficulty was to present the material from which the "experience" itself was formulated, although this of course only set the problem back a stage. Blancanus described his astronomical "observations" as being provided by "experiments"—that is, developed from the comparison of observational data on the positions of celestial bodies. Thus the presentation of the observational data itself could render more credible the observations resting upon them. Astronomical tables were also equivalent to observations, to the extent that they derived from predictive models that themselves depended on observational data. They could thus serve a similar purpose: the gradual realization of the tables' predictions provided a continuing opportunity for experiencing the planetary motions embodied in the models.[92] In optics, Scheiner similarly used experiments as the underpinnings of his scientific experiences, allowing the detailed description of experiments—accounts of what happens when certain situations are created—to render more immediate the properties of nature purportedly determined by them.

A common technique in optics for rendering an experiment immediate was, as we have seen, to present it in the form of a recipe. The reader is instructed to perform a series of operations and then told what outcome will result, or else (more revealingly) the series of operations is presented in the subjunctive mood of a geometrical *problema*.[93]

91. On *petitiones*, see Wallace, *Galileo and His Sources*, p. 113.
92. Cf. chapter 4, section I, below, for more on this point.
93. Cf. above, text to n. 82.

This literary device afforded to experiential claims a status that transcended what appears to us as an obvious difference between geometrical constructions and novel empirical findings: the latter's unpredictability. A geometrical construction is transparent, because in following its steps one sees the outcome generated inevitably before one's eyes; even recourse to compass and ruler is unnecessary. To render evident the proper result of a contrived empirical situation, by contrast, one would need to manipulate specialized apparatus with appropriate skill. However, the geometrical literary structure did in fact serve to accord a sort of transferred transparency to described experimental procedures. Procedure and outcome appeared formally inseparable: optical experiments, for example, detailed procedures for generating those conditions under which particular effects would be manifest— such as a specific change in the apparent position of a pointer in refraction. As constructional techniques, empirical descriptions were justified by their intended outcomes; there appeared to be no question of the outcome differing from the one presented. Just as constructions in geometry were generated from postulates that expressed conceded possibilities, so the use of a geometrical paradigm served to re-create unfamiliar experience by generating it from familiar experience—that is, easily picturable operations.[94] Steven Shapin has described the reading of Robert Boyle's detailed experimental narratives as "virtual witnessing";[95] a broadly similar characterization applies here, but with one crucial qualification. There is indeed a real difference between geometrical construction and novel empirical outcome, but these contrived situations were not intended to appear *novel*.

Indeed, the avoidance of apparent novelty, and hence disputability, in the presentation of empirical suppositions even allowed Aguilonius, in his work on vision, to dispense with the kind of detailed protocols that Scheiner provided.[96] By the same token, Blancanus described the use of an apparatus to show the rarefaction and condensation of the air that was eminently presentable precisely because it was not crucial. He notes that although this behavior of air "is established by many experiences" (which he regards as unnecessary to recount), "it is however agreeable to bring forward now a more beautiful as well as most evident one: let a glass flask be constructed, as you see in the figure,"

---

94. See, for more on this point in regard to mathematical postulates, chapter 8, section II, below.

95. Shapin, "Pump and Circumstance" (1984); see also Shapin and Schaffer, *Leviathan and the Air-Pump* (1985), chap. 2.

96. Cf. n. 84, above.

and so forth.[97] If claimed results were neither especially surprising nor overly recondite, inherent plausibility reinforced the persuasive effect of the overall presentation. The efficacy of this technique was enhanced if no controversial issue rested on the claimed results: as long as not too much rested on any particular procedure, the geometrical form sufficed.

Jesuit mathematicians in the early seventeenth century, then, employed techniques designed to incorporate recondite, constructed experiences into properly accredited knowledge about the natural world. The norms of scientific procedure to which they adhered derived from the Aristotelian model employed by Jesuit natural philosophers and logicians, and their incentive was the attempt, initiated by Clavius, to raise the status of the mathematical disciplines and their practitioners in the Jesuit academic system. The traditions of practice in astronomy and optics had exploited particular ways of generating and using experiential data, ways that created problems for the characterization and presentation of those subjects as sciences according to Jesuit philosophical criteria. Blancanus and Scheiner make clear the point that much of the experiential basis of astronomy and of optics was manufactured by expert practitioners and could not easily be transformed into the evident, universal experience that would provide adequate principles for a true science. Scheiner's approach to optics differed somewhat from Aguilonius's, although the latter shared in many ways Scheiner's Aristotelian methodological ideals. Aguilonius's lack of concern with some of the methodological specificities involved in converting the Latinized optics of Alhazen into a science that could meet the objections of certain Jesuit philosophers seems to have stemmed from his disregard for the dimension of *discovery*. The foundations from which his science develops are fairly fixed, whereas Scheiner's leaves open the possibility of making, and rendering "evident," new empirical truths—not just new formulations or new inferences.

Thus began a reformulation of the criteria by which specialized empirical suppositions—in Blancanus's astronomical terms "observations"—in the disciplines of mixed mathematics could be judged evident. Astronomical data and accounts of instrumental techniques, or procedures detailed as instructions or quasi-geometrical constructions,

---

97. Blancanus, *Sphaera mundi*, p. 111: "quod etsi multis constet experientijs, libet tamen pulcherrimam nunc aeque ac evidentissimam afferre: construatur, vitrea ampulla, uti in figura vides." Note that Blancanus refers to the kinds of procedures that Scheiner calls "experiments" as "experiences"; clearly the terminological usage cannot be taken as a clear or invariable indicator of epistemological status.

went some way towards making private knowledge-claims into public, self-evident truths. The closer that such presentations came to appearing like thought experiments—procedures that will, with correct presentation, create conviction of the truth of an untried outcome—the closer to proper scientific suppositions became their conclusions. Controversy, however, or the threat of controversy, demanded more radical measures. New means of employing experience then led towards a characteristically seventeenth-century—and modern—notion of the experiment as historical event.

Three

EXPERTISE, NOVEL
CLAIMS, AND
EXPERIMENTAL EVENTS

I. Mathematical Form, Witnesses, and Novelty

The previous chapter showed how Jesuit academic culture at the beginning of the seventeenth century included a place for mathematical sciences that involved a unique methodological conceptualization of the place of experience in the making of scientific knowledge. The crucial notion may be characterized as that of expertise, whereby the experience of a specialist acted as a legitimate surrogate for common experience, which the dominant Aristotelian model required for the creation of the evident premises of true scientific demonstration. Jesuit mathematicians formulated the recondite, specialized experiences of astronomy and optics by using geometrical forms and operational instructions to the reader to expand the available stock of common experience in a way analogous to the role of the traditional commentary form in natural philosophy; their ability to do this rested on their credibility as competent practitioners of their specialties. The location of a statement of experience in a text bearing the marks of authority and laying claim to a standard place in the body of literature on its subject allowed the unquestioned acceptance of the experience, but only if it was not seen to *depend* on a specific textual citation. That is, the forms considered in chapter 2 for establishing statements of experience could not function successfully in the event of challenges.

New experiences of nature were proposed at a much faster rate in the seventeenth century than before—the Torricellian experiment, the moons of Jupiter, sun spots, the circulation of the blood, the differential refrangibility of light—and this novelty generated problems for the establishment of knowledge. New phenomena, by definition, had been unavailable to common experience, and using them to make knowledge within a scholastic idiom therefore required much work in restructuring that experience. Niccolò Cabeo's work on magnetism provides an illuminating example of the handling of experience in one of these areas of novelty. His technique may be located squarely within the generic constraints of mixed mathematics.

Cabeo, a Jesuit teacher of natural philosophy and of mathematics, approached the textual constitution of experience through an explicit borrowing of the techniques of the mathematicians.[1] In the *Philosophia magnetica* of 1629,[2] he insists that he is doing physics—natural philosophy—rather than mathematics; nonetheless, he quite deliberately borrows the mathematicians' rhetorical resources for his own ends.[3] Book III of his treatise, on the directional property of magnets, opens with a chapter discussing "the method and manner of approaching this book."[4] It notes that the proper way to investigate the nature of magnets is to use Aristotle's demonstrative regress.[5] The previous books, Cabeo says, had proceeded *ex signis*, that is, from the effects or appearances; they also included a posteriori arguments that attempted to work backwards from such appearances to causal explanations. Now that material would be integrated.

> This, furthermore, will be the method of explication: I will present as a title the effect itself by the clearest, briefest words I can; generally in that way whereby mathematicians are wont to present their propositions, which they then assemble in demonstrative form: next, I will begin to explicate the matter with additional figures where needed, and in an uncomplicated manner. By a clear method, and with confirmed experience, I will attempt to deliver the reason of the effect from the proper principles of the magnetic philosophy.[6]

---

1. On Cabeo and his book, see Thorndike, *A History of Magic and Experimental Science*, vol. 7 (1958), pp. 267–269; Pumfrey, "Neo-Aristotelianism and the Magnetic Philosophy" (1990), on pp. 181–183; and biographical sketch in Sommervogel, *Bibliothèque de la Compagnie de Jésus* (1890–1932/1960), q.v.
2. Cabeo, *Philosophia magnetica* (1629).
3. For the emphasis on physics, see e.g. ibid., pp. 69, 77.
4. Ibid., pp. 197: "Ratio, & forma addendi hunc librum."
5. Ibid., p. 198.
6. Ibid.: "Ea poro erit explicandi ratio: proponam quasi titulum ipsum effectum verbis quantum potero clarissimis in sua brevitate; ea ferè ratione, qua solent mathematici suas propositiones praemittere, quas deinde colligunt habita demonstratione: mox rem incipiam explicare additis figuris, ubi opus fuerit, & expressis rerum formis. Explicate ratione, & probata experientia, eius rationem reddere tentabo ex proprijs magneticae philosophiae principijs."

The value of geometrical form in natural philosophy was as a means of borrowing the supposed clarity of geometrical demonstration. The "proper principles of the magnetic philosophy" are summaries of experience that the demonstrative regress helped to warrant, but which the geometrical exposition made openly accessible.[7]

Indeed, as a scholastic physical treatise Cabeo's work is somewhat heterodox in format, being a systematic topical treatise rather than a commentary.[8] Cabeo sought properly physical explanations for effects, but he was not averse to mathematical reasoning, and he accepted that it was peculiarly conclusive. After discussing electrical attraction and presenting his effluvial explanation of it, for instance, he recognized that there were still some difficulties with his ideas. If anyone should come up with a better explanation, he would willingly change his own—with the exception of whatever was demonstrated "mathematically."[9] Cabeo respected the solidity of mathematical demonstrations even though they failed to get at the more difficult matters of physics, and the mixed mathematical sciences provided him with a paradigm for the use of experience in scientific demonstration.

Like the electrical effect, magnetic phenomena (except in their crud-

---

7. On the demonstrative regress, see chapter 1, section IV, above. Cabeo's talk of "proper principles" refers to the standard Aristotelian requirement that each science rest on its own unique principles so as to avoid the logical difficulties that would result from the mixing of categorically different subject matters: see chapter 2, section I, above, and Funkenstein, *Theology and the Scientific Imagination* (1986), pp. 35–37, 303–307. Cabeo's invocation of the mathematical model of demonstration in discussing magnetism had been anticipated by William Gilbert, *De magnete* (1600), preface, 2d p.: "And even as geometry rises from certain slight and readily understood foundations to the highest and most difficult things, whereby the ingenious mind ascends above the aether: so does our magnetic doctrine and science in due order first show forth certain things of a less rare occurrence; from these proceed more extraordinary ones; at length, in a sort of series, are revealed things most secret and privy in the earth, and the causes are recognized of things that, in the ignorance of those of old or through the heedlessness of the moderns, were unnoticed or disregarded" (adapted from Gilbert, *On the Loadstone* [1952], p. 1): "Et veluti geometria à minimis quibusdam & facilimis fundamentis, ad maxima & difficillima assurgit; quibus mens ingeniosa, supra aethera scandit: ita doctrina nostra & scientia magnetica, ordine convenienti, quaedam primùm ostendit minùs rara; ab illis magis praeclara emergunt, tandemque serie quâdam, globi telluris arcana maximè, & abdita reserantur, & eorum causae agnoscuntur, quae vel priscorum ignorantiâ, vel recentiorum negligentiâ, incognita & praetermissa fuerunt."

8. Cf. on this point Baroncini, "L'insegnamento della filosofia naturale nei collegi italiani dei Gesuiti" (1981), esp. pp. 170–174; also Heilbron, *Electricity in the 17th and 18th Centuries* (1979), pp. 109–110.

9. Cabeo, *Philosophia magnetica*, p. 195: "ista, ut mathematice demonstrata profero." On the effluvial theory of electricity see Heilbron, *Electricity in the 17th and 18th Centuries*, pp. 180–183.

est manifestations) were by no means a part of everyday experience; they required precise experimental contrivance. More obviously than in most cases, therefore, the evident character of the phenomena was compromised and the authority of the reporter brought into prominence. Cabeo tries to deal with the difficulties in his "Praefatio ad Lectorem":

> I have delayed the reader in the very antechamber of the treatise, moreover, to advise him that I am about to bring forward nothing in these magnetical disputations which, so far as I have been able, I did not confirm by experiments again and again, even with others gathered to watch, so that not only would I have many witnesses of the truth, but I would remove myself from suspicion of error, as long as many most attentive people also observe the experiment.[10]

In this passage Cabeo is referring chiefly to effects that he wishes to explain through appropriate physical principles, using the demonstrative regress, rather than to the establishment of the principles themselves. However, his attitude towards experience implicated the same issues of evidentness: he could not make demonstrative knowledge of effects if their reality was in question; the demonstrative regress itself depended on it.

First, Cabeo says that he repeated his experiments many times, in effect creating an experience from individual sense impressions (Scheiner's "experiments") in the proper Aristotelian, and Jesuit mathematical, fashion. Second, he vitiates the apparently private nature of that experience by calling on witnesses whose implied assent helped to constitute it as shared or public. Third, Aguilonius's requirement that experiences be based on many instances because the senses sometimes err was additionally met by the multiplication of sensory impressions from numerous observers. Cabeo thus uses testimony to warrant discrete experimental events as integral parts of the methodological grounding for his "magnetical philosophy." That development was made possible by the prior use, in the mathematical sciences, of accounts of privileged experience (the description of experimental procedures, for example, or the citation of astronomical data) to ground the less-than-evident claims associated with the constructed universal ex-

---

10. Cabeo, *Philosophia magnetica*, "Praefatio," 3d p.: "In ipso autem tractationis vestibulo Lectorem admonendum duxi, me nihil allaturum in hisce magneticis disputationibus, quod, quantum per me licuit, experimentis semel atque iterum non comprobarim, alijs etiam ad spectaculum convocatis, ut non solum plures haberem veritatis testes, sed me ipsum ab erroris suspicione subducerem; dum curiosius plures idem spectant experimentum."

periences routinely used in those disciplines—Blancanus's "observations" and Scheiner's "experiences." The scientific imperative that experiential claims be evident—and be *seen* to be evident—thus led to the development of new techniques of justification centered on historical accounts of legitimatory events. An area that came to receive particular attention both because of its relative novelty and its apparent close association with the subject matter of the mathematical sciences concerned falling bodies—one of Galileo's "new sciences."

II. Arriaga and Cabeo on Falling Bodies

The novelty of Galileo's claims about freely falling bodies depended for its recognition on the preexistence of a piece of learned common experience, perceived through the structure of Aristotelian physics, whereby "everyone knew" that heavier bodies fall faster than lighter ones. In 1632, independently of Galileo's work but clearly responding to the increased currency of related ideas, there appeared a discussion of falling bodies in the *Cursus philosophicus* by the Spanish Jesuit Roderigo de Arriaga, a teacher at the college in Prague. This work became one of the most widely known scholastic textbooks of the seventeenth century and was often reprinted, going through several editions.[11]

The proposition that "all heavy bodies fall downwards by themselves equally" appears as a subheading to a set of "disputations" on Aristotle's *On Coming-to-Be and Passing-Away*.[12] The difficulty of Arriaga's task appears from his admission that the proposition flies in the face of received opinion:

> Hitherto, nothing has been so generally accepted among philosophers and other men as that bodies that are heavier consequently descend faster to the earth; to doubt concerning which was once a heinous thing. Hence, again, it was eminently fixed in the minds of all that a stone moves faster at about the middle [of its flight] than in its initial motion.[13]

---

11. Sommervogel, *Bibliothèque de la Compagnie de Jésus*, q.v., is as always a useful reference; see also Thorndike, *A History of Magic and Experimental Science*, vol. 7, p. 399; Schmitt, "Galileo and the Seventeenth-Century Text-book Tradition" (1984), on p. 223.

12. Arriaga, *Cursus philosophicus* (1632), p. 582 col. I: "Omnia gravia aequaliter per se cadunt deorsum." Aristotle's treatise bears the by-then usual humanistic Latin title *De ortu et interitu* in place of the medieval translation *De generatione et corruptione*.

13. Arriaga, *Cursus philosophicus*, p. 582 col. I: "Nihil hucusque inter Philosophos ac reliquos homines ita receptum, quàm corpora, quò sunt graviora, eò velociùs descendere ad terram, de quâ re dubitare, olim fuisset nefas. Hinc rursus altè omnium animis insitum fuit, velociùs lapidem circa centrum moveri, quàm in initio motus."

68   Chapter Three

In stressing the universality of these convictions, Arriaga attempts to dramatize his own dissent from them rather than to establish their solidity. This approach is the opposite of what one might expect from normal scholastic practice. Instead of its being a weighty argument for the true state of affairs embodied in common experience, Arriaga presents the general opinion as a kind of natural curiosity: "It is wonderful, inasmuch as this matter can be so easily tested by experience, that never in all the centuries have [philosophers] tested it even once, as they would have discovered thereby the falsity of their opinion."[14]

Like Cabeo, Arriaga now proceeds to accredit his assertions by providing a dialectical context: "A few years ago some people doubted that two globes equal in shape and size fall equally from a certain distance even if the [lighter] one is unconnected to the heavier, and likewise for any other shape." But now, Arriaga says, "I not only find this in [bodies of] equal shape and size, but in [all] bodies whatever, however much unequal in form, shape, or size, so that one dry piece of bread crust of two inches fell equally speedily from a very high place as a stone that I was able only with difficulty to hold with my hands." Arriaga makes his universal experiential claim ("I find this") and only subsequently adduces a specific example by way of illustration. Even there, the establishment of a kind of universality is implied for the specific bread/stone experience: Arriaga states that he has tried this "not once or twice, but often," so that his account is explicitly subsumed to a normal state of affairs known to him through experiential familiarity. The requisite universality is warranted by an extension even beyond Arriaga's personal experience itself: not only has he tried the phenomenon often, but also, like Cabeo's work, "in the presence of many people." It has thus become quasi-public. Arriaga concludes:

> The same [holds] even more for a small stone together with any rock: indeed, a thick piece of paper, compacted from six or seven others into a flat and round shape and very small, falls as fast as a big stone. What philosopher (if experience and eyes were not witnesses greater than any objection) could say this without being considered foolish? Perhaps the same would happen in many received opinions, if we could examine them by experience itself.[15]

---

14. Ibid.: "Et mirandum est, quòd cùm res haec experientiâ tam facilè comprobari possit, numquam per tot saecula vel semel id experti sint, ut inde suae opinionis falsitatem deprehenderent."

15. Ibid.: "Dubitarunt [*sic:* dubitaverunt?] hac de re aliqui ante aliquot annos in duobus globis aequalis figurae & magnitudinis, qui ex quacumque distantiâ aequaliter cadunt, licet sit unus decuplo altero gravior; & idem est de qualibet aliâ figurâ: ego verò

Arriaga thus attempts to invest his claims with credibility by stressing the frequency with which he has experienced the relevant behavior, as well as its sharing with others. His strategy suits well the Aristotelian methodological context in which it is embedded. Arriaga's discussion is designed to create conviction, or at least a willingness to lend credence, in readers who have certain expectations about how experience ought properly to be construed and utilized in the establishment of a knowledge-claim. Above all, Arriaga does not find it necessary to provide a narrative of a discrete historical event.

When, in 1646, Cabeo came to address precisely the same question in his wide-ranging commentary on Aristotle's *Meteorology*, he presented his experience in a form similar to that of the *Philosophia magnetica* of almost twenty years earlier and very close to that used by Arriaga. The full title of Cabeo's commentary promises to consider not just meteorology "first from the sayings of the ancients, then chiefly from experiments of singular things," but also "almost the whole of experimental philosophy." [16] Thus, in book I, he asks "whether the speed of all falling bodies be equal." [17] Specifying that this does not include bodies that get blown about in the air, like feathers or paper, Cabeo sets about justifying an affirmative answer.

> Let there be, first of all, two weights of the same kind, such as two pieces of lead, either having exactly the same figure (such as both being spherical) or not, which drop simultaneously from a high place; I say that, physically, they reach the ground from no matter what altitude at the same time: both I myself and others also have tried this

---

non solùm in aequali figurâ & in magnitudine, sed in quibuscumque corporibus, quantumvis formâ, figurâ, magnitudine inaequalibus, idem deprehendo, ita ut unum siccum corticem panis duorum digitorum, cum saxo quod manibus vix tenere poteram, ex alto valdè loco aequè citò cadere, non semel aut bis, sed saepè expertus coram multis sim. Idem à fortiori de exiguo lapillo cum quocumque saxo: imò chartam crassam, ex sex vel septem aliis compactam in figurâ planâ & rotundâ ac valdè exiguâ, tam citò cadere quàm magnum lapidem. Quis hoc Philosophorum diceret (si experientia & oculi non essent testes omni exceptione maiores) qui non putaretur stultus? Ita fortè in multis receptis opinionibus contingeret, si eas experientiâ ipsâ examinare possemus." This kind of presentation should not be regarded as original with Arriaga: for late sixteenth-century examples see Wallace, "Traditional Natural Philosophy" (1988), in notes on pp. 222, 223: Giuseppe Moleti ("fatta la prova, non una volte, ma molte"); Girolamo Borro ("non semel, sed saepenumero"); Galileo, from "De motu" of c. 1590 ("de hoc saepe periculum feci"). All three are discussing falling bodies.

16. Cabeo, *In quatuor libros Meteorologicorum Aristotelis commentaria* (1646); the title's continuation includes: "Quibus non solum meteorologica, tum ex antiquorum dictis, tum maxime ex singularum rerum experimentis explicantur sed etiam universa fere experimentalis philosophia exponitur."

17. Ibid., Lib. 1, p. 97 col. I: "An omnium cadentium velocitas sit aequalis."

by many experiments, and I have always observed [the weights] to fall in an exactly equal time, even if one be of one ounce and the other fifty.[18]

Like Arriaga, Cabeo stresses that other people too have tried the experiment many times, not just he alone. He also avoids narrating specific trials, framing his claims as already universalized statements of experience: "they reach the ground"; "I have always observed," and so on. The scientific "experience" subsumes the frequently repeated individual instances in the usual Aristotelian fashion. Cabeo takes care, however, to make clear the precise physical conditions underlying his assertion. He acknowledges that, so far, "perhaps this experiment might give someone well-founded doubt," because it only concerns two bodies of the same material—scarcely a sufficient ground on which to support a proposition that compares the speeds of *all* falling bodies. That insufficiency would not apply to an experiment "in which one might see not only two balls of lead greatly different in weight, but also balls greatly different in material, such as lead and wood, and disparate in figure, such as squared, or pyramidal, and rotund." Then, Cabeo suggests, no doubt could be entertained if, from a high place, and falling through a tranquil sky, "however much the difference in weight, it were impossible to note a sensible difference in the time at which they struck the ground."[19] Cabeo goes on to give the reader instructions on how best to carry out such experiments, describing the appropriate mode of procedure in a way long familiar in astronomy (classically, Ptolemy's descriptions in the *Almagest* of astronomical instruments and their use) and in optics.[20] Finally, Cabeo adduces a rational explanation as an auxiliary to his empirical claims. At least for the equality of fall of homogeneous bodies, one could con-

---

18. Ibid., p. 97 cols. I–II: "Sint primo duo gravia eiusdem rationis, ut duo plumbea, sive omninò similem habeant figuram, ut quod ambo sint sphaerica, sive non, quae simul ex edito loco decidant; dico simul physicè ex quaecunque altitudine ad terram pervenire: hoc multis experimentis, & ego ipse sum expertus, & alij etiam experti sunt, & semper omninò aequali tempore descendere depraehendi, etiam si unum esset unius unciae, alterum quinquaginta."
19. Ibid., p. 97 col. II: "Verum fortasse alicui suspitionem afferret, hoc experimentum, ex eo, quod videat non solum duos globos plumbeos, valde impares in pondere. Sed etiam globos valde impares in materia, ut plumbeum, & ligneum, & dispares in figura, ut quadratum, seu piramidale, & rotundum, si simul ex edito loco, tranquillo Coelo cadant, ambo simul ad terram pervenire, ita ut quantumcunque sit discrimen ponderis, non possit notari sensibile discrimen temporis, quo ad terram allidunt."
20. See chapter 2, section III, above.

sider the point that dropping one, two, or ten identical balls together would make no difference to the rate of fall of each; therefore one should expect no difference if the balls were consolidated together to make a single large one. But still, this reason by itself would not be persuasive in the absence of the experiment.[21] The experiment yields a "constant and certain" perception of the equality of time of fall.[22] Notice, however, that Cabeo, in the passage quoted earlier, specified that bodies of differing weight reach the ground at the same time "physically"—not necessarily "mathematically."

Neither the activities of Arriaga and his associates nor those of Cabeo and his were in themselves capable of creating scientific knowledge. Lived experience and formal knowledge met through the medium of language, and the language by which Arriaga and Cabeo formulated their knowledge made appeal to the familiarity of a scholastic rhetoric of experience. Mere utilization of the appropriate rhetoric was not, of course, in itself sufficient to guarantee acceptance of their claims. The work involved in creating self-evidence, so as to tap the reader's sources of conviction, could always be undermined by a determined opponent.

III. Riccioli versus Arriaga and Cabeo

In 1651, in his great work of astronomy and cosmography grandly titled the *Almagestum novum*, the Jesuit astronomer Giovanni Battista Riccioli rejected the assertions of Arriaga and Cabeo that the speed of falling bodies is independent of weight. Undaunted by the growing number of supporters of the proposition (much as Arriaga had been undaunted by the previous uniformity of its denial), Riccioli attempted to discredit the experiences on which they relied. He starts with Cabeo, paraphrasing him in this manner:

> ... he affirms most emphatically from his own often repeated experiments, that if two balls are dropped at the same time from the same height, one of one ounce and the other of ten pounds or whatever greater weight, either both being of lead, or one lead and the other stone or wood; providing that the air is still, and that the lighter one is not of so small a weight that it is not forceful enough to overcome the resistance of the air, or tosses about in the breeze (like a feather

---

21. Ibid., p. 97 col. II; the reason is a variant on the familiar Benedetti-Galileo argument.
22. Ibid., p. 98 col. I: "perceptum certum, & constans."

or piece of paper), it will happen that both reach the ground at the same moment, and no difference can be detected by the senses in the fall: from which he infers the speed of all falling bodies to be equal among themselves.[23]

Now comes Riccioli's counterattack. Cabeo has, quite appropriately, stressed that he tried these things often; Riccioli wishes to weaken that powerful commonplace by suggesting that Cabeo's resultant expertise was not, in fact, suited to settling the question at issue. "However," he says, "I do not know from what height he released those balls." This was to be a telling point. But its full effect is achieved through the most powerful stroke at Riccioli's command: to insert himself into Cabeo's account and replace Cabeo's narrative voice with his own. He means to say that one cannot tell *from Cabeo's own account* what the height of release was. His own account will be more reliable than Cabeo's: he explains that "when we were at Ferrara at the same time in 1634" he participated in these very experiments.

Riccioli proceeds to detail the experience that he had gleaned from the work on which Cabeo apparently based his own, differing experience:

> ... besides wooden balls, we released stones of diverse weights from the tower of our chapel of the Society of Jesus, with once a bronze basin, at another time a wooden board having been put by me under [the drop], so that from the different sounds I would distinguish better which one reached the ground faster.

He remembers certainly "that I noticed that the heavier one reached [the ground] a little bit more quickly." Now the matter of the altitude of the drop comes into play, and Cabeo's authority is crushingly demoted: "But because that difference [in time of fall] was tiny—for the tower at that place, from which they were released, did not exceed eighty feet—for that reason he [Cabeo] could never be persuaded to

---

23. Riccioli, *Almagestum novum* (1651), part 2, Lib. IX, sect. IV, cap. XVI, p. 382 col. II: "... asseverantissimè affirmat ex proprijs saepiusque iteratis experimentis, si globi duo simul dimittantur ex eadem altitudine, unus unciae unius, alter decem librarum, vel cuiuslibet maioris ponderis, sive ambo sint plumbei, sive unus plumbeus sit, alter vel lapidus vel ligneus; dummodo & aër sit tranquillus, & illud quod levius est, non sit tantulae gravitatis, ut non valens vincere resistentiam aëris aut aurae fluctuet in aëre, (cuiusmodi esse plumam vel chartam) fore ut ambo eodem momento ad terram perveniant, nullumque sensibile discrimen in casu notari possit: ex quo infert, omnium cadentium velocitatem per se aequalem esse." Cabeo enters the discussion as one who agrees with Baliani.

admit any inequality or difference in the descent."[24] That constitutes Riccioli's final word on the worth of Cabeo's claims.

Arriaga presented Riccioli with a different problem. Riccioli could not claim any privileged authority in rewriting Arriaga's text; instead, he had to rely on exploiting features of the text itself. He mentions that Arriaga's position is shared by Bartholomaeus Mastrius and Bonaventura Bellutus, authors of a recent set of disputations on Aristotle's *De caelo* and *Meteorologica*, but takes care to stress that they only follow Arriaga and should not be seen as independent authorities.[25] Riccioli accurately cites the assertion as Arriaga propounded it: "any two heavy bodies, whether of the same or of different species [i.e., material composition], or bulk, or figure, however much the difference in weight, descend from the same height in the same time, and fall down by themselves with equal speed." He takes care to note that "truly, the foundation of so great an assertion is the experiment of Arriaga," so as to emphasize that other statements of the disputed phenomenon are merely derivative of Arriaga's. Riccioli then proceeds to weaken Arriaga's crucial experiment itself:

... he says that he has often released from the same table simultaneously a piece of dry bread crust of two inches in size, and a pen, with which he wrote, and a great stone, which he could only with difficulty

---

24. Ibid.: "Nescio autem ex quanta altitudine globos illos dimiserit, hoc tamen certò memini, cùm essemus simul Ferrariae Anno 1634. & ex turri nostri templi Societatis IESU demitteremus lapides diversae gravitatis, nec non globos ligneos, me subiecta uni pelui aenea, alteri tabula lignea, ut ex diversitate sonitûs meliùs distinguerem, uter citiùs ad terram perveniret, advertisse id quod gravius erat aliquantulò citiùs pervenire. Sed quia illud discrimen exiguum erat, neque enim turris ille locus, ex quo demittebantur, excedebat 80. pedes, idcirco ille nunquam adduci potuit, ut eam vel ullam inaequalitatem admitteret, aut discrimen in lapsu." Note how Riccioli demotes Cabeo's claims by shifting them from statements about the world to statements about Cabeo's own stubbornness: see the discussion of "modalities" in Latour, *Science in Action* (1987), pp. 22–26 (an idea first presented in Latour and Woolgar, *Laboratory Life* [1979, 1986]).

25. Riccioli, *Almagestum novum*, pt. 2, p. 382 col. II; the reference is to Mastrius and Bellutus, *Disputationes in libros de celo et metheoris* (1640). Their *quaestio* on the subject is headed: "Explicantur proprietates quaedam motus naturalis gravium, & levium, praecipuè quò ad velocitatem"; see esp. pp. 175 cols. I–II, with attempted explanations of the supposed phenomenon to p. 179 col. I. These explanations are always tested against Arriaga's assertions about the phenomenon itself. Riccioli does not seem to mention Hugo Sempilius on this subject, although Sempilius had underwritten the same position. Sempilius, *De mathematicis disciplinis libri duodecim* (1635), p. 15 col. I remarks on the explanatory virtues of mechanics; one issue is why a ball of lead and one of wool fall from the same height to the ground in the same time: "Cur globus plumbeus & laneus aequalis magnitudinis ex eadem altitudine cadentes, eodem sensibili tempore ad terram perveniant."

hold in his hands, and he has noticed these simultaneously to strike the pavement at the same moment, and to move equally fast, from which he exclaims and bemoans that in so easy a thing none of the philosophers has ever established [the characteristics of] descent by experience, but almost all have persisted in relying on their predecessors.[26]

Arriaga had, of course, claimed to release objects "from a very great height," but Riccioli takes advantage of his lack of specificity to reduce the dignity and persuasiveness of Arriaga's experience. He uses the presence of bread crust in the account to reduce the experiment to a matter of dropping things from a dining table, as if it were a matter of casual observation under inadequate circumstances. Rubbing salt into the newly created wound, Riccioli's final presentation of Arriaga's criticism of "the philosophers" contrasts the poverty of Arriaga's experience with his arrogance in assailing others.

Riccioli closes his review of previous work on falling bodies by describing the contrasting nature of his own subsequent presentation. "Hitherto," he says, "I have not spoken about the *opinions* of others, but about their *errors*, because the opposing of them by us and by others who have been present as witnesses to our observations, shortly to be reported, is indeed not just probable, but evident and certain."[27] Cabeo, Arriaga, and the rest were unreliable because they were incompetent; either their trials were misconceived or they presented them inadequately. Somehow, by contrast, Riccioli's accounts of his own experimental work would be entirely reliable—indeed, infallible.

The challenge ought to have been enormous. In keeping with the dominant scholastic conventions that constituted scientific experience in this period, Riccioli wanted experiences that were not only certain,

---

26. Riccioli, *Almagestum novum*, pt. 2, p. 382 col. II: "... asserunt, duo quaecumque corpora gravia sive eiusdem, sive diversae speciei, aut molis, aut figurae, quantumcumque differentia in gravitate, eodem tempore ex eâdem altitudine descendere, & aequali per se velocitate labi: Tantae verò assertionis fundamentum est experimentum Arriagae, quo ait se saepius ex mensa eâdem dimisisse simul siccum panis corticem duorum digitorum, & calamum, quo scribebat, & saxum ingens, quod vix manibus sustinere poterat, & advertisse haec simul eodem momento pavimentum percutere, atque aequè velociter moveri, ex quo exclamat & conqueritur, in re tam facili nullum unquam philosophorum experientia hunc descensum comprobasse, sed omnes ferè in fide parentum permansisse." Arriaga's mention of his pen comes a little after his main account of the experiments: Arriaga, *Cursus philosophicus*, p. 583 col. I.

27. Riccioli, *Almagestum novum*, pt. 2, p. 383 col. I: "Hactenus de aliorum non dico opinionibus, sed erroribus, quia illarum oppositum nobis & alijs, qui testes adfuerunt observationibus nostris, mox referendis, iam non est probabile tantùm, sed evidens ac certum."

but also *evident*. Indeed, for scientific purposes the two attributes were inseparable: an experience could never be established as certain without first being established as evident, since an experience that was not evident could always be doubted. But what did it mean for an experience to be evident?

We have already seen some of the practical means by which philosophical and mathematical texts created evidentness in the early seventeenth century. At root, these textual means relied on the existence of well-formed expectations and habits of cognition among their readership, expectations and habits reinforced by orthodox pedagogy. Experiences attested in authoritative texts (like Aristotle's account of the extended illumination of Mount Caucasus, used by Galileo's friend Mazzoni as the basis for a philosophical argument),[28] or common wisdom (such as the idea that the bodies of murder victims bleed on the approach of the murderer, which even the sceptical Descartes accepted as true)[29] routinely stood in for the reader's own experience; in a similar way, the aura of certainty surrounding mathematical demonstration allowed geometrically structured descriptions of contrived experiences—experimental procedures—to re-create conviction. When Cabeo, and Arriaga, recited the indeterminate frequency with which they had conducted their experimental tests and stressed the presence of witnesses, they were tacitly appealing to the sorts of criteria that Aguilonius had formulated in terms of explicit methodology, rooted in standard Aristotelian pedagogy. But Riccioli's attempts at undermining their claims dramatize the epistemological ambiguities and opportunities of these rhetorical techniques. Novel experiences, not novel experiments, were the real issue: the outcomes of experimental procedures were not philosophical knowledge, and hence could not be contested, until they had been worked up into true scientific experiences that had philosophical consequences.

The extension of private experience to others through the medium of the text amounted to inciting unquestioning acceptance of the universal nature of the author's own experience. Riccioli's handling of the work of opponents involved attempts at investing himself with superior authority, derived from a purportedly greater understanding of how to make legitimate scientific experience. Riccioli tried to convince his readers that he possessed the right kind of *expertise*; if he did it successfully his proclaimed experiences would receive the public war-

---

28. See Shea, "Galileo Galilei: An Astronomer at Work" (1990), p. 52.
29. Descartes, *Œuvres* (1964–76), vol. 9, p. 309; this is only in the French version (1647) of the *Principia philosophiae*, pt. 4, section 187.

rant necessary for them to count as properly scientific—as "evident and certain."[30] The discussions of the behavior of pendulums and falling bodies in the *Almagestum novum* serve as useful illustrations of the available techniques.

IV. Riccioli's Experiments to Establish Expertise

Since it was a work not just of astronomy but also, more generally, of cosmography, Riccioli devotes book II of the *Almagestum novum* to the "sphere of the elements," the terrestrial realm; two of its chapters deal with the behavior of pendulums and falling bodies.[31] The first may be seen as a prolegomenon to the second insofar as its goal seems to be the establishment of instrumental techniques rather than the discovery of new aspects of nature: it is headed "Concerning the oscillations of a pendulum suitable to measuring other motions and times, as much in elementary and compound bodies as in the stars."[32] Riccioli nonetheless signals the comparative novelty of the enterprise, presenting the "Occasion for exploring the oscillations of the pendulum":

> When I was professor of philosophy at Parma, and wished to learn by what increment of velocity heavy bodies descend faster and faster towards their end, and saw that it required a very precise measure of time, and usually of its smallest sensible parts; the occasion at length offered itself to me to measure it.[33]

30. *Expertus* to mean "experienced" is classical Latin usage, although "expert" as a noun, like "expertise," is apparently recent usage in both English and French (the *Oxford English Dictionary*'s earliest example of the noun "expert" in English is from 1825). Sargent, "Scientific Experiment and Legal Expertise" (1989), p. 28, stresses the nature of "experience" in the English common-law tradition as a matter of "expertise" on the part of judge and jury, in a sense quite similar to that used here, although Sargent never shows precisely how this applies to English experimental philosophy. It would seem to help more, however, in understanding Francis Bacon than Robert Boyle, Sargent's chief subject of examination: cf. ibid., p. 20 n. 6; and especially Martin, *Francis Bacon* (1992), esp. pp. 164–171.

31. Riccioli, *Almagestum novum*, pt. 1, "Liber secundus: De sphaera elementari," chaps. 20, 21. Some of this material is employed, in concert with relevant passages from part 2, in Koyré, "An Experiment in Measurement" (1968), on pp. 102–108. Koyré's account is not always reliable, and sometimes combines different passages in questionable ways.

32. Riccioli, *Almagestum novum*, pt. 1, p. 84 col. I: "De Perpendiculi Oscillationibus ad motus alios & tempora mensuranda idoneis, tam in Elementis & mixtis, quàm in syderibus."

33. Ibid.: "Cum profiterer Parmae Philosophiam publicè, optaremque experiri quo incremento velocitatis Gravia descenderent velociùs, ac velociùs versus finem, videremque ad id requiri mensuram temporis subtilissimam, & minimarum fermè particularum eius sensibilium; Oblata est mihi tandem occasio huius mensurae."

He found that other people who had written about it (including Galileo) had not actually measured the periods of oscillation. He therefore set about rectifying that omission, drawing on the work of himself and his colleagues.

Riccioli is concerned with presenting the distilled wisdom of his own experience of the behavior of pendulums, not with recording individual experimental trials. Accordingly, he presents a diagram of a pendulum apparatus together with its description and definitions of terms. The description is an abstract account of a *type* of apparatus rather than a report of an actual piece of equipment; it takes the form of statements such as: "Let AB in the present diagram be an iron rod."[34] Riccioli continues with a number of "propositions" mimicking geometrical presentation, the confirmation of which consists of statements of "experiments" (*experimenta*). Again, these experiments are not given as reports of discrete trials. Thus the first proposition, stating the isochrony of pendulums, is justified first of all by a set of instructions. "Make yourself, then," says Riccioli, "a clock . . . , for example of one semiquadrant of an hour, or of ten minutes, so that there is less tedium at the beginning of the experiments" and so on, concluding: "If, however, you were not diligent in doing this, there might be a difference of one or two oscillations between the vibrations of the first and second [periods of] time."[35] This first experiment leaves unquestioned the truth of the universal experience that the instructions purportedly enable the reader to realize; in effect, the instructions act as an elaborate statement of the "proposition" itself. The second experiment, however, adduces discrete events—but only as instances of how to check the reliability of what is now an *instrument*.

It commences, much like the first, with instructions on how to establish the apparatus: "arrange in the plane of the meridian two threads suspended from above from the same point," proceeding to describe the construction of a sighting instrument. The idea is to have a pendulum the swings of which are counted to mark the interval between the transits of particular stars across the meridian. The number of swings for the same transit interval on different nights are compared, "for if all things have been carried out exactly, and the pendulum moves unvaryingly through its plane, you will find either no disparity in the

---

34. Ibid.: "Sit in praesenti diagrammate Regula ferrea AB"; entire description through p. 84 col. II.

35. Ibid., p. 85 col. I: "Fac enim tibi horologium ex pulvere stanneo breve, putà unius semiquadrantis horae, aut 10 minutorum, ut minus sit taedium in principio experimentorum"; "Si tamen haud ita diligens fueris, unius aut alterius vibrationis differentia esse poterit, inter primi & secundi temporis vibrationes."

number of oscillations, or else a contemptible one." Once again, the crucial claim is given as an assured universal truth. Riccioli now illustrates his claim with a historical narration designed to show how the matter should be handled in practice:

> Thus when we wanted to consider whether the axis of the usual pendulum used for the measuring of astronomical times had suffered any sensible difference by attrition, as of holes E, I [holes housing the horizontal bar that supports the pendulum], or by something else, we counted for three nights the number of oscillations from the transit of Spica to the transit of Arcturus through the same meridian, with witnesses and judges Fathers Franciscus Maria Grimaldus and Franciscus Zenus, who are most practiced in this matter, actually on 19 and 28 May and 2 June, and twice we found simply 3,212 oscillations, and once 3,214, having practiced our custom of putting single counters into a dish after each group of thirty oscillations.[36]

Once this account of how the device had been calibrated is over, the remainder of the chapter largely abandons the citation of participants and witnesses to specific trials. Henceforth all but one of the propositions appear without comment or only with very brief corroboration, such as proposition VIII, concerning the relation of pendulum length to frequency (length inversely proportional to the square of the frequency): "Thus by many experiences from Galileo, Baliani, and us."[37]

Riccioli's basic approach to establishing the propositions he presents in this chapter thus amounts to little more than assertion of his own reliable experience of their truth. The only exceptions involve actual measurement of the period of a pendulum, and these are matters that do not form the subject of a proposition—the account of the transit measurements occurs in a proposition concerning isochrony, not actual numbers, and it describes checking the reliability of the apparatus. The establishment of isochrony in the face of possible frictional impedi-

---

36. Ibid.: "Secundò colloca in plano Meridiani duo fila ex eodem puncto supernè suspensa"; "nam si exactè peracta sunt omnia, & perpendiculum per idem sui planum incessit, aut nullam in numero vibrationum disparitatem deprehendes, aut contemptibilem. Sic nos cùm perpendiculum, ad mensuranda tempora Astronomica, usitatum expendere vellemus num attritione axis, & foraminum E, I, aut aliunde aliquam sensibilem diversitatem passum esset, testibus & adiutoribus PP. Francisco Maria Grimaldo & Francisco Zeno, in hac re exercitatissimis, numeravimus tribus noctibus, nempe Maij 19. & 28. & Iunij 2. vibrationes à transitu Spiçe ad transitum Arcturi per eundem Meridianum, & bis deprehendimus vibrationes simplices 3212. & semel 3214. usi more nostro singulis calculis, in vas post tricennas quasque vibrationes compositas iniectis."

37. Ibid., p. 86 col. I: "Ita ex Galilaeo, Baliano nostrisque pluribus experimentis." The exception is proposition XI, again involving exact measurement of the period of a pendulum, and using a historical account of work done with F. M. Grimaldi.

ment, together with the uncontested period/length relationship independently corroborated by Galileo and Baliani, was of central importance in rendering the pendulum suitable as a time-measuring instrument. Riccioli's historical accounts describe an exercise in calibration as an illustration of procedure, rather than representing discrete experiments the findings of which were to contribute to an understanding of nature.

The exercise is turned to account in the succeeding chapter, where the accurate measurement of time is essential for overturning what Riccioli argues is a false but increasingly prevalent belief: that all heavy bodies accelerate at the same rate regardless of their weights. The chapter is headed "On the speed of heavy bodies descending with natural motion, and the proportion of increase of their speed."[38] Riccioli claims to have tried to discover the proportion, using pendulums, before the discussions of Galileo (in the *Dialogue* of 1632) and Baliani (in his *De motu naturali gravium* of 1638) ever appeared. He promises to recount his own findings as well as his agreements and disagreements with those two writers.

Proposition I directly challenges the Galilean view: "Of two bodies of equal size [*molis*], that which is heavier descends faster with natural motion from the same terminus to the same terminus, than that which is lighter." He substantiates the assertion with an account of dropping twelve balls made from solid clay, each weighing twenty ounces, together with twelve others made from compressed paper, in pairs from the tower of the Asinelli in Bologna. The heavier would land first, hitting the ground three pendulum swings before the other, "which," he stresses, "was confirmed by experiment repeated twelve times." He is aware that this result contradicts the widely known claims of Galileo and Baliani; the reason for the discrepancy, he asserts, is that they examined the matter using altitudes of only perhaps 50 or 100 feet, whereas this tower is 312 feet high. Furthermore, he continues, "Witnesses and judges to this and the following experiments included, although not always everyone simultaneously, Fathers Stephanus Ghisonus, Camillus Roderigus, Iacobus Maria Pallavicinus, Vincentius Maria Grimaldus, Franciscus Maria Grimaldus, Franciscus Zenus, Georgius Cassianus, and others."[39]

Proposition II asserts the same thing for two bodies of different

---

38. Ibid., p. 89 col. I: "De Velocitate Gravium Naturali motu descendentium, & Proportione incrementi velocitatis eorum."

39. Ibid.: "Corporum duorum aequalis molis illud quod gravius est, naturali motu citiùs descendit ex eodem termino ad eundem terminum, quam illud quod est levius"; "... quod repetito duodecies experimento confirmatum est"; "His & sequentibus experi-

weight but identical composition. Again a historical presentation is used: "For we dropped, August 4, 1645, from the tower of the Asinelli, many chalk balls . . . and we always observed the heavier one to strike the pavement" about four feet ahead of the lighter. Riccioli then presents the differences between rates of fall of balls of differing composition (wax, chalk, wood, lead, iron). Proposition III observes that these findings show that Aristotle was nonetheless wrong in asserting (if this is what he really meant) that rate of fall is directly proportional to weight.[40]

Proposition IV maintains the odd-number rule for free fall, credits Galileo and Baliani with its prior statement, and observes that it has been "often confirmed by our experiments." Riccioli and F. M. Grimaldi, it seems, dropped "many chalk balls" from "diverse towers or the windows of buildings" chiefly in Bologna—Riccioli names five of them together with their altitudes.[41] The two Jesuits used a pair of very short pendulums for accurate measurement of time, a technique accredited by the previous chapter. Now Riccioli presents actual numbers in a telling way: "Moreover, among many experiments, I set forth in the following table two most select ones, the most certain of all, written below."[42] These are not "raw data," then, but didactically effective examples—and even they are not to be read as outcomes of individual trials, as Riccioli's exposition makes clear: "Thus in the first experiment, when we had observed the aforementioned ball, by careful procedure many times repeated, travel from a height of ten feet to the pavement in only five vibrations of the aforementioned pendulum. . . ."[43] The imprecision of the phrase "many times repeated" continues to play as important a role for Riccioli as it had for other methodological Aristotelians.[44]

Riccioli's tables in the two chapters epitomize his use of experience

---

mentis testes, & adiutores interfuere, licet non semper omnes simul, PP. Stephanus Ghisonus," etc.

40. Ibid.; ibid., p. 89 col. II: "Demisimus enim Anno 1645. Augusti 4. ex Turri Asinellorum plurimos cretaceos globos . . . & semper observavimus graviorem percutere pavimentum."

41. Ibid., pp. 89 col. II through 90 col. I: "nostrisque experimentis saepius comprobata"; "plurimos globos cretaceos"; "ex diversarum Turrium aut aedium fenestris."

42. Ibid., p. 90 col. II (above table): "Inter multa autem Experimenta, duo infrascripta selectissima, & omnium certissima . . . propono in tabella sequenti."

43. Ibid., p. 90 col. I (below table): "In primo itaque Experimento, cùm ex altitudine pedum 10. globum praedictum iterata saepius operatione observassemus ad pavimentum pervenire tantum vibrationibus quinque praedicti perpendiculi. . . ."

44. See for more on this point chapter 5, below.

in establishing these mathematically dressed propositions. The first of the tables encapsulates the length/frequency relationship asserted by chapter XX's proposition VIII.[45] It coordinates the two variables through an arrangement into columns, in a form that might best be characterized as a "ready-reckoner." It is thus akin to that other seventeenth-century innovation, logarithm tables, and resembles most of all the much older model of astronomical tables for representing the motions of celestial bodies. Astronomical tables, such as the thirteenth-century Alfonsine Tables, the sixteenth-century Prutenic Tables, and the seventeenth-century Rudolphine Tables, were not, it should be remembered, tables of astronomical data. Instead, they were generated by geometrical models of celestial motions. Similar layouts appear occasionally in texts of the other major classical mathematical science of nature, geometrical optics, as with Ptolemy's and Witelo's tables of refraction. It is now well known that those tables were generated using a simple algorithm to interpolate values between a scattering of empirically determined calibration points; they were not tables recording raw measurements.[46] Riccioli's pendulum table is of just this kind: the listed values coordinate with empirical determinations through the choice of "typical" touchstone measurements, but their rationale is the exposition of the length/frequency relationship, which they are calculated to exemplify. Similar tables appear on succeeding pages, listing the number of oscillations of pendulums of particular lengths against astronomical time. The ideal nature of the tables becomes even clearer in light of the presentation, immediately following, of a series of *Problemata* (the standard term for geometrical constructional problems) the solutions to which rely on application of the relationship displayed in the tables.[47]

The tables in chapter XXI have the same characteristics. The information on differing rates of fall of balls of different composition appears in the form of a tabular comparison between the balls of each pair, noting which outstrips which, and by how far, on a standard drop. The figures are straightforwardly empirical insofar as there is no underlying relationship to be established; nonetheless, each number appears

45. Riccioli, *Almagestum novum*, pt. 1, p. 86 col. I.
46. On Ptolemy, Lejeune, "Recherches sur la catoptrique grecque" (1957–58), esp. pp. 152–166; idem, *L'Optique de Claude Ptolémée* (1989); Smith, "Ptolemy's Search for a Law of Refraction" (1982); on Witelo, Crombie, *Robert Grosseteste* (1953), pp. 219–225; Lindberg, *Theories of Vision* (1976), pp. 118–121.
47. Riccioli, *Almagestum novum*, pt. 1, pp. 87, 88 col. I. Cf. Shapin, *A Social History of Truth* (1994), p. 326.

Experimenta.	Vibrationes simplices Perpendiculi.	Tempus Vibrationibus congruens.	Spatium confectum à globo cretaceo vnciarū 8. in fine tēporis.	Spatiū ergo seorsim confectū singulis tēporibus æqualibus.	Proportio incremēti velocitatis in simplicibus numeris expressa.
		"	'''	Pedes Romani.	Pedes Romani.	Num.parit.impares
I	5	0	50	10	10	1
	10	1	40	40	30	3
	15	2	30	90	50	5
	20	3	20	160	70	7
	25	4	10	250	90	9
II	6	1	0	15	15	1
	12	2	0	60	45	3
	18	3	0	135	75	5
	24	4	0	240	105	7
	ferè 26	4	ferè 20	280		

Figure 1. Table comparing height and duration of fall for heavy bodies, from G. B. Riccioli, *Almagestum novum* (1651), pt. 1, p. 90. Courtesy of Division of Rare and Manuscript Collections, Carl A. Kroch Library, Cornell University.

as an unconditional value rather than as the result of a specific trial or even as the determined average of a specified number of trials.[48] The table justifying the odd-number rule is itself exactly what one would expect from the pendulum examples (see figure 1). It shows distances fallen from rest in times measured by the oscillations of a given pendulum, with the final column reducing the figures to display a perfect adherence to the Galilean progression of distance with time. The table is followed by a calculation of how long it would take one of these chalk balls to fall from the moon to the earth (a problem borrowed from Galileo's *Dialogo*).[49] The table requires no concept of experimental error, because it does not tabulate discrete experimental events. It tabulates the lessons of "experience," and does so according to a well-understood form.

Later in the work, as a prologue to a discussion of the motion of the earth, there is another, much fuller study of fall, which nonetheless follows a similar pattern.[50] It commences with the undermining of Cabeo, Arriaga, and others discussed above; it continues with sections each presenting a "group of experiments" (*classis experimentorum*) that

---

48. Ibid., p. 89 col. II.
49. Ibid., pp. 90 cols. I–II; 91 col. I.
50. On Riccioli's discussion of the earth's motion, see Koyré, "A Documentary History of the Problem of Fall" (1955), pp. 349–354; Grant, "In Defense of the Earth's Centrality and Immobility" (1984), pp. 51–54; Russell, "Catholic Astronomers and the Copernican System" (1989).

purportedly establish a particular conclusion. Riccioli had advertised these conclusions as "certain."[51] The movement between recipe-like or instructional accounts of contrived experiences, necessarily universal in form although springing from Riccioli's expertise, and historical accounts of individual trials or sets of trials again occurs with the implied epistemological priorities just examined—although the greater prolixity of the historical accounts serves to establish even more powerfully that Riccioli knows whereof he speaks. Even some of the non-historical universal experiences that Riccioli presents are clearly marked as warrants of the truth of the assertions to which they are attached, rather than as the available grounds for making the assertions in the first place. At the end of three *experimenta* of this kind, for instance, Riccioli writes: "And it is evident by other innumerable experiments of this kind that a heavy body falling naturally from a higher place has acquired ever greater impetus at the end of [its] motion."[52]

As before, the historical narrations are themselves set in a context of justifying more general experiential claims. The chief example, wherein Riccioli and numerous Jesuit colleagues dropped balls from various towers in Bologna,[53] is introduced by saying: "The fifth experiment, taken up by us very frequently, has been the distance that any heavy body traverses in equal times in natural fall."[54] The succeeding account includes a diagram of the Asinelli tower where most of the trials were conducted,[55] but the reader is never allowed to forget that these details should not be taken to restrict Riccioli's warrant for his claims: specifying investigations that he and his collaborators undertook in May of 1640, he adds "and then at other times"—Riccioli's expertise can be seen in his historical narrations, but it is not dependent on them. Riccioli is not adducing evidence for his claims; he is presenting tokens of his experiential knowledge.

In the detailed historical reports found in the fourth section a number of specific dates and named witnesses are presented, and results,

---

51. Riccioli, *Almagestum novum*, pt. 2, p. 383, col. I: "... ab experimentis sequentibus ad conclusiones certas procedemus."

52. Ibid, p. 384 col. II: "Et alijs innumerabilibus experimentis huiusmodi evidens sit, maiorem semper ac maiorem impetum in fine motûs acquisivisse corpus grave naturaliter cadens ex altiore loco."

53. See above for Riccioli's earlier presentation of this material; also Koyré, "An Experiment in Measurement."

54. Riccioli, *Almagestum novum*, pt. 2, p. 385 col. I: "*Quintum* igitur Experimentum sumptum à nobis saepissimè, fuit spatij dimensio, quod grave quodpiam, aequalibus temporibus naturali descensu conficit."

55. Ibid., p. 385 col. II.

here sometimes of single rather than multiple trials, appear in a table of actual data for the different descents of paired balls of different weights or compositions. These figures are, however, adduced to show only "the unequal descent from the same height through the air of two heavy bodies of diverse weights," with some "corollaries" drawn from them that show that heavier bodies fall faster except when there is a big counteracting difference in density.[56] There are no correlating measurements of the specific weights of the balls; the actual figures themselves, therefore, rather than the overall trends, have no special significance—they serve only as marks of the sort of thing one finds.

Throughout the sections of the chapter, the reader is bombarded with assurances of the frequency with which the experiments were made ("frequently repeated," "From all these and many other experiments it is evident to us," etc.).[57] The tables of data are not treated as complete and foundational; the description of a table of results for bodies falling through water, for example, is described as having been compiled "ex selectis experimentis."[58] Finally, following the individual propositions and their *experimenta*, there appear "theorems selected from the foregoing experiments." These are the items of natural knowledge, such as an argument for the reality of absolute levity, that Riccioli wishes to have accepted on the strength of his expert testimony. Derivation of these scientific conclusions, the distilled "certainties" that he has promised, apparently qualify as "theorems" because they are justified by reference to the previously established experiences (which make the conclusions "manifest," "agree" with them, "show" them, and so forth), much as geometrical theorems are deduced from uncontested premises.[59] Last of all, to complete the mathematical packaging, Riccioli gives "problems selected from the foregoing theorems," showing how the conclusions may be put to work for such tasks as comparing rates of fall of bodies in air and in water.[60]

Riccioli, then, uses experience to establish expertise, a quality that he as the author of a text can then employ to undermine the knowledge-claims of others, such as Galileo, Arriaga, Cabeo, and Baliani (he repeats his contradictions of them as he lays out his concluding "theorems"). The witnesses he sometimes cites serve to increase the

---

56. Ibid., p. 387 col. I: "pro Duorum Gravium diversi ponderis Descensu Inaequali ex eadem altitudine per Aerem"; corollaries on pp. 388 col. I to 389 col. I.
57. ". . . [O]bservationem . . . saepiusque repetitam"; "Ex omnibus his & alijs plurimis experimentis nobis evidens est": ibid., pp. 390 col. I and 391 col. I.
58. Ibid., p. 390 col. II.
59. Ibid., pp. 394 col. I to 396 col. I.
60. Ibid., pp. 396 col. I to 397 col. II.

credibility of the experiences described, and thus the authenticity of his own competence to speak. Citation of witnesses would have been irrelevant if Riccioli had expected his readers to perform these experiments for themselves. The citations enhance the persuasive efficacy of the text, not the accessibility of nature.

In 1669, in the fifth edition of his *Cursus philosophicus*, Arriaga responded to unnamed critics. Specifying with nice imprecision, as had Cabeo, the appropriate restriction of his claims to heavy bodies that fall directly down rather than float in the manner of paper or feathers, he reasserts the "truth of experience" which obviates any need for "profound speculation." The recent writers who oppose his conclusion assert the "contrary experience" out of bitterness and will not accept Arriaga's refutations gracefully. Arriaga professes astonishment that others, judging the matter under the same circumstances, could deny "so many and such clear experiences in my favor"; nonetheless, so fair is he, that in the face of the "exclamations" against him, "I have made the same experience again very often, even with some of them themselves in person, and just as I have taught, I have perceived the thing to be so."[61] These protestations, however, are now backed up by historical descriptions of specific trials, introduced in this way:

> Recently, from the top of the cupola of the Prague Cathedral, which is very high, I released from my hands at the same time a small stone and another more than twenty times heavier, and still both touched ground at precisely the same time. The same thing was done from Karlstejn Castle which is even higher, and they fell in exactly the same way, even if bodies of unequal weight were dropped.[62]

Riccioli, a prominent "recent writer" who denied Arriaga's claim, had ridiculed Arriaga's justification of it by exploiting the imprecisions and vaguenesses of his account. The fifth edition, with reports of actual trials and their locations, moved to fill the rhetorical breach by following a protocol similar to one exploited by Riccioli himself.

V. Grassi, Galileo, and Historical Narration

So far in this chapter, we have seen Riccioli and Arriaga adduce, on rare occasions, specific times, places, and witnesses as instances of instrumental calibration or tokens of credibility, but we have not seen an

---

61. Arriaga, *Cursus philosophicus* 5th ed. (1669), p. 691, original passage quoted in Schmitt, "Galileo and the Seventeenth-Century Text-book Tradition," p. 224 n.; my translations. The sections on falling bodies in the fourth edition of the *Cursus philosophicus* (Paris, 1647) appear to be identical to those of 1632.

62. Ibid., p. 224 and n.; trans. Schmitt.

individual experimental event constituted as grounds for belief in a universal statement. The methodological objections that such a move would have called forth within the dominant academic culture should be clear from chapter 2. An examination of a more famous controversy, this time between the two mathematical scientists Orazio Grassi and Galileo Galilei over the nature of comets, allows us to see the deployment of another historically reported event and the ways in which, its appearances to the modern eye notwithstanding, the event did not, because it could not, serve as a "crucial experiment" in which nature gave an unequivocal answer to a question.[63] A singular event could not be treated as a universal natural behavior when physical properties were involved; the truth of its historical report, by contrast, could be used as a vindication of the competence or expertise of the reporter.

Grassi held the chair of mathematics at the Collegio Romano, previously occupied by Clavius, from 1616 to 1624 and again from 1626 to 1628.[64] Galileo operated free of disciplinary shackles as Tuscan court philosopher and mathematician, and sinecure holder of the University of Pisa. Grassi's *Libra astronomica* of 1619, published under the pseudonym of Lothario Sarsi, formed part of a celebrated dispute with Galileo over the comets of 1618, a dispute that was to produce Galileo's famous *Assayer* in 1623. The core of the controversy was the question of whether comets were bodies traversing the heavens or, as Galileo claimed, refractive phenomena in vaporous exhalations rising up from the earth. Although ostensibly a matter of optics and astronomy, the arguments ranged widely, and in the third part of the *Libra astronomica* Grassi attempted to undermine one of Galileo's attacks on Aristotle. Grassi did not support the particular Aristotelian thesis at issue, but he sought to catch Galileo in an error.

Galileo had identified an apparent contradiction between Aristotle's dictum that the celestial spheres are perfectly smooth and his claim that air can be kindled by the motion of the lunar sphere to produce shooting stars and comets. The empirical principle that Galileo used to produce his demonstration of a contradiction—the statement that functioned as the equivalent of the major premise of a syllogism—was

---

63. For accounts of varying aspects of this controversy see Shea, *Galileo's Intellectual Revolution* (1972), chap. 4; Moss, *Novelties in the Heavens* (1993), chap. 8; Biagioli, *Galileo, Courtier* (1993), chap. 5, which is much to be preferred over the rather breathless account in Redondi, *Galileo Heretic* (1987), chap. 6.

64. On Grassi, see, in addition to references in previous note, Stillman Drake, "Introduction" to Drake and O'Malley, *The Controversy on the Comets* (1960), esp. p. xv; Costantini, *Baliani e i Gesuiti* (1969); letters between Grassi and Baliani are also printed in Moscovici, *L'expérience du mouvement* (1967), pp. 230–262.

that "neither air nor fire adheres to polished and smooth bodies," as the moon's sphere is said to be. Thus it cannot be kindled by the latter's motion. Galileo justified his empirical principle with an appeal to experience:

> [l'esperienza] shows this, for if we cause a concave circular vessel of very smooth surface to revolve about its center with as great a velocity as we wish, the air contained within it will remain at rest. This may be clearly demonstrated by the tiny flame of a little, lighted candle held inside the hollow of the vessel; the flame will not only fail to be extinguished by the air contiguous to the surface of the vessel but it will not even be bent, though if the air were moving so swiftly it ought to put out a much larger flame. Now if air does not participate in this motion, still less would some other lighter and more subtle substance acquire it.[65]

Galileo thus seeks experiential support for a premise required in his demonstration of Aristotle's fallacy. The experience is presented as a general unproblematic statement of how things behave, glossing over the constructed character of the conditions. In similar circumstances Grassi would probably have done the same thing, but in the context of the controversy he decides instead to call into question the experience itself. In doing so, he raises the stakes involved in the dispute, and makes it much harder for Galileo to argue with him. He could not merely counterclaim, since this would not have given him any argumentative edge over Galileo. So he introduces a quite different technique, translating the question of whether or not a rotating smooth bowl will disturb a candle flame into a test of personal probity.

Contrary to Galileo's assertion, says Grassi, smooth dishes when rotated about their axes do carry their contents, whether water or air, around with them. To justify this claim, he first provides a detailed description of apparatus—a procedure long familiar in optics and astronomy (see figure 2). But he then details a discrete historical event:

> Lest anyone believe that we tested this negligently and carelessly, we obtained a brass hemispherical vessel skillfully hollowed out on a lathe, and we undertook that the axis would pass through its center as a spherical axis—if it were prolonged. Furthermore, we constructed a fitfitfitfitfitfirm and stable foot lest it be agitated by the motion of the vessel.[66]

---

65. Galileo, *Opere* (1890–1909), vol. 6, pp. 53–54; trans. Drake in Drake and O'Malley, *The Controversy on the Comets*, pp. 29–30. This episode is briefly discussed in Shea, *Galileo's Intellectual Revolution*, pp. 92–93.

66. Grassi, *Libra astronomica* (1619), in Galileo, *Opere* 6, p. 156; trans. O'Malley in Drake and O'Malley, *The Controversy on the Comets*, p. 110.

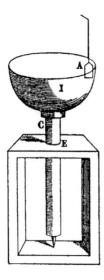

Figure 2. Orazio Grassi's apparatus for detecting the motion of air in an axially rotated hemisphere, from Galileo, *Opere* 6, p. 156.

The outcome of the experiment, like the procedural details, appears in the same form: "But not only water was carried around by the motion of the vessel, but the air itself, from which especially Galileo took his example."[67] Most remarkable of all, however, is Grassi's means of enhancing the account's credibility:

> I have no few witnesses to the fact that I say this not more surely than truly; first, many fathers of the Collegio Romano—however, many others were willing to recognize this on the authority of my teacher [Grassi is writing under the guise of one of his own pupils]—and many others as well. I ought not be silent about that one among them whose name is very well known to me not more by birth than by his singular erudition, and who can verify my activities and substantiate my remarks. I speak of Virginio Cesarini, who was astonished that a thing that many today hold certain could ever be proved false; and yet he saw what was done, which many denied could be done.[68]

Grassi, in order to vindicate his claims for a particular historical occurrence, cites witnesses. In effect, the requirement in scientific demonstration that empirical premises be *evident* is here replaced by a surrogate technique whereby what is to be made evident is Grassi's competence to speak on the matter; the event represents a sort of trial of that competence which Grassi is seen to pass. Having made the issue of testimony explicit, Grassi multiplies witnesses and exploits

67. Ibid.
68. Ibid., p. 157; trans. O'Malley, p. 111.

the authority of one "whose name is very well known to me not more by birth than by his singular erudition"; Cesarini had the peculiar advantage, given the context and the situated character of the dispute, of being one of Galileo's followers.[69] Grassi thus responds to Galileo's general empirical assertion by citing events the gainsaying of which would put Galileo in the position of questioning his word and that of others.[70]

Nonetheless, singular, constructed trials remained an undesirable expedient, for all the techniques designed to accommodate them. Grassi therefore supplemented the effect of his historical narrative with independent sources of credibility. He makes a point of saying that "many others were willing to recognize" the truth of the claimed results "on the authority of my teacher," Grassi himself. The behavior of the "many others" thereby put them on a par with anyone who might accept astronomical data on the authority of an appropriate observer; Grassi appears to expect that his authority ought under most circumstances to be enough. Grassi also goes on to say: "These things are certain from experiment; yet if this experiment were lacking, reason itself would also teach them," and he accordingly attempts to justify his results with arguments concerning the humidity and hence adhesiveness of water and air.[71] He soon returns, however, to more refined narratives of the original experiment, again presented as a report in the first person singular of his construction and manipulation of special apparatus.[72] The report ends by noting: "Those same ones who were mentioned by me earlier saw these final experiments, but I do not consider it necessary that they testify again."[73] By this means, of course, they are already seen to have done so.

Galileo attempted to circumvent Grassi's denial of his assertion in the *Assayer*. He did so not by narrating a historical event that yielded an outcome the opposite of Grassi's, but by appealing once again to an already universalized experience founded on many instances. Galileo addresses himself to Grassi according to the charade that a student, Sarsi, had written Grassi's text:

> Now to counter all Sarsi's [experiences], let your Excellency make a bowl rotatable upon its axis, and in order to make sure whether the air which it contains goes along when it is spun rapidly, take two

---

69. On these client-patron relationships see Biagioli, *Galileo, Courtier*, esp. chap. 1; on Cesarini, see further refs. in ibid., p. 296 n. 92.
70. Cf. Shapin, *A Social History of Truth*, chap. 3, on "giving the lie."
71. Grassi, *Libra astronomica*, in Galileo, *Opere* 6, p. 157; trans. O'Malley in Drake and O'Malley, *The Controversy on the Comets*, p. 111.
72. Grassi, *Libra astronomica*, in Galileo, *Opere* 6, pp. 157–159.
73. Ibid., p. 159; trans. O'Malley, p. 115.

lighted candles; attach one inside the bowl an inch or two from its top, and hold the other in your hand inside the bowl at a like distance from the top. Then make the bowl rotate rapidly; if after a while the air also goes round with it, then since the bowl doubtless moves the contained air and the attached candle with equal speed, the little flame of the first candle will not bend at all, but will remain as if the whole were stationary. This is what happens when a man runs with a lantern which encloses a flame—this is not blown out or even bent, inasmuch as the air which surrounds it moves with equal speed. The same effect is seen even more clearly in a rapidly sailing ship, for lights which are placed below decks make no movement whatever, but remain in the same state as if the ship were at rest.[74]

Galileo's familiar rhetorical tactic of appealing to common experience worked to establish his empirical assertions, appropriately, as things that "everyone knows." Examples abound in his writings, most famously in the later *Dialogue Concerning the Two Chief World Systems* of 1632, which includes another use of the rapidly moving ship as an argument for a kind of relativity of motion.[75] Its invocation here, however, serves his purpose somewhat less effectively. He continues:

But the little candle which is held fixed will give a sign of the circulation of the air which, blowing upon it, will cause it to bend. If the result is the opposite—that is, if the air does not follow the motion of the bowl—the fixed flame will remain straight and quiet while the other one, which is carried rapidly by the bowl, will bend by being pushed against the quiet air. Now whenever I have seen the [experience], it has always happened that the little stationary candle has remained burning with a straight flame while the one attached to the bowl has always been much bent and frequently blown out. And surely your excellency will see the same, or anyone else who wishes to put it to the proof. Thus you may judge what must be said about the action of the air.[76]

Grassi responded to the *Assayer* with his book *Ratio ponderum* in 1626, and in the section dealing with this passage he points out that Galileo has simply falsified his position: Grassi's point was that the air in the bowl did not rotate at the same speed as the bowl itself, whereas Galileo made it appear as if Grassi's claim required all-or-nothing—either the air was not moved at all, or it moved with the same rotation

---

74. Galileo, *Il Saggiatore*, in Galileo, *Opere* 6, p. 328; trans. Drake in Drake and O'Malley, *The Controversy on the Comets*, pp. 288–289.
75. Galileo, *Opere* 7, pp. 169–175.
76. Ibid. 6, pp. 328–329; trans. Drake, pp. 289.

as the bowl.⁷⁷ Grassi also denies that the flame of the candle held stationary by the hand will be unaffected by the bowl's motion. Galileo (who had implied universality with the phrase "whenever I have seen the experience") might have failed to see the disturbance occur, he says:

> however, it happens for those who, more open-mindedly trying it in my presence, have wanted themselves personally to hold the candle clutched in their hand and to set the bowl in motion—most illustrious and erudite men, among whom are numbered (most dear to Galileo) Virginio Cesarini and Giovanni Ciampoli.⁷⁸

Grassi again appeals to the testimony of witnesses, implicating Galileo's arguments in a network of loyalties that invite him to question the word of his own protégés. Grassi's establishment of his empirical claim on the foundation of testimony tacitly appeals to that earlier historical narration of a particular event in which his witnesses had first played their parts. It also makes clearer than before that what was at issue was his own credibility as an authority on this matter.

Extraordinary circumstances—a controversy with Galileo—had thus called forth extraordinary resources in scientific argumentation. Those resources were potentially available within the Jesuit tradition of mathematical sciences due to the acute relevance for Jesuit disciplinary politics of the methodological issues involved in using experience in the formulation of a true science. As we have seen in this and the previous chapter, personal authority—a quality of "expertise"—was always implicitly invoked when astronomers or opticians presented the results of their expertly contrived manipulations. In a context of controversy, Riccioli found it expedient to enforce the credibility of his claims concerning falling bodies with circumstantial details involving witnesses; however, he did not resort to reports of historical events in such a way as to make the truth of his foundational claims flow directly from the truth of the reports. Similarly, the Jesuit mathematician Grassi's citation of witnesses represented a raising of the stakes when Galileo employed a quite usual form of deploying experience: Grassi presented a historical event as a test of his own credibility, his recognized expertise.⁷⁹

77. Grassi, *Ratio ponderum* (1626), in Galileo, *Opere* 6, p. 473.
78. Ibid., p. 474.
79. One might note the similarity with Galen's famous demonstrations in the public culture of second-century Rome: see Scarborough, *Roman Medicine* (1969), chap. 8, esp. p. 116. On the crucial epistemic differences between "experiments" and "demonstrations" in modern science, see Collins, "Public Experiments and Displays of Virtuosity" (1988).

Galileo's position in the dispute was in many ways typical of his philosophical outlook. His understanding of demonstrative scientific methodology and its practical management shows a close kinship with the Aristotelianism of the Jesuits and others.[80] That kinship extended to the nuances of the use of experience found in his celebrated work of astronomical observation, where his attempts at justifying novelty required all the resources that a received view of philosophical demonstration might be made to yield.

80. See especially, among studies cited in chapter 2, above, Wallace, *Galileo and His Sources* (1984); idem, *Galileo's Logic of Discovery and Proof* (1992).

Four

APOSTOLIC SUCCESSION, ASTRONOMICAL KNOWLEDGE, AND SCIENTIFIC TRADITIONS

I. Tradition and Time

The epistemic challenge offered to methodological Aristotelians by the discipline of astronomy in the seventeenth century issued, as we saw in chapter 2, from its reliance on discrete data—observational records made at particular times and places by particular people. The creation of universal knowledge about the heavens from such historically situated testimony called upon certain requisite assumptions relating to the admissibility of inductive inferences from sometimes *unrepeatable* singulars. Validation of the precession of the equinoxes, for example, relied on observations that, for all practical purposes, would never be replicated (until precession had completed 360 degrees, tens of thousands of years later). A long time frame, and the attendant necessity for trustworthy records, therefore played a central part in astronomical experience. Niccolò Cabeo addressed precisely this point in his commentary on Aristotle's *Meteorology*, under the general heading: "On the truth of histories, and their use for philosophizing."[1]

Histories, says Cabeo, allow us to overcome the brevity of our own lives.

> For just as we write in good faith what happens in our own time, and so to speak paint the state of our affairs, so that we put it on view to those who come after, and their successors, they also have done the same thing who were before us, from which it comes about by the common association of men that by combined endeavors our life be-

[1]. Cabeo, *In quatuor libros Meteorologicorum Aristotelis commentaria* (1646), Lib. 1, p. 399 col. 1: "De historiarum veritate, & utilitate ad philosophandum."

comes as it were more prolonged, and fuller, and we are able to pronounce on the state of things as if we had been present at them all.²

Cabeo's cooperative, socialized understanding of historical knowledge makes it clear why Descartes, the apostle of the individual knower, had such a low estimation of history. But, as Cabeo goes on to argue, the point applies not just to reports of the affairs of men:

> This is shown by similar things in astronomy, for there are many movements of the heavens, or of the slowest stars, which cannot be observed by one man in the brief span of his life; and whereas one man alone cannot do it, many people continue in concert together, until by continued observations it is brought about that the life of many becomes as it were a single very long expanse of life of hundreds and thousands of years.³

No sense of methodological impropriety attaching to the necessary reliance on testimony and human records is betrayed by Cabeo's historicized, community-based (one might say "conservative") characterization of astronomical science. The reason would seem to be the apparently self-validating character of the universalized "experiences" that result from this accumulated data. Cabeo notes that, as a result of this long process of astronomical endeavor, there have emerged "from the power of those observations laws and canons of celestial motions which correspond best to things."⁴ It does not matter that the observational data derive from non-evident historical testimony, because data could *never* be evident; they were not in themselves candidates for the empirical principles of scientific demonstrations because they were not in themselves universals. As we have seen above, in scholastic-Aristotelian philosophy, a scientific demonstration took the form of logical deduction from universal generalizations or principles the truth

---

2. Ibid., p. 399 col. 2: "Sicut enim nos fide bona scribimus, quae succedunt nostra tempestate, & quasi pingimus statum rerum nostrarum, ut posteris, ac successoribus proponamus: hoc idem fecerunt etiam illi, qui fuerunt ante nos, ex qua communi hominum societate fit, ut collatis studijs, vita nostra fiat quasi productior, & amplior, & de statu rerum pronunciare possimus, ac si omnibus praesentes fuissemus."

3. Ibid.: "Probatur hoc à simili in Astrologia sunt enim multi caelorum motus, seu astrorum tardissimi, qui non possunt ab uno homine brevi vitae tempore observari; & quod unus solus homo praestare non potest, multi simul convenientes obtinent, dum observationibus continuatis efficitur, ut plurimorum vita, unicus sit quasi vitae tractus longissimus 100. & 1000 annorum."

4. Ibid.: "Ex vi istarum observationum ... leges, & canones motuum caelestium, qui optimè rebus respondent." On the term "laws of motion" in astronomy (apparently a sixteenth-century development) see N. Jardine, *The Birth of History and Philosophy of Science* (1984), p. 240

of which was taken to be evident and unquestionable. The establishment of that evidentness necessarily stood outside the procedures of scientific demonstration proper because, by their very nature, fundamental principles could not be deductively demonstrated. From a practical point of view, the only requirement was that they be *accepted* as evident.[5] In the case of astronomy, as Cabeo goes on to indicate, that acceptance was generated by the routine, and ongoing, fulfillment of predictions. The "laws and canons" of motion that constituted the principles of astronomy allowed, he said, both the finding of "the definite position of particular stars" and the certain prediction of "eclipses and other configurations of the planets."[6] The fallibility and singularity of historical testimony were therefore irrelevant to astronomy's scientific status. The principles of astronomy were not justified by the already accumulated data; such a justification would have required such matters as evaluation of the reliability of testimony. Instead, the acceptance of the principles rested on an ongoing familiarity with their verisimilitude as guides to current and future appearances.[7] On this view, then, the legitimacy of the knowledge claimed by astronomy depended on the discipline's continuing practice, rather than on any modern ideal of "replication."

Cabeo stresses the point that, for astronomy, "the observation of one person does not suffice." If these "laws and canons" of celestial motion had been derived from the observations of one man, they "would have needed his life to last a thousand years." Cabeo goes on to address the cooperative aspect of knowledge-creation more generally:

> Therefore, just as from many observations of different men, made in different and successive times, true knowledge has arisen about the motion of the heavens that cannot be observed except over a very long period of time, so also in other matters, both geographical and concerning other parts of philosophy, we are able to profit from the observations of others.[8]

5. As discussed in chapter 1, section V, and chapter 2, section I, above.
6. Cabeo, *Commentaria*, Lib. 1, p. 399 col. 2: "... quibus invenitur certus locus singulorum astrorum, & praedicuntur certo eventu eclypses, & aliae configurationes planetarum."
7. Cf. chapter 2, text to n. 91, above.
8. Cabeo, *Commentaria*, Lib. 1, p. 399 col. 2: "Observatio unius non sufficit"; "si debuissent unius hominis observationibus haberi, debuisset durare vita ad mille annos"; "Sicut ergo ex multis diversorum hominum observationibus, diversis, sibique succedentibus temporibus factis, extitit vera cognitio de motu coeli, qui non potest nisi longissimo tempore observari: ita etiam alijs in rebus, & geographicis, & aliarum philosophiae partium, ex aliorum observationibus proficere possumus." Geography, like astronomy, was of course a mixed mathematical science.

Crucially, the kind of science of which Cabeo speaks is one rooted in *tradition*. The conventional image of the Scientific Revolution as a new beginning that swept clean the philosophies of the past has no place for such a conception, just as it has no place for such schoolmen as Cabeo. That image is well on its way towards historiographical rejection, but a nuanced understanding of its difficulties—an understanding that also recognizes its strengths—requires the incorporation of time itself into the making of natural knowledge in the seventeenth century.[9]

Thomas Kuhn has emphasized the "essential tension" between innovation and tradition in science through his examination of the place of conservative as well as revolutionary behaviors in the structure of scientific endeavor.[10] The cultural origins of this "tension" have recently been considered by the sociologist Joseph Mali, who traces its development to the confrontation between Protestant and Catholic polemicists during the Reformation.[11] Mali identifies a view of tradition (one frequently associated with the eighteenth-century philologist Giambattista Vico)[12] that, rather than emphasizing unchanging customary praxis, stresses tradition's creative malleability as a constitutive part of any human society. The application of the idea to Kuhnian scientific communities follows immediately, as a way of overcoming the apparent paradox of an activity largely governed by routine "normal science" that nonetheless values innovation as its highest achievement: innovation becomes the highest expression of creative continuity.

Mali's discussion may, however, be pushed further. This kind of tradition need not merely be seen as flexible enough to permit the ongoing reorientation (on this view, strictly "nonrevolutionary") of its concepts and practices. The role of an *explicit* idea of tradition in the life of a community may also positively imply the *sanctioning* of change. Counterintuitively, one may say that the idea of a tradition will play no justificatory role in a society that has no need of absorbing change, because such a society will justify its adherence to the status quo through reference to equally stable concepts—an ideal envisioned by upholders of an "original intent" reading of the U.S. Constitution, for example. Appeal to tradition, by contrast, acknowledges a temporal dimension in human affairs and tries to control it by perceptions of

---

9. For an attempt to relate empiricist science in the seventeenth century to differing notions of the past, see Pumfrey, "The History of Science and the Renaissance Science of History" (1991).
10. Kuhn, "The Essential Tension" (1977).
11. Mali, "Science, Tradition, and the Science of Tradition" (1989).
12. See on this Mali, *The Rehabilitation of Myth* (1992).

both similarity and difference between past and present. It should go without saying, of course, that any particular tradition is something that is always actively created by its present users.[13]

II. Phenomena, Observations, and the Temporal Constructedness of Jovian Moons

Cabeo pointed to the essential role of tradition and the community in the creation of natural knowledge. There is something authentically Aristotelian about that viewpoint. It generated problems in the handling of experiential *novelty*, however, as Cabeo's discussion of falling bodies considered in chapter 3 has already illustrated. In effect, the proper constitution of scientific experience called for the integration of novel phenomena into a recognized tradition that routinely served to validate the science's empirical content. However, work by sociologists such as David Bloor and H. M. Collins, drawing on Wittgenstein's later philosophy, has stressed the indeterminacy of methodological rules in science. Any simple techniques for identifying the character of something claimed as new, so as to determine its place in the existing scheme of knowledge, are always, in principle, open to unlimited interpretations. Which interpretation is deemed by the relevant community to be the proper one, and hence to be the correct application of the rules, is a matter of social contingency.[14] But even explicitly traditional forms of knowledge of the kind under examination here have their rules. Rather than constituting the formal structure of accreditation for the knowledge that results from their successful application, however, these rules serve as a guide in determining whether any particular new piece of knowledge is a legitimate development of the tradition or an illegitimate departure from it. They are, in this sense, meta-rules, which accredit a particular development of a tradition instead of directly accrediting new knowledge-claims themselves.[15]

As we saw in chapter 2, Blancanus had justified the publication of his comprehensive astronomical textbook *Sphaera mundi* (1620) in part through the necessity of including the new telescopic discoveries made by Galileo; his mentor Christopher Clavius had not seriously addressed them in a similar work some years earlier.[16] But in 1611 Clavius

13. Cf. Hobsbawm and Ranger, *The Invention of Tradition* (1983), esp. Hobsbawm, "Introduction: Inventing Traditions," pp. 1–14. Shils, *Tradition* (1981), addresses his topic from a sociological perspective; for a historical investigation of possible uses of "tradition," see Pocock, "Time, Institutions and Action" (1971).
14. See especially Bloor, *Wittgenstein* (1983); Collins, *Changing Order* (1985, 1991).
15. This matter is addressed in explicitly historical terms in section VI, below.
16. See chapter 2, n. 50, above.

himself had described Galileo's novelties as kinds of "monstrosity."[17] Galileo had thrown into the arena of astronomical practice novelties of an unprecedented kind, ones that violated ordinary expectations of what the astronomer's job should look like. They were monstrous because they were deviations from the normal. Galileo's new astronomical claims therefore offer especially valuable insights into the making of scientific experience in the early seventeenth century.

In his *Sphaera mundi*, Blancanus had provided appropriate metarules to structure the articulation of his tradition of practice: these were his categorizations of the various forms of empirical knowledge in astronomy discussed above in chapter 2. A "phenomenon" was a celestial behavior or appearance that was immediately obvious even to the casual observer. An "observation," by contrast, was a piece of empirical knowledge that could only be created through the use of instruments and calculatory techniques that were the province of the expert astronomer.[18] Observations were thus the property of a fraternity of astronomical observers who participated in an authenticated astronomical tradition. Galileo's monsters had therefore to be assimilated into that tradition by displaying the hallmarks of genuine "observations."

The monsters first appeared in 1610, in Galileo's little book *Sidereus nuncius*. While some attention has of late been given to Galileo's representations there of the moon,[19] little has been paid to the depictions in the same text of the satellites of Jupiter.[20] For pages at a time Galileo presents formalized diagrams of the stars apparently accompanying Jupiter as it traverses the heavens; these diagrams frequently track the changing appearance of the planet and its companions not just from night to night, but over the course of a single night. Galileo intends thereby to demonstrate that the stars are in fact moons, what Kepler was soon to dub "satellites," orbiting Jupiter in analogous fashion to our own moon around the earth.

If this material were to convince Galileo's readers, those readers needed to bring to it particular expectations that, nonetheless, Galileo did not in the main regard as requiring articulation. This is remarkable, in that not only were Galileo's diagrams purported representations of

---

17. See chapter 1, text to n. 26, above.
18. See chapter 2, section II, above.
19. See, e.g., Winkler and Van Helden, "Representing the Heavens" (1992); Edgerton, *The Heritage of Giotto's Geometry* (1991), chap. 7.
20. Parker, "Galileo and Optical Illusion" (1986), stands as a partial exception, but it is not about representational conventions. Van Helden, "Telescopes and Authority" (1994), however, notes on p. 10 the novelty of Galileo's sequential diagrammatic representation.

things seen using an instrument possessed by no one else, but they also required even then a considerable amount of effective interpretive glossing. Each diagram represents the disk of Jupiter with noteworthy neighboring stars; Galileo's first task is to show that these stars are not fixed—merely forming the backdrop against which Jupiter's proper motion may be tracked—but are associated with the planet itself (a known fixed star serves as a reference in later diagrams). He then wishes to show that the companions orbit the planet in regular, quite short periods and are four in number. But the views depicted are, of course, static; even the most charitable observer peering through a Galilean telescope would only see Jupiter with some stars in its vicinity, much as Galileo represents in each of his diagrams.

The knowledge that Galileo wishes to make, therefore, resides primarily not in the momentary appearance of Jupiter but in the juxtaposition of many such appearances across time; his readers are not to find such a procedure worthy of critical scrutiny. Galileo's final goal, not yet achieved in the *Sidereus nuncius*, is to determine with exactitude the periods of all four moons, but the conclusion that there *are* four moons circling Jupiter is itself considerable. The issue is the same, on a smaller scale, as that later discussed by Cabeo, but rather than relying upon several generations of observers to construct the "laws and canons" of these particular celestial motions, Galileo's readers only receive testimony from one observer regarding several nights' worth of observational records. The records themselves are then the raw material from which the knowledge-claims about Jovian satellites are constructed.

There is nothing remarkable about this aspect of Galileo's procedure in the context of contemporary astronomical practice as codified by Blancanus. Galileo creates his determination that Jupiter has four moons, each with at least a roughly estimable orbital period, through use of instruments and expert calculatory techniques deployed across a temporal space represented by his sequence of diagrams; he is making "observations" in Blancanus's sense. (Of course, Blancanus's account of proper astronomical method implicitly required that the practitioner possess generally recognized credentials for doing this sort of thing, but Galileo was a professor of mathematics at Padua and presumably had no worries on that score.)[21] There might, however, appear to be some doubt as to the proper cognitive category for the individual views through a telescope represented by each of Galileo's

21. Galileo had no real credentials as a working *astronomer,* of course, and in the usual sense of that term he never did.

diagrams taken separately. Such sights could be witnessed without expert gloss, but they nonetheless required recondite instrumentation and some skill in using it. Galileo does not make clear how he understands their cognitive status. But a very similar case occurring not long afterwards prompted him to address this problem directly.

III. Sunspots as "Phenomena"

In late 1611 the German Jesuit Christopher Scheiner saw dark spots against the surface of the sun. His account of the discovery took the form of three letters written to Mark Welser, a prominent merchant of Augsburg, who fulfilled his duty by promptly publishing them. Galileo, who had also seen such spots (as his correspondence attests), reacted vigorously when he saw the danger of losing priority, and himself wrote to Welser criticizing Scheiner's interpretation of the spots as orbiting bodies rather than properties of the solar surface. Scheiner responded with a "More Accurate Disquisition" (*Accuratior disquisitio*) regarding his position, and Galileo again assaulted his arguments.[22]

Galileo's position in this well-known exchange is usually explained by reference to his anti-Aristotelianism, a view supported by his use of sunspots as a hammer for scholastics both in the present case and, most famously, in the *Dialogo*: Scheiner's attempts to represent the spots as orbiting the sun offended Galileo's perception of them as evidence of generation and corruption in the heavens, contrary to Aristotle's teachings.[23] But the mutability and fluidity of the heavens were not unheard-of propositions in academic circles by this time, nor scandalous ones (albeit not generally favored).[24] More crucial, perhaps, was

---

22. These texts are all reprinted in Galileo, *Opere* (1890–1909), vol. 5. See for overviews of the dispute Drake, *Discoveries and Opinions of Galileo* (1957), pp. 81–85; Shea, *Galileo's Intellectual Revolution* (1972), chap. 3; Moss, *Novelties in the Heavens* (1993), chap. 4.

23. See Galileo, *Opere* 7, pp. 76–78. This interpretation is emphasized, for example, in Drake, *Discoveries and Opinions of Galileo*, p. 83, and in Shea, *Galileo's Intellectual Revolution*, esp. pp. 49–55; see for a different argument (more compatible with my own) Pitt, "The Heavens and Earth" (1991), esp. p. 137.

24. The most notable late sixteenth-century exponent of this view is perhaps Robert, later Cardinal, Bellarmine: Baldini and Coyne, *The Louvain Lectures* (1984), publishes the lectures from the early 1570s in which he expounded these ideas. See also Baldini, *Legem impone subactis* (1992), chaps. 8, 9; Blackwell, *Galileo, Bellarmine, and the Bible* (1991), pp. 40–45; Pitt, "The Heavens and Earth." Lattis, "Christoph Clavius and the *Sphere* of Sacrobosco" (1989), pp. 149–161, 326–337, discusses both Bellarmine's views and the opinions of other Jesuits during the first three decades or so of the seventeenth century both in support of the fluidity of the heavens and on the Tychonic system. See also Wallace, "The Certitude of Science" (1986), for examples from Jesuit commentaries on Aristotle's *De caelo*; Krayer, *Mathematik im Studienplan der Jesuiten* (1991), Pt. II, chaps. 6, 7, on Catten-

Galileo's wish to use the discovery as support for the proposition that natural philosophy could and should learn from the mathematical sciences. If an astronomical discovery could directly inform physical speculation, then the disciplinary hierarchy would be at least leveled, if not inverted, and this was the preeminent vocational goal of the Tuscan grand duke's new "philosopher and mathematician." So Galileo was concerned to portray the point of contention between himself and Scheiner as one resting on the solidity of deductive mathematical argument. Ironically, Scheiner, the Jesuit mathematician, seems to have been in considerable sympathy with this position, although his agreement played no part in the substance of the current debate.[25]

Galileo explained how he wanted his arguments to be understood in one of his letters to Welser. The meta-rules found in Blancanus clearly inform Galileo's thinking. He wrote:

> The different densities and degrees of darkness, their variations [*mutazioni*] of shape, their agglomerations [*accozzamenti*] and separations [*separazioni*] are directly evident to our sight without any need of reasoning, as a glance at the diagrams I am enclosing will show. But that the spots are contiguous to [the surface of] the sun and are carried around by its rotation has to be deduced and concluded by reasoning from certain particular features [*certi particolari accidenti*] which our sensory observations yield.[26]

---

ius's support for the Tychonic system and its fluid heavens. Baldini and Coyne, *The Louvain Lectures*, pp. 3–4, discuss Scheiner's apparently approving citation of Bellarmine's views in Scheiner, *Rosa ursina* (1630). (Scheiner's book took a long time from the commencement of its printing to its completion; its title page therefore leads to the frequent dating of it as 1626–1630.) Scheiner discusses at considerable length this issue of the nature of the heavens and cites many authorities besides Bellarmine concerning the proposition that the heavens are corruptible (*Rosa ursina*, pp. 658–698), and that they are either fluid or igneous (ibid., pp. 699–746)—which latter position in particular Scheiner supports (e.g., ibid., p. 624 col. 2).

25. On the disciplinary hierarchy, including some later views of Scheiner, see chapter 6, section III, below; Westman, "The Astronomer's Role in the Sixteenth Century" (1980); Biagioli, "The Social Status of Italian Mathematicians" (1989); idem, *Galileo, Courtier* (1993), pp. 218–227. An opponent of Galileo's in his controversy over bodies in water in 1612, Vincenzo di Grazia, criticized "those who wish to prove natural facts by means of mathematical reasoning, among whom, if I am not mistaken, is Galileo. All the sciences and all the arts have their own principles and their own causes by means of which they demonstrate the special properties of their own subject. It follows that we are not allowed to use the principles of one science to prove the properties of another." This conventional position was, of course, vigorously rejected by Galileo. See Shea, *Galileo's Intellectual Revolution*, pp. 34–35, quote on p. 34.

26. Galileo, *Opere* 5, p. 117; translation slightly modified from Shea, *Galileo's Intellectual Revolution*, pp. 54–55.

Galileo here characterizes those features of sunspots that can be depicted on a diagram (several of which are presented with his published letters) as tantamount to Blancanus's "phenomena." They are "directly evident to our sight without any need of reasoning."[27] But claims for those properties of sunspots constituted by time as well as position—that is, characteristic motions and those things that can be inferred from them—must be "deduced and concluded by reasoning." These include the arguments for the contiguity of the spots with the solar surface. This second category, then, is epistemologically equivalent to Blancanus's "observations," being constructed by expert astronomers.

In the writings on sunspots that so offended Galileo, Scheiner used a similar terminological scheme to that of Blancanus—just as he was to do, in a more formalistic way, in his optical treatise *Oculus* in 1619.[28] Scheiner initially introduced the matter in a way that, unlike the remarks of Galileo quoted above, displays the peculiarity of the present case deriving from the involvement of special instruments in mere individual sensory perceptions. Those special instruments might appear to make such perceptions the privileged property of what Scheiner called in his *Oculus* the *peculiari empirici*, those people who would in the present case be the astronomers who make so-called "observations." Yet at the beginning of the first letter to Welser, in which he claims to be the first to see the sunspots, Scheiner speaks of the "phenomena that I have observed around the sun" as "new and almost incredible." He and friends of his have brought forth into the clear light of truth (*in clarissimam veritatis lucem*) new things hitherto unknown to, or even denied by, astronomers.[29] Once Scheiner has outlined his basic claims, he again uses the words *phaenomenon* and *phaenomena*, now in connection with the refutation of possible explanations of the *phaenomena* different from that associating them with the body of the sun. So Scheiner describes how the sunspots appear to different eyes, and through different telescopes (the term is *tubus opticus*),[30] and how they do not move even when the telescope is rotated about its axis.[31] This, he says, could not happen if the *phaenomenon* were brought about by the telescope itself.[32] Similarly, it cannot be something existing in the air, because it does not exhibit diurnal change or parallax. Throughout

---

27. See also, on related issues, Wisan, "Galileo's Scientific Method" (1978), pp. 20–24.
28. See above, chapter 2, section II.
29. "Phaenomena, quae circa Solem observavi, petenti affero, mi Velsere, nova et paene incredibilia." Galileo, *Opere* 5, p. 25.
30. Rosen, *The Naming of the Telescope* (1947).
31. Galileo, *Opere* 5, pp. 25–26.
32. Ibid., p. 26.

a number of further exploded objections, Scheiner's reference term (although not explicitly used in every case) apparently remains *phaenomenon*.[33]

What is of interest here is the fact that Scheiner speaks of a *phaenomenon* that is "new and almost incredible." A *phaenomenon* for Blancanus had been something evident to, and known by, everyone, as against the recondite *observationes* of the skilled astronomer. Galileo had spoken of those appearances of sunspots which could be depicted on a diagram as "directly evident to our sight," in contradistinction to things known by reasoning from certain particular features yielded by sensory *observations*—these latter being the province of the mathematician (but also "philosophical astronomer")[34] Galileo. The difference between the two categories in the case of sunspots was, however, much less sharp than Galileo apparently wished to make it, and Scheiner's initial stress on the novelty of these *phaenomena* indicates the problem. Something new and, indeed, "almost incredible" scarcely looks like a natural candidate for an *evident* "phenomenon." And yet that is how Scheiner describes it and how Galileo treats it.[35]

When Scheiner switches terminology, and speaks of "the observations themselves,"[36] he refers to the small diagrams accompanying his first letter. After discussing the diagrams and their shortcomings, he then proceeds to explain how, and under what circumstances, he had made them. He does so because, as he says, astronomers (*mathematici*) "prefer to believe their own eyes in such a matter than those of others."[37] Scheiner's diagrams tend to underline this point, by being small and rather schematic.[38] If his fellow astronomers were going to investigate for themselves, Scheiner's claims did not need to rest on his own observations. Galileo, by contrast, ornamented his own letters on sunspots with large and detailed diagrams, copied from telescopic projec-

---

33. Ibid.

34. Galileo talks of "philosophical astronomers" in his first letter to Welser, ibid., p. 102.

35. In his *Accuratior disquisitio,* however, although Scheiner speaks of the telescopic phenomena as having been recently discovered, he notes that they are in themselves "most ancient" (*antiquissima*): ibid., p. 69.

36. Ibid., pp. 26–27.

37. Ibid., p. 27: "Et quoniam memini, te aliquando quaerere, quinam essent isti aquilarum pulli, qui Solem recta auderent intueri; compendia etiam, quae mathematici, qui propriis in tanta causa oculis quam alienis credere malent, tuto sequantur, expertus monstrabo."

38. Ibid., pp. 32–33. The diagrams are not much better (or bigger) in the *Accuratior disquisitio.* Cf. Winkler and Van Helden, "Representing the Heavens," on maps/diagrams of the moon in the seventeenth century.

tions.³⁹ In the *Accuratior disquisitio*, Scheiner remarks at one point that "were I also to show the solar spots to be seen without any telescope, by the eye of anyone, why will anyone raise objections (whoever does so), if he is honest?"⁴⁰ He then proceeds to explain how the spots may be glimpsed without use of a telescope by a sort of pinhole camera arrangement, as well as by use of a mirror to create a similar image on a sheet of paper. "And I have received this way of observing, long having been sought for in vain, from a certain excellent friend of mine." It is, he says, "free from blemish."⁴¹

A way of viewing sunspots that does not require the offices of a telescope thus achieved perfection precisely by virtue of its freedom from artificial visual assistance. Because sunspots can now be observed, said Scheiner, without a telescope, they are accessible to "the eye of anyone." In this way, one might say, they truly become "phenomena" in Blancanus's sense. The distinction that Galileo made was somewhat different, of course: he was not, as was Scheiner in the material we have just considered, concerned simply to establish the brute phenomenon of spots on the sun; he wanted to go beyond it to what might be learned through the application of the methods of the mathematician. At this point, telescope or not, we are beyond the reach of phenomena and incontestably in the realm of observations made by *mathematici*.

Scheiner's procedure for creating credibility resembles features of Riccioli's work with falling bodies and pendulums—he adduces witnesses.⁴² The second sentence of his first letter to Welser makes the point that the new and almost incredible phenomenon had produced admiration not only in him, but also in his friends.⁴³ In the *Accuratior disquisitio*, immediately after the discussion of how sunspots can be made visible even without a telescope, Scheiner makes a point of listing the names of a good many illustrious and learned scholars and mathematicians (not all of them his coreligionists) who confirm or assent to his general assertions about the reality of the sunspots. The last of these are Jesuit brothers of Scheiner's, Christopher Grienberger,

---

39. Galileo, *Opere* 5, pp. 145–182, with one large picture to a page. On their production see Kemp, *The Science of Art* (1990), pp. 94–96.

40. Galileo, *Opere* 5, p. 61: "Iam si ostendero, maculas Solares etiam videri sine ullo tubo, oculo homine cuiusvis, quid opponet, quisquis opponit, ut non imponat?"

41. Ibid., p. 62: "Et hunc observandi modum, diu frustra quaesitum, accepi ab optimo quodam amico meo. Quae maculas indagandi ratio omni etiam prorsus errandi labe caret."

42. See chapter 3, section IV, above.

43. Galileo, *Opere* 5, p. 25.

Clavius's successor at the Collegio Romano, and Paul Guldin, both of whom are said to have seen the spots on specified dates. Guldin's observations, presented in the form of small diagrams, are given to show how they agree with ones independently made by Scheiner himself.[44] A good number of Scheiner's cited supporters, however, had not actually witnessed the spots themselves, as he cheerfully acknowledges. They are instead character witnesses.

The astronomical phenomena of which Cabeo was to write were underwritten by "principles" that expressed a supposed periodicity. This phenomenon, however, of spots that appeared and disappeared with no principles yet found to order them, seemed to consist of genuinely historical—that is, unrepeatable—events. Only reliable testimony could establish assertions of specific appearances at specific times. But Scheiner's intention was not to establish specific appearances for their own sake; it was to establish universal or generic properties of sunspots. The specifics acted towards that end, especially in the case of those observations made concurrently with Guldin, insofar as they established the truth of the general claims; that was their purpose, in exactly the way that Scheiner cited the names of various illustrious people who had not even seen the phenomenon themselves but who were prepared to assent to its reality.[45] The payoff came in the list of universals that Scheiner felt warranted in presenting after all the particulars had been laid out.

> *The consequences of these observations are these:*
> 1. These appearances are not merely in the eye.
> 2. They are not a flaw of the glass.
> 3. They are not a sport of the air, but have to do neither with it nor with any heaven much lower than the sun.
> 4. They move around the sun.
> 5. They are close to the sun, because with other cases, such as the moon, Venus, and Mercury, [these latter bodies] seemed to have shown clearly a greater removal [than that of the spots].
> 6. They are of very flat or thin bodies, in that, as a consequence, their diameter is diminished in longitude, but conserved in latitude (that is, because they become slender close to the limit of the solar perimeter).

44. Ibid., pp. 62–65 (diagrams on p. 63). A similar procedure involving independent drawings (although not this time in different locations) was used by Kepler in his attempts to confirm Galileo's observations of the Jovian moons: Van Helden, "Telescopes and Authority," p. 12. Cf. similar remarks on the establishment of the reality and character of sunspots by the Roman Jesuit professor of mathematics Franciscus Eschinardus in 1689, as quoted in Feldhay and Heyd, "The Discourse of Pious Science" (1989), on p. 115.

45. Galileo, *Opere* 5, p. 62.

7. They are not to be received into the number of the stars,

    1, because they are of irregular figure;

    2, because they likewise vary;

    3, because they all undergo equal motion, and, since they depart little from the sun, it ought already to have returned them several times, contrary to what has happened;

    4, because they repeatedly arise in the middle of the sun, which will have escaped the keenness of the eyes at first glance;

    5, because occasionally some of them disappear before the completion of their course.[46]

Thus could Scheiner's observations be made into universal science, formalized in a way that Galileo eschewed.

Once again, however, Galileo's procedures closely resembled Scheiner's in practice even while they differed from them in appearance.[47] In his second letter to Welser on sunspots (of which Scheiner remained ignorant in the *Accuratior disquisitio*), Galileo notices a means of con-

---

46. Ibid., p. 65:

"*Consequentiae harum observationum sunt hae:*

(1) Has apparitiones non esse tantum in oculo.

(2) Non esse vitri vitium.

(3) Non aëris ludibrium, sed neque in ipso, neque in aliquo caelo versari, quod sit Sole multo inferius.

(4) Moveri circa Solem.

(5) A Sole prope distare, quod alias in longa ab ipso remotione illustratae viderentur, ut Luna, Venus et Mercurius.

(6) Esse corpora multum plana sive tenuia, propterea quod in longitudine sphaerae diminuatur ipsarum diameter, at in latitudine conservetur (hoc est, quod gracilescant iuxta perimetri solaris extensionem).

(7) Non esse in numerum stellarum recipiendas,

    (1) quia sint figurae irregularis;

    (2) quia eandem varient;

    (3) quia aequalem omnes subeant motum, et, cum parum absint a Sole, oportebat eas iam aliquoties rediisse, contra quam factum;

    (4) quia subinde in medio Sole oriantur, quae sub ingressum oculorum aciem effugerint;

    (5) quia nonnunquam dispareant aliquae ante absolutum cursum."

A similar numbered list of general conclusions, but necessarily less comprehensive, appeared at the conclusion of Scheiner's original letters on the sunspots, his *Tres epistolae* (Galileo, *Opere* 5, p. 31); they are there given the good mathematical title *Corollaria*.

47. This is not always an applicable point to make of Galileo's writings, however. Galileo's *Il saggiatore*, written in 1623 as a polemic against Grassi, has all the baroque features of Renaissance polemic deployed in Grassi's own writings in the debate. But Galileo, most famously in the *Dialogo* but also in most of the *Discorsi*, often avoided the formalized mathematical exposition of his arguments in favor of more accessible, informal arguments. His letters on sunspots frequently fall into that category, as contrasted with Scheiner's.

firming the reality of the new phenomenon as evidence of mutation in the heavens. It is, he says, easy:

> Anyone is capable of procuring drawings made in distant places, and comparing them with those he has made himself on the same days. I have already received some made in Brussels by Sig. Daniello Antonini for the 11th, 12th, 13th, 14th, 20th and 21st of July which fit to a hair those made by me, and others sent to me from Rome by Sig. Lodovico Cigoli, the famous painter and artist. This argument should be enough to persuade anybody that such spots are a long way above the moon.[48]

This is, of course, exactly what Scheiner had done in regard to the observations of Guldin, in that case to confirm the reality of the phenomenon itself. It is important to note that Galileo wants to stress that these are "easy" matters that "anyone" can accomplish. It remains less certain that the same characterization could have been applied to the Medicean stars.

IV. From Points of Light to Orbiting Bodies: The Sociology of Observation

Galileo's account in the *Sidereus nuncius* of his discovery and investigation of the moons of Jupiter has him quickly reaching the conclusion, "entirely beyond doubt, that in the heavens there are three stars wandering around Jupiter like Venus and Mercury around the Sun. This was at length seen clear as day in many subsequent observations, and also that there are not only three, but four wandering stars making their revolutions about Jupiter."[49] The initial conclusion had been formed, however, on the basis of just four discrete views of Jupiter, each presented schematically in the account.

Galileo's lengthy subsequent parade of depictions formed the basis not only of the conclusion that there were four stars accompanying Jupiter on his course, but also of a determination, imprecise and incomplete as Galileo admitted it to be, of their orbital periods.[50] The reader is given little time in the course of the narrative to question the initial conclusion, the most spectacular, that Jupiter has orbiting companions. The rather lengthy material following the initial conclu-

---

48. Galileo, *Opere* 5, p. 140; translation slightly modified from Drake, *Discoveries and Opinions of Galileo*, p. 119.

49. Galileo, *Opere* 3, pt. I, p. 81; trans. Albert Van Helden, in Galileo, *Sidereus nuncius* (1989), p. 66.

50. Drake, "Galileo and Satellite Prediction" (1979). For a valuable parallel, see the account of the discovery of pulsars in Woolgar, "Discovery" (1980).

sion no doubt helped to make it all seem more plausible, by lending evidential weight to the claim, but Galileo made the initial claim itself appear quite unproblematic. This is despite the fact that, as Stillman Drake has persuasively argued, Galileo's own initial interpretation of what he saw was of bodies moving to and fro in a straight line, successively approaching and moving away from Jupiter.[51] Kinematically, this viewpoint need never have been abandoned. In the letter of 1611 written (in Italian) by Clavius and his mathematical colleagues to Cardinal Bellarmine, answering the cardinal's queries about Galileo's observational claims, the writers commit themselves only to the proposition that there are four stars, different from the fixed stars, that move back and forth in what appears to be a straight line (*in linea quasi retta*) with ever-changing distances from Jupiter.[52] No doubt Galileo wanted the depictions that he used to support his assertions to be like the sunspots, "directly evident to our sight," and thus, "phenomena," but he could not count absolutely on others seeing things in the same way.

The crucial feature of Galileo's little book, seen as an attempt to establish new truths about the natural world, is that when it was published the observations on which it was based could not be made by anyone but Galileo, because only he possessed a good enough telescope.[53] Nonetheless, Galileo published, presumably with the expectation of being believed, and named (with permission, of course) the four new planets going around Jupiter the "Medicean stars," after the family of his soon-to-be patron, the Grand Duke of Tuscany. Galileo's initial campaign to win general acceptance of the stars took a remarkable form, one that shows in very concrete terms what was at stake in creating shared experience in the philosophy of nature. Neither tradition nor an adequate observational community provided Galileo with the mechanisms through which scientific experience was usually con-

---

51. Drake, "Galileo and Satellite Prediction," p. 86; idem, "Galileo's First Telescopic Observations" (1976), pp. 161–163. See also Shea, "Galileo Galilei" (1990), esp. pp. 59–70. The cognitive model of a heliocentric Mercury and Venus, well known from antique suggestions, must have played a more than didactic role in Galileo's development of the orbital interpretation. The lack of absolute inevitability of the orbital conception shown by Galileo's initial alternative reading demands comparison with the long-drawn-out story of the strange appearance of Saturn: cf. Van Helden, "Saturn and His Anses" (1974); idem, "'*Annulo cingitur'*" (1974); idem, "The Accademia del Cimento and Saturn's Ring" (1973).

52. Galileo, *Opere* 11, p. 93 (letter of 24 April 1611).

53. Thomas Harriot represents a partial, and isolated, exception: see Shirley, *Thomas Harriot* (1983), pp. 397–398, on Harriot's lunar observations of 26 July (Old Style) 1609, and Hetherington, *Science and Objectivity* (1988), pp. 11–17, comparing Harriot's drawings with Galileo's.

stituted. Galileo's initial campaign involved sending to recipients all over Europe telescopes that he had made himself, so that there would be others to see the new planets. The telescopes went only indirectly to astronomers such as Kepler; they went directly to prominent noblemen who had astronomers as clients.[54] The important thing was that credible astronomers—their credibility signaled by powerful patronage—should confirm Galileo's claims. But the gaining of their support required that they first be made, by the provision of the telescopes, into a new kind of astronomer. In effect, Galileo was at this stage not merely making new claims about the world; he was also creating, by means of providing the technical wherewithal, the very observational community that he needed to validate those claims—a telescopic community. That was how he tried to make his astronomical experience universal.

For some months, however, the moons of Jupiter remained inhabitants of Galileo's book alone. And yet, much like the sunspots the existence of which Scheiner reported as having been credited by illustrious referees who had, nonetheless, never seen them, Galileo's assertions met with a considerable (although not uniform) degree of acceptance.[55] Clearly Galileo had handled things appropriately for much of the audience he wished to convince. In one notable case, there exists explicit testimony concerning the grounds on which Galileo's assertions met with belief.

Very shortly after the *Sidereus nuncius* appeared, Kepler published a pamphlet responding to it, in the form of a letter addressed to Galileo.[56] At the time Kepler had still not obtained access to a telescope adequate to the task of observing the moons of Jupiter for himself. In that situation, he regards it as necessary to apologize for believing Galileo's claims. "I may perhaps seem rash," he says to his addressee, "in accepting your claims so readily with no support from my own experience." Kepler recognized that the proper kind of scientific experience was always *shared*. He then proceeds to explain why, nonetheless, it may be legitimate to believe Galileo.

First of all, there are social grounds for Galileo's credibility. "Why should I disparage him, a gentleman of Florence?" asks Kepler; "would it be a trifling matter for him to mock the family of the Grand Dukes

54. Van Helden, "Conclusion" to Galileo, *Sidereus nuncius*, on pp. 91–92; see also Westfall, "Science and Patronage" (1985); Biagioli, *Galileo, Courtier*, pp. 56, 133–134.
55. See Drake, "Galileo and Satellite Prediction"; Van Helden, "Conclusion" to Galileo, *Sidereus nuncius*, for discussions.
56. Kepler, *Dissertatio cum nuncio sidereo*. The text is discussed in Moss, *Novelties in the Heavens*, pp. 85–96.

of Tuscany, and to attach the name of the Medici to figments of his imagination?" These are reasons based on the sort of person Galileo was—a gentleman—and on the imputed social meaning of his claims, to do with his relationship to the Medici.[57] These are essential, constitutive features of the knowledge that Galileo was making, as further witnessed by the fact that he was then in the business of sending telescopes to notable patrons rather than directly to their astronomer-clients. But Kepler also gives reasons for belief that appeal to Galileo's disciplinary status and provide further evidence concerning what could be taken as appropriate in making certain kinds of new knowledge about nature. Indeed, the very first of Kepler's defenses of his credulity speaks directly to this issue: "Why should I not believe a most learned mathematician, whose very style attests the soundness of his judgment?" he asks. Kepler thus acknowledges that the way in which Galileo presents his arguments lends them support; and that way is that of the mathematical scientist, the astronomer—the *mathematicus*.[58]

Kepler provides one other salient reason for believing Galileo: "Shall he with his equipment of optical instruments be disparaged by me, who must use my naked eyes because I lack these aids? Shall I not have confidence in him, when he invites everybody to see the same sights, and what is of supreme importance, even offers his own instrument to gain belief from people's eyes?"[59] Kepler's words invoke an interesting double set of assumptions that lay bare part of the mecha-

---

57. Kepler, *Dissertatio*, p. 107; trans. Rosen, *Kepler's Conversation* (1965), pp. 12–13. See also Dear, "From Truth to Disinterestedness" (1992), esp. pp. 626–627. On Galileo's patronage relationship to the Medici, see above all Biagioli, *Galileo, Courtier*, esp. chap. 2; also Westfall, "Science and Patronage." The relationship between social status and credibility in the seventeenth century is the central theme in Shapin, *A Social History of Truth* (1994). Kepler's association of the likelihood of Galileo's empirical claims with the status of his dedicatee appears to have been a standard one: compare the dedicatory practices of the sixteenth-century writers on America, Oviedo and Las Casas, who each said that "the majesty of the person to whom his work was addressed (Charles V in the case of the *Historia general*, Philip II in that of the *Brevissima relación*) was a pledge of their accuracy. Who, asked Oviedo, would lie to Caesar?" Pagden, *European Encounters with the New World* (1993), p. 58.

58. Kepler, *Dissertatio*, p. 107; trans. Rosen, *Kepler's Conversation*, p. 13. The *Sidereus nuncius* deploys mathematical arguments at several crucial junctures, perhaps most notably in Galileo's discussion of the heights of the purported lunar mountains: see Galileo, *Opere* 3, pt. 1, pp. 71–72. Galileo's material on the Jovian companions reproduces at the level of "raw data" the early stages of the kinds of astronomical procedures (not routinely seen in published form) that went into the development of any planetary model and thereby reinforced the idea that similar regularities applied here.

59. Kepler, *Dissertatio*, p. 107; trans. Rosen, *Kepler's Conversation*, p. 13.

nism at work in the kinds of science produced by many mathematical writers besides Galileo. On the one hand, Kepler regards it as grounds for belief that Galileo invites others to check his claims by making their own observations; but on the other hand he denies his own right to cast doubt on Galileo's claims precisely because he himself has no instrumental basis for doing so. But he does not therefore suspend belief; he enthusiastically endorses Galileo's discoveries.[60]

These discoveries cannot be regarded as entirely trouble free, however. Although the Jesuit astronomers of the Collegio Romano, led by Clavius, quickly confirmed Galileo's chief assertions (unlike Kepler, on the basis of their own observations), the most belligerent doubter of the Jovian moons, Martin Horky, ridiculed the entire edifice of Galileo's arguments, rooted in the optical instrument itself.[61] Scheiner, more moderately and more credibly, at least saw interpretive options besides that chosen by Galileo. These he exploited in his work on sunspots some two years later.

V. Moons and Maculae

Scheiner clearly did not regard the sunspots as an isolated curiosity of the heavens, and he does not appear, despite Galileo's insinuations, to have been motivated solely by the intention to reduce them to orbital motion that would save the doctrine of the immutability of the heavens (indeed, had that been the case there would have been little incentive for him to have publicized them). Instead, as his original three letters to Welser show, he saw in their properties and behavior a possible explanatory model for a wider range of new telescopic phenomena in the heavens. In particular, he suggested that they might provide clues to the more puzzling aspects of the behavior of Jupiter's companions and the strange appearance of Saturn.

In the list of *corollaria* that comes at the end of the *Tres epistolae*, Scheiner lays out, in the same general form used in the *Accuratior disquisitio*, the principal conclusions that could be drawn from the obser-

---

60. This may be compared with Riccioli's later procedures (see chapter 3, above): the establishment of expertise through appeal to purportedly public (or publicly available) demonstrations that thus act as moral ratification. On Kepler's subsequent endeavors to validate Galileo's observations, see Van Helden, "Telescopes and Authority," pp. 12–13.

61. See Van Helden, "Conclusion," in Galileo, *Sidereus nuncius*, pp. 100–102. Horky's *Contra Nuncium Sidereum* is in *Opere* 3, pt. 1, pp. 127–145; one of his complaints is that the precision of Galileo's claimed positional measurements for the satellites as represented in his diagrams is too great for the capabilities of Galileo's instrument (ibid., p. 144).

112   Chapter Four

vations he had presented. These include the claim that the sunspots often appear like moons, showing phases:

> 4. If through the splendor of the sun the illuminated parts of them can be distinguished from the parts not illuminated, [we will see] around the sun many horned, gibbous, new, and even perhaps full moon shapes.[62]

No doubt the variety of shapes displayed by sunspots made this an easy claim to support, one way or another. He then continues:

> 5. It is perhaps the same cause with the illumination of other stars.
> 6. Hence it is also reasonable that the companions of Jupiter, as to their motion and location, be of a not disparate nature [from that of the sunspots]: whence we hold for certain as a general matter that they [the Jovian companions] are not merely four [in number] but many, and not stretched around Jupiter in only one circle [i.e., not occupying just one circular plane], but in many. Given which, it is easy to reply to certain objections, and also many contradictions about them regarding [their] motions are resolved; for they appear to move inclined to Jupiter sometimes towards the south, sometimes towards the north.[63]

Scheiner's idea here is that the sunspots are a kind of celestial phenomenon, clearly different from ordinary planets or ordinary stars, of which the companions of Jupiter recently discovered by Galileo merely represented another example. The peculiarities of sunspots were not, therefore, grounds for regarding them as being other than orbiting bodies of some sort; instead, these peculiarities were to be seen as

---

62. Galileo, *Opere* 5, p. 31; cf. diagram in ibid., p. 30.

"4. Si per splendorem Solis liceret partes illarum collustratas a non collustratis discernere, visuras nos plurimas circa Solem lunulas cornutas, gibbas, novas, et fortasse etiam plenas." Scheiner had changed his mind on this particular point by the time of the *Accuratior disquisitio*, as item 6 of his list of conclusions quoted in section III, above, shows. He accordingly altered the appropriate characteristics of stars (*stellae*), which now no longer have themselves generically to be round; they thus remained congruent with sunspots (ibid., p. 53).

63. Ibid., p. 31; I thank Albert Van Helden for his suggestions regarding the latter part of this translation:

"5. Eandèm fortassis esse rationem, quo ad sui illustrationem, aliorum astrorum.

"6. Consentaneum hinc etiam esse, Ioviales comites, quoad motum et situm, haud disparis esse naturae: unde nos ferme pro certo tenemus, illos non tantum esse quatuor, sed plures, neque in unico tantum circulo latos circa Iovem, sed pluribus. Quo dato, facile respondeatur ad quasdam obiectiones, et multae etiam circa illos in motibus diversitates solvantur; apparent enim ii ad Iovem aliquando in austrum, aliquando in boream inclinati."

typical of a certain newly found kind of celestial phenomenon.⁶⁴ Not only this, but:

> 7. Neither do I entirely fear to conjecture something similar around Saturn: for why does it sometimes show an oblong appearance, while sometimes there appears a retinue with two stars touching its sides? But for the time being I keep this to myself.⁶⁵

Thus the system of sun and planets was well endowed with "sunspots," both of the original variety and of other kinds cognitively modeled upon them. Their characteristics were that they moved around their parent body; that they were of variable aspect, as if appearing and vanishing unexpectedly; and that they moved to a certain extent to both north and south of the equator of the parent body, just as the prototypical sunspots seemed to be confined to a band stretching to each side of the sun's equator.

Galileo's response to these points in his first letter to Welser was characteristically ironic. "It grieves me," he wrote, "to see Apelles [Scheiner's pen name] enumerate the companions of Jupiter in this company (I believe that he means the four Medicean planets)." He continues:

> These show themselves constant, like any other star, and they are always light except when they run into the shadow of Jupiter, at which times they are eclipsed just as is the moon in the earth's shadow. They have their orderly periods, which differ among the four, and which have been already exactly determined by me. Nor do they move in a single orbit, as Apelles either believes or thinks that others have believed; they have distinct orbits about Jupiter as their center, the various sizes of which I have likewise discovered.⁶⁶ I have also detected the reasons for which one or another of them occasionally

---

64. Shea, *Galileo's Intellectual Revolution*, p. 51, describes Scheiner's position in terms of a prejudiced determination that the sunspots were orbiting bodies and fails to explain why Scheiner should then have treated the Jovian companions as he did.

65. Galileo, *Opere* 5, p. 31:

"7. Neque omnino vereor suspicari simile quid circa Saturnum: quare enim modo oblonga specie, modo duabus stellis latera tegentibus comitatus, apparet? Sed hic adhuc me contineo."

66. As noted above in the translated passage from Scheiner's item 6 on this subject, Scheiner probably meant that the companions of Jupiter move in the same circular *plane*, rather than that they all move in the same circular orbital path. There would then be no suggestion that they circled all at the same distance from the planet. The words "neque in unico tantum circulo latos circa Iovem, sed pluribus" no doubt could also be read with Galileo's interpretation, but Scheiner would surely not have made so elementary an error.

tilts northward or southward in relation to Jupiter, and the times when this happens.[67]

At the time of composition of Scheiner's letters, Galileo's further work on the orbits of the Jovian moons had not been published.[68] The determinations to which Galileo refers are evidently those published in his *Discourse on Bodies in Water*, which came out in 1612 and had only recently been seen by Welser in October of that year, after the publication not only of Scheiner's letters but also of his *Accuratior disquisitio*.[69] Galileo makes clear, in effect, that Scheiner's characterization of the moons as "appearing and disappearing" rather like sunspots was not unreasonable; it was the same phenomenon as the one explained by Galileo in terms of periodic eclipsing of the moons by Jupiter's shadow.

> And perhaps I shall have replies to the objections that Apelles hints at concerning these things, whenever he gets round to specifying them. That there may be more of these planets than the four hitherto observed, as Apelles says he holds for certain, may possibly be true; such positiveness on the part of a person who is (so far as I know) very well-informed makes me believe that he must have very good grounds for his assertion which I lack. Hence I should not like to say anything definite about this, lest I might have to take it back later.[70]

Scheiner had not been persuaded by the *Sidereus nuncius*, the only source available to him regarding Galileo's ideas concerning Jupiter, that the simple scheme (however complicated in quantitative determination) of the four moons was the correct way to see Jupiter's unexpected appearance. Scheiner explicated his alternative approach further in the *Accuratior disquisitio*, written after having seen Galileo's first letter to Welser from which the above extracts are taken. In a short section of that publication dated 14 April 1612 (a relevant time of year in the Christian calendar), Scheiner presents diagrams to illustrate aspects of some "remarkable things . . . which don't feed the stomach,

---

67. Galileo's first letter, in Galileo, *Opere* 5, p. 109; trans. Drake, *Discoveries and Opinions of Galileo*, p. 101, with very minor alterations.

68. This was work carried out chiefly in 1611–1612; see Drake, "Galileo and Satellite Prediction."

69. See fourth letter of Welser to Galileo in Galileo, *Opere* 5, pp. 184–185; Drake, *Discoveries and Opinions of Galileo*, pp. 121–122. The second edition of Galileo's *Discourse on Bodies in Water* shows a change in Galileo's views compared with the first edition of March 1612. In March, Galileo seems not to have been fully convinced that the spots *were* on the surface, whereas in the second edition, at the end of the year, he was (I thank Richard S. Westfall for having originally drawn my attention to this point): cf. Drake, *Cause, Experiment and Science* (1981), p. 20.

70. Galileo, *Opere* 5, pp. 109–110; trans. Drake, *Discoveries and Opinions of Galileo*, p. 101.

but delight the eyes."⁷¹ The Lenten feast that he presents involves successive telescopic views of what he calls the "Jovian stars," illustrated by the usual sorts of diagrams. This report is determinedly independent of Galileo's existing claims: Scheiner even includes a star as a candidate companion to Jupiter that seems to be moving in a path quite far removed from the plane of motion of the others. Scheiner is not prepared to concede that Galileo has already resolved everything.

Scheiner subsequently changed his mind on all these issues. By 1614 he had adopted a view much closer to that later presented in his great astronomical work of 1630, the *Rosa ursina*, which deals especially with the sun and sunspots: the sunspots have turned into features at no discernible distance from the sun's surface ("whether they be stars is hitherto disputed," he could still say in 1614, but this was clearly answered in the negative by the time of the *Rosa ursina*), while the Jovian companions are now four in number and just as Galileo had asserted.⁷² But the fact remains that in the first couple of years after Galileo had introduced telescopic astronomy, Scheiner felt entirely justified in reading Galileo's claims, and Galileo's diagrams of Jupiter's appearance in the *Sidereus nuncius*, in ways significantly different from those sought by Galileo himself. There was nothing cut-and-dried about Galileo's presentation of astronomical experience in the *Sidereus nuncius*. As Blancanus said when he laid out his meta-rules, astronomers make observations; the content of those observations therefore depended crucially on the assumptions and procedures of astronomers themselves. Astronomical experience among Galileo's astronomical peers was something made through time. Consequently, Galileo could not fully control the ways in which other astronomers would respond to those aspects of his work that were *monstrous*—literally, unprecedented. His monsters had to be built into an ongoing astronomical enterprise, to be naturalized. Scheiner himself attempted to accomplish such a naturalization along different lines from those favored by Galileo, but each represented a potentially viable continuation of an existing tradition of astronomical practice. Traditions are always open-ended; as means of legitimation they are never preprogrammed.

VI. Apostolic Succession, Humanism, and Method

A deeper understanding of the knowledge-making traditions involved in early-modern philosophical enterprises such as those of Galileo and Scheiner may be had through further consideration of sixteenth- and

71. Galileo, *Opere* 5, p. 54: "spectabiles ... qui pascant non ventrem, sed oculos delectent."
72. Scheiner, *Disquisitiones mathematicae*, p. 66 (sunspots), pp. 78–87 (Jovian moons).

seventeenth-century conceptions of authority and tradition in religion. It must be stressed that (for the time being) the purpose of this consideration is only to set up an analogy to the situations hitherto discussed; it is not to suggest that religious practices structured, conditioned, or reflected those found in the making of natural knowledge. Such might still be the case, of course.

The Catholic Church, in the face of the Protestant challenge, appealed to the sanction of tradition to justify its claims to possessing authentic apostolic authority. It did so precisely because of the undisputed changes in doctrine and practice that had occurred throughout the history of the Church. The Protestants saw these changes as corruptions of the original, authoritative model of the early Church, whereas the Catholics claimed them as the unfolding of God's will.[73] The conflict did not arise over the propriety of a historically sensitive view of Christianity—both Catholics and Protestants agreed that the true Church should look back to its origins. The crucial difference concerned the correct characterization of the changes that had occurred during the history of the Church, and it involved not only disagreement regarding the status of specific events, but also regarding a more general question: what keeps a changing tradition authentic, and what constitutes its subversion or corruption? The Protestant challenge shows that the legitimating function of tradition was not universally seen as self-evident.

Catholics claimed that their tradition was authentic because the development of the Church had been guided by the Holy Spirit, evidence for which lay in the continued occurrence of miracles throughout Church history. The Protestants maintained that the medieval Church had subverted the practices of the original, true Church, and that only a reformation could undo the corruption; they tended to deny the occurrence of true miracles subsequent to the period of the early Church. For Catholics, innovations introduced since the time of the early Church fathers were authentic changes within a tradition; for Protestants, the early Church was the true model, and all there was to do was to emulate it. The latter position did more than simply deny the validity of the Catholic tradition; it denied the legitimacy of any significant change in Christianity and therefore the legitimating potential of any kind of tradition.[74]

At first glance, it appears as though the earlier stages of the Scientific

---

73. See, for a useful discussion of these and related issues, Roberts, "The Politics of Interpretation" (1981).

74. There is an element of oversimplification in this account; there were, of course, important differences between different Protestant sects on this as on other issues. How-

Revolution should be seen as broadly similar in these respects to the Protestant Reformation. Prominent figures such as Copernicus, Vesalius, and François Viète denied legitimacy to existing traditions of astronomy, anatomy, and mathematics; they each criticized the existing state of affairs and wanted to renovate it. However, their intended renovation resembles the Catholic stress on legitimatory tradition more than it does the Protestant view of the restoration of pristine, authentic doctrine. In attempting to restore the ancient achievements in their fields, they stressed the importance of recovering the activities that had enabled those accomplishments.

Their ambition typifies the attitude of the Renaissance humanist, who regarded classical culture as the pinnacle of human achievement and wished to restore it in place of the decadent dark ages that had swept it away. In understanding the meaning of tradition in this period, then, the paradigmatic cultural role is that of the Renaissance humanist.[75] Humanist restoration centered above all on imitation, which meant the learning of a practice. Ancient practitioners of the arts and sciences were clearly good at what they did; their products, whether orations or theorems, showed that clearly. The humanist wanted to recover those ancient abilities, not to produce fake orations or mimic Euclidean proofs as an act of homage. In the study of nature as in other spheres, the humanist enterprise appealed to textual authority, but that authority resided not in what the text said about nature; it resided in how the text was produced. Renaissance humanists wanted to learn how to compose orations like Cicero or to compose histories like Tacitus or Livy; but they also wanted to learn how to compose astronomical models like Ptolemy, to compose anatomies like Galen, to compose theorems like Euclid, or to compose mechanical demonstrations like Archimedes.

Far from seeking *prisca sapientia*, however, humanist practitioners could see themselves as carrying on an authentic ancient tradition even when they disagreed on particular issues with its preeminent ancient exponent.[76] The ancient authority to whom Vesalius looked was Galen,

---

ever, the version given here serves to make the intended general point; it approximates well to Luther's position: cf. Roberts, "The Politics of Interpretation," esp. pp. 19–20.

75. Kristeller, "The Humanist Movement" (1961); Gray, "Renaissance Humanism" (1963). Cf. Grafton, *Defenders of the Text* (1991), chap. 7.

76. We *do* of course have, in this period, interest in restoring pristine ancient knowledge—doctrines—rather than reestablishing ancient enterprises as live traditions. Apart from the hermetic matters stressed by Frances Yates (Yates, *Giordano Bruno* [1964]; idem, "The Hermetic Tradition in Renaissance Science" [1965/1984]), Francis Bacon also appealed to the "wisdom of the ancients." Bacon thought that the pre-Socratics knew what they were talking about because they had enjoyed a closer connection to things them-

and yet Vesalius claimed to be able to do better than Galen because Galen's human anatomy was based largely on apes whereas Vesalius's was based on dissection of actual human cadavers. Copernicus looked to Ptolemy as his model, even though he disagreed both with Ptolemy's adherence to geocentrism and with his use of equants. Just as Galen represented the Greek enterprise of anatomy, so Ptolemy represented the Greek enterprise of astronomy, and just as Vesalius wanted to be a better Greek anatomist than Galen, Copernicus wanted to be a better Greek astronomer than Ptolemy.[77] They were thus attempting to restore interrupted, lost, or corrupted traditions—ancient enterprises the restoration of which still allowed room for change and development.[78]

One of the difficulties that such humanist endeavors faced was that of identifying the essential features of ancient practices. Certainly, Cicero and Quintilian gave helpful information on how to compose an oration, so that it was not necessary solely to rely on the finished products themselves. Galen talked about how to do anatomy; one did not have to divine it solely from the results of his own anatomical work. But Euclid did not write about how to invent theorems; he simply provided finished ones, which is why Viète and other mathematicians tried to reconstruct an "art of analysis" that they believed the ancients must have had but chose to keep secret. Copernicus had to reconstruct what he took to be the authentic rules underlying Greek astronomy; part of his task involved determining which were essential and which

---

selves: see Rossi, *Francis Bacon* (1968), chap. 3. Joy, *Gassendi the Atomist* (1987), discusses Gassendi's attempts to revive Epicurean atomism on a doctrinal level and within the context of a historical critique; see also Joy, "Humanism and the Problem of Traditions" (1993).

77. See, of the more recent literature, Rose, *The Italian Renaissance of Mathematics* (1975); Edelstein, "Vesalius, the Humanist" (1943); Morse, "The Reception of Diophantus' *Arithmetic*" (1981); Long, "Humanism and Science" (1988); Cochrane, "Science and Humanism in the Italian Renaissance" (1976); Vasoli, "The Contribution of Humanism to the Birth of Modern Science" (1979); Westman, "Humanism and Scientific Roles" (1980); Laird, "Archimedes among the Humanists" (1991); Biagioli, "The Social Status of Italian Mathematicians"; Reeds, "Renaissance Humanism and Botany" (1976); Nutton, "Humanistic Surgery" (1985); idem, "Greek Science in the Sixteenth-Century Renaissance" (1993). Rosen, "Renaissance Science" (1961), reviewed Burckhardt's position and Pierre Duhem's opposing views, concluding that Burckhardt's perspective held up in the light of more recent work; such seems to remain the case.

78. The "time" element is a little difficult here; attempts at "restoration" implied a kind of subsuming of the present to the past or a juxtaposing of the two that transcended linear chronology. "Tradition" did not always lie along a simple temporal progression. Cf. Wilcox, *The Measure of Times Past* (1987), esp. chap. 6.

extraneous. Thus, in the dedicatory letter to the pope at the beginning of *De revolutionibus*, he reports on how he looked for evidence that a central, stationary earth was not a necessary element of the Greek astronomical tradition. Similarly, his rejection of Ptolemy's equant device involved putting a particular construal on how the "rule" about uniform circular motion ought to be interpreted. Copernicus took the Greek astronomical enterprise as being independent of the specific doctrines of geocentrism or heliocentrism.[79]

Ancient precedent could therefore be used in two distinct ways: although it might be used as a source of doctrine, it could also, and much more importantly, function as methodological exemplar. Thus while Pierre Gassendi sought to restore the specific doctrine of Epicurean atomism, Viète sought to restore the ancient "art of analysis."[80] The former endeavor was more straightforward than the latter, because a claim to be engaged in the same enterprise as were the authors of its ancient exemplars, while at the same time admitting deviations from their actual doctrines, required that the characterization of their enterprise contain some extra ingredient. One, if partial, solution was to construct a historical narrative. Galileo's early patron, the Archimedean mathematician and engineer Guidobaldo del Monte, felt obliged to reconstruct a heroic "history of mechanics," with Archimedes as its high point, as a way of defining the ancient enterprise to which he appealed—even though Guidobaldo disagreed with certain aspects of Archimedes' work.[81] Vesalius took advantage of Galen's criticisms of his own contemporaries: Galen had represented the state of affairs in his time as the end product of a long period of decline in the physician's art; Vesalius simply borrowed Galen's characterizations as a way

---

79. Copernicus, *De revolutionibus* (1543), dedicatory preface; for a convenient annotated translation, see Copernicus, *On the Revolutions* (1992), pp. 3–6. See Westman, "Proof, Poetics, and Patronage" (1990); also Rose, *The Italian Renaissance of Mathematics*, chap. 5. Oddly, this last point is very close to the way in which Lakatos and Zahar read Copernicus: Lakatos and Zahar, "Why did Copernicus's Research Programme Supersede Ptolemy's?" (1978).

80. This can never, of course, be a strict separation; in addition to renovating specific doctrines, Gassendi's program also involved modifications of Epicurean atomism together with the implicit injunction to carry on a progressive natural philosophical enterprise on its basis. The distinction between the two dimensions of the process are nonetheless evident.

81. Rose, *The Italian Renaissance of Mathematics*, chap. 10, esp. p. 234; Bertoloni Meli, "Guidobaldo dal Monte and the Archimedean Revival" (1992); see also Laird, "Archimedes among the Humanists"; idem, "The Scope of Renaissance Mechanics" (1986). Also of relevance is Biagioli, "The Social Status of Italian Mathematicians."

of revealing the true tradition that Vesalius, following Galen, wished to restore.[82]

A striking example of such an approach in the seventeenth century may be found in Jesuit astronomical texts. Cabeo had described the establishment over time of the "laws and canons" of astronomy. Thus his account necessarily relied on the prior establishment of a canon of accepted astronomical authorities, which were both sources for the data and exemplars of developing astronomical practice. That is why some of Cabeo's Jesuit colleagues who were authors of major treatises on astronomy adopted the device of drawing up chronological listings of the chief astronomers and mathematicians from antiquity to their own day.

Blancanus included with his *De mathematicarum natura dissertatio* of 1615 more than twenty-five pages of a "clarorum mathematicorum chronologia," heavily populated by astronomers. (The first entry for the section covering the seventeenth century is, of course, Clavius, whom Blancanus acknowledges as *praeceptor meus*.)[83] Hugo Sempilius, in his *De mathematicis disciplinis libri* of 1635, devoted forty-nine double-column pages to an "Index auctorum,"[84] giving a separate, exhaustive listing of both ancient and modern writers for each of the mathematical disciplines. Another extensive listing appeared five years after Cabeo's treatise, in Riccioli's *Almagestum novum*. As part of his introductory apparatus, Riccioli presents a chronology of astronomers, astrologers, and cosmographers that runs from Zoroaster through to the latest authority, Giovanni Battista Riccioli. It is followed by an alphabetical listing containing descriptions of each figure and his accomplishments.[85] Throughout their treatises, these Jesuit mathematicians cite other astronomers and mathematicians liberally, contemporaries as well as predecessors—a habit typical of Jesuit scholarship, itself notoriously humanist in tone.[86] It created the sense of a well-established community of practitioners engaged in a continuing enterprise.

---

82. Vesalius, *De humani corporis fabrica* (1543), prefatory dedication to Charles V; translated in O'Malley, *Andreas Vesalius* (1964), pp. 317–324.
83. Blancanus, *De mathematicarum natura dissertatio* (1615), pp. 37–64; quote p. 62.
84. Sempilius, *De mathematicis disciplinis libri duodecim* (1635), pp. 262–310.
85. Riccioli, *Almagestum novum* (1651), Pt. 1, pp. XXVI col. 1 to XXVIII col. 2 (chronology), pp. XXVIII col. 1 to XLVII col. 2 (alphabetical). See also, for a couple of additional such items, Thorndike, *A History of Magic and Experimental Science*, vol. 7 (1958), p. 49.
86. See references on Jesuit education in chapter 2, n. 1, above, as well as Dear, *Mersenne* (1988), chap. 1; also N. Jardine, *The Birth of History and Philosophy of Science*, chap.

However, the establishment of a particular tradition's historical existence still left the problem of justifying that tradition—just as, despite unquestioned chronicles of popes and Church councils that established a genuine ecclesiastical historical development, Protestants would not accept the authenticity of Catholic tradition. Humanist natural philosophers and mathematicians wished to participate in ancient traditions not merely because of the antiquity and historical verifiability of the latter, but because of a belief that they represented privileged ways of doing things. Catholics held their tradition to be justified, and thus capable of legitimating innovations in doctrine, by the continual, behind-the-scenes guidance of the Holy Spirit. Just so, humanistically informed philosophers had their own functional equivalent of the Holy Spirit. It was something of consuming interest by the end of the sixteenth century: method.

The work of Neal Gilbert and others has shown how the topic of method became a commonplace of logic texts and textbook prefaces in many disciplines. Petrus Ramus is, no doubt, the most famous of the sixteenth-century "methodists," but it is crucial to remember that discussion of method throughout this period usually occurred with reference to classical sources, especially Aristotle and Galen.[87] "Method" served an analogous function to that of the Holy Spirit in the Catholic tradition because it identified the hidden source of a tradition's legitimacy.

When Viète sought the "art of analysis" by finding evidence of its possession by the ancients, he wanted a guarantee that he was doing true Greek mathematics. There is a similar dimension to some of Descartes's writings on the same subject. In the fourth of his *Rules for the Direction of the Mind*, written in the years around 1620, Descartes asserts that there is a method that will guarantee the reliability of human knowledge. He goes on to comment:

> I can readily believe that the great minds of the past were to some extent aware of it, guided to it even by nature alone. For the human mind has within it a sort of spark of the divine, in which the first seeds of useful ways of thinking are sown, seeds which, however neglected and stifled by studies which impede them, often bear fruit of their own accord. This is our experience in the simplest of sciences, arithmetic and geometry: we are well aware that the geometers of

---

8, on disciplinary histories in astronomy; Grafton, "Humanism, Magic and Science" (1990), esp. pp. 115–116, on Pico's historical reconstruction of astronomy and astrology.

87. Gilbert, *Renaissance Concepts of Method* (1961); see Dear, "Method in the Study of Nature" (forthcoming), for an overview and bibliography.

antiquity employed a sort of analysis which they went on to apply to the solution of every problem, though they begrudged revealing it to posterity.[88]

A little later in the same text he refers to the "true mathematics" of the ancients, saying: "Indeed, one can even see some traces of this true mathematics [*Mathesis*], I think, in Pappus and Diophantus who, though not of that earliest antiquity, lived many centuries before our time. But I have come to think that these writers themselves, with a kind of pernicious cunning, later suppressed" their knowledge—so as to conceal their method. Instead, they merely exhibited what Descartes calls "barren truths"; that is, true doctrines, but not true methods productive of those doctrines.[89]

Talk of method, then, was a way of reifying the vaguer notion of an ancient enterprise or endeavor. A modern practitioner's claim that he was continuing the work of the ancients received its meaning from the concept of a method that defined what the ancients had done. Only by codifying the essential character of an ancient discipline's practices could new work be presented with all the marks of authenticity to justify its novelty. Method is here playing the role of meta-rules discussed above in section II.

But if method was the humanist philosopher's Holy Spirit, he still needed some equivalent of miracles to confirm its presence. Miracles were the manifest signs of divine superintendence over the history of the Church. For philosophers such as Viète, technical achievements played the same role: they provided the evidence that a method had guided the sciences of the ancients, while their modern accomplishment guaranteed that this method, and hence that tradition, had been successfully restored.

In his *Geometry* of 1637, Descartes had come to adopt a new position. He said that because he could now solve problems that had thwarted the ancients, his symbolic algebra must be something new: instead of Viète's goal, the ancient method of analysis, it represented his own, superior method.[90] Had the ancients commanded Descartes's approach, "they would not have put so much labor into writing so many

---

88. Descartes, *Œuvres* (1964–1976), vol. 10, p. 373; trans. Dugald Murdoch in Descartes, *The Philosophical Writings*, vol. 1 (1985), p. 17. For relevant commentary by Jean-Luc Marion, see Descartes, *Regles utiles et claires pour la direction de l'esprit* (1977), pp. 137–152.

89. Descartes, *Œuvres*, vol. 10, pp. 376–377; trans. Dugald Murdoch in Descartes, *The Philosophical Writings*, vol. 1, pp. 18–19.

90. Descartes, *La géométrie*, book I (in Descartes, *Œuvres*, vol. 6). Not everyone agreed: as late as 1657 John Wallis, himself a noted mathematician, regretted that "most of the

books in which the very sequence of the propositions shows that they did not have a sure method of finding them all [i.e., roots—in effect, of quadratic equations], but rather gathered together those propositions on which they had happened by accident."[91] Descartes's argument that the ancients had lacked his form of algebra was aimed at repudiating the humanist enterprise itself, and it did so by repudiating the legitimatory role of tradition. Such was only fitting for someone who regarded the individual, not the community, as the proper locus of cognitive activity. But it would have been a considerable blow to Cabeo and the astronomical enterprise that he described to have abandoned the past in quite so resolute a way; time was its true object of study.

Galileo attempted to portray the moons of Jupiter as inevitable products of what amounted to ordinary astronomical practice; like the sunspots, they were made according to the normal procedures of astronomical science and were as solid as a mathematical science could make them. Scheiner characterized his own work in much the same way, and argued for different conclusions. The meta-rules, or astronomical method, that informed the presentations of both disputants allowed divergent results; by the same token, those rules were adequate to validate each vision of a future astronomy as a legitimate continuation of astronomical tradition. But there is considerable irony in Galileo's presence in this story: Galileo did not want to admit to the reality of the tradition. In one of his sunspot letters he ridicules reliance upon authorities in the making of knowledge (implicitly censuring Scheiner himself), saying, much like Descartes, that "in the sciences the authority of thousands of opinions is not worth as much as one tiny spark of reason in an individual man."[92] There thus lies a major rhetorical, and substantive, difference between Scheiner and his adversary: despite their similarities, Scheiner recognizes that there are more difficulties in making natural knowledge than Galileo is prepared to admit.

Ancients have hidden their Analytics (since there is no doubt that they possessed such a thing)": see De Gandt, "Cavalieri's Indivisibles and Euclid's Canons" (1991), p. 159.

91. Descartes, Œuvres, vol. 6, p. 376; trans. slightly modified from Descartes, *The Geometry* (1952), p. 17.

92. Galileo, *Opere* 5, pp. 200–201; trans. Drake, *Discoveries and Opinions of Galileo*, p. 134. There is a more famous expression of this idea in the *Dialogo*: Galileo, *Opere* 7, p. 78.

Five

THE USES OF EXPERIENCE

I. An Evident Science of Motion

Galileo's work on falling bodies has long stood as a classic case of early-modern physical science, but its character has been contested ever since Alexandre Koyré's *Études galiléennes* of 1939. Against the then-prevailing view of Galileo as a great experimentalist, Koyré wanted to portray him as a metaphysician. According to Koyré, Galileo did not actually perform the experiments that he describes in his published writings; instead, the importance of Galileo's work lay in his neo-Platonic conceptual schema.[1] Nowadays, however, there is general agreement that Galileo did indeed develop his ideas on fall in concert with physical apparatus (although according to what precise relationship is less clear).[2] Nonetheless, Galileo undoubtedly included a considerable number of "thought experiments" in his work, wherein the outcome of a contrived trial was deduced from more general considerations typically not themselves relying on deliberate test. The most accurate conclusion to draw, perhaps (in keeping with the material that we have examined in previous chapters), is that there is a much smaller difference between Galileo's thought experiments and what are usually considered his "real" experimental work than is often allowed. One feature that they share is the way in which the experience appears as part of Galileo's finished argumentation.

"Experiments," as hallmarks of modern experimental science, are

---

1. See especially Koyré, "Galileo and Plato" (1943/1968); idem, *Galileo Studies* (trans. 1978; original 1939).
2. See n. 4, below.

constituted textually as historical accounts of events that act as evidence, in a quasi-legal sense, for the truth of a universal knowledge-claim.[3] "Experiences" in the Aristotelian sense were, as we have seen, usually constituted as statements of *how things happen* in nature, not as statements of how something *had happened* on a particular occasion. So when Galileo rolled balls down inclined planes and shot them from the edges of tables to measure their distances of travel, he was not performing "experiments" in the modern sense unless he did these things as underpinning for their formal presentation in his writings as discrete historical events. But such is not the case.[4] His famous description of inclined planes, adduced to justify his law of fall in the *Discourses and Demonstrations Concerning Two New Sciences* of 1638, now appears more problematic than was recognized in the days of Koyréan and anti-Koyréan jousting over Galileo's credentials as an experimental scientist. Instead of describing a specific experiment or set of experiments carried out at a particular time, together with a detailed quantitative record of the outcomes, Galileo merely says that, with apparatus of a certain sort, he found the results to agree exactly with his theoretical assumptions—having, he says, repeated the trials "a full hundred times." He had shortly before claimed to have done this "often."[5] Both phrases are just ways of saying, in effect, "again and again as much as you like." Galileo thus establishes the authenticity of the experience that falling bodies do behave as he asserts by basing it on the memory of *many instances*—a multiplicity of unspecified instances adding up to experiential conviction.

Recent research has shown that Galileo aimed at developing scientific knowledge, whether of moving bodies or of the motion of the earth, according to the Aristotelian (or Archimedean) deductive formal

---

3. On apparent legal connections, see B. Shapiro, *Probability and Certainty in Seventeenth-Century England* (1983), chap. 2; Sargent, "Scientific Experiment and Legal Expertise" (1989); Martin, *Francis Bacon* (1992); and cf. chapter 3, n. 30, above. The universality of a philosophical or scientific knowledge-claim is, of course, a crucial difference from the specificity of a factual determination in the law.

4. There is a fairly considerable literature on this: a fundamental reference is Drake, *Galileo's Notes on Motion* (1979); see also, for useful discussion and further references, Naylor, "Galileo's Experimental Discourse" (1989). Inferences from his manuscript notes that Galileo did indeed shoot balls from tables, measure actual distances, and so forth are quite solid; the point is to consider how those things were made into universal knowledge of nature.

5. Galileo, *Opere* 8 (1890–1909), p. 212: "per esperienze ben cento volte replicate"; p. 213: "molte volte." Trans. Drake in Galileo, *Two New Sciences* (1974), pp. 170, 169. Cf. Galileo's use, in the *Sidereus nuncius*, of *sexcentis* to mean something like "innumerable": Galileo, *Opere* 3, p. 75.

structure of the mixed mathematical sciences.⁶ His literary practice regarding experience bears out that finding. Galileo is a long way from the literary construction of an experiment in the sense of a reported, singular historical event. He did not provide narratives of what he had done and seen; instead, he told his reader *what happens*. The effect of naturalness attaching to this form of assertion is what renders Galileo's use of experience tantamount to the invocation of thought experiments: the reader is reassured that the world's working in a particular way is entirely to be expected, entirely consonant with ordinary events. There was nothing contentious or novel about such a construal of experience; Galileo could even allow it to be exhibited in the earlier *Dialogo* by Simplicio, his Aristotelian straw man, with no sense of danger to himself. In the course of the famous exchange concerning the dropping of weights from the mast of a moving ship, Salviati asserts that the outcome can be known without resort to experience. Simplicio retorts incredulously: "So you have not made a hundred tests, or even one? And yet you so freely declare it to be certain?"⁷ The Galilean figure of "a hundred times" thus appears even in the construal of proper scientific experience by a notional opponent. A little later, Galileo has Salviati himself refer to the determination of the rate of acceleration of an iron ball by tests "many times repeated."⁸ For Galileo, the proper con-

---

6. Outstanding studies of Galileo's methodological ideals and practice include Wisan, "Galileo's Scientific Method" (1978); McMullin, "The Conception of Science in Galileo's Work" (1978); Wisan, "Galileo and the Emergence of a New Scientific Style" (1981); Lennox, "Aristotle, Galileo, and 'Mixed Sciences'" (1986); and among the many important studies by William A. Wallace, the following are particularly important on methodological matters: Wallace, *Galileo and His Sources* (1984); idem, "The Problem of Causality in Galileo's Science" (1983); idem, "Randall *Redivivus*" (1988); idem, *Galileo's Logic of Discovery and Proof* (1992). Wallace's general approach is followed up in an interesting article by Hemmendinger, "Galileo and the Phenomena" (1984). Also important in elaborating these methodological points are Crombie, "Sources of Galileo's Early Natural Philosophy" (1975); idem, "Mathematics and Platonism" (1977); Carugo and Crombie, "The Jesuits and Galileo's Ideas of Science and of Nature" (1983); Carugo, "Les Jésuites et la philosophie naturelle de Galilée" (1987). Although Charles Schmitt identified systematic differences in the handling of experience between Galileo's early writings and those of the Paduan Aristotelian Zabarella, those differences exemplify the disciplinary differences between natural philosophy and the mathematical sciences and tend to vanish when comparing Galileo with Jesuit mathematical authors: Schmitt, "Experience and Experiment" (1969); cf. Baroncini, *Forme di esperianza e rivoluzione scientifica* (1992), chap. 2, for other criticisms of Schmitt that tend to reduce the differences.
7. "Che dunque voi non n'avete fatte cento, non che una prova, e l'affermate così francamente per sicura?": Galileo, *Opere* 7, p. 171; trans. Drake in Galileo, *Dialogue Concerning the Two Chief World Systems* (1967), p. 145.
8. Galileo, *Opere* 7, p. 249.

strual of experience was not an issue; this was simply what scientific experience *was*.

Galileo's discussion of fall along inclined planes appears in the Third Day of the *Two New Sciences*, and the science of motion proper is constituted by a geometrical treatise in Latin. The presentation of the Latin treatise is interrupted periodically by discussion, in Italian, between the three participants in the dialogue, and the account of experiential support for the assertions of the formal treatise appears in one of those interruptions. Thus Galileo's literary presentation of experience is sharply distinguished from the strict deductive structure of his formal mathematical science of motion. Galileo tried as much as he could to make the basic assumption of that science appear intuitively obvious (namely, that the distances covered by a freely falling heavy body in uniform periods of time from rest follow the sequence 1, 3, 5, 7, and so on), but the Italian gloss bears witness to his failure. It tries to bolster the Latin treatise with an appeal to universal experience, but the constructed, and essentially recondite, character of that experience is precisely the reason why it could not appear in the formal science. Galileo's problem was that a true science had to rely on evident and universally acceptable premises; in having to adduce specialized, contrived experiences, Galileo admitted failure.[9]

One of Galileo's most striking results was his derivation of a parabolic path for projectile motion. This appears in Day Four of the *Two New Sciences* as a mathematical conclusion deduced from the prior principles of uniform acceleration in free fall and a quasi-inertial principle of uniform horizontal motion (an approximation of circular motion around the center of the earth), justified by an argument denying any mutual interference between the two. At no point does Galileo even consider testing the result empirically; there would be no scientific meaning to such a test, strictly speaking.[10] Others were not so sanguine about it. The Jesuit Niccolò Cabeo, who was a natural philosopher rather than a mathematician (although we have already seen his

---

9. Ibid., vol. 8, pp. 190–213; trans. Drake in Galileo, *Discourses and Demonstrations* (1974), pp. 147–170. And unlike astronomers, for example, Galileo could not appeal to a preexisting community of specialists to underwrite his assertions.

10. Galileo, *Opere* 8, pp. 274–275, discusses the inaccuracies of the parabolic path demonstration arising not only from the effect of the medium but also from the lack of strict parallelism of the lines of descent (Guidobaldo del Monte's objection to Archimedes' demonstration of the law of the lever). Guidobaldo had investigated the path of a ball rolled along a transversely inclined surface: see Drake's remarks in Galileo, *Two New Sciences*, p. 143 n. 138.

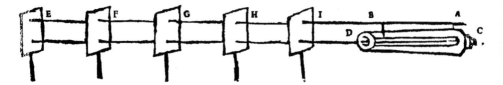

Figure 3. Culverin and screens to track the paths of projectiles, from Niccolò Cabeo, *In quatuor libros Meteorologicorum Aristotelis commentaria* (1646), Lib. 3, p. 41.

sympathy to mathematical procedure), conducted an elaborate attempt at empirical investigation of the question a few years after Galileo's treatise was published (see figure 3). He arranged a set of paper screens equally spaced, one behind the next, in front of the muzzle of a cannon. When the cannon was fired, the height of the cannonball as it passed through each screen would then be marked by the position of the hole in the paper, and these results could be used to reconstruct the actual path.

Cabeo describes the setup in some detail in his commentary on Aristotle's *Meteorology* of 1646, concluding:

> Thus, with matters so disposed, fire is applied to the cannon or culverin; the ball will fly towards E, and will perforate all those interposed sheets, and it does not burn, as experience shows—not at a distance of ten feet from the culverin—but it will only perforate. Note therefore on the individual sheets the distance from the marked point of the axis where the ball perforates the sheet, and because those sheets are equally spaced, you will have exactly the path that the ball, having been carried from the mouth of the culverin to the final mark E, traces out.[11]

Cabeo then claims to have experienced this "not once, nor twice," but repeatedly—the standard formula. Echoing the procedure that we saw in the previous chapter, he also specifies that he has not experienced

11. Cabeo, *In quatuor libros Meteorologicorum Aristotelis commentaria* (1646), Lib. 3, p. 42 col. 1: "Sic rebus dispositis applicetur ignis bombardae. vel sclopo; volabit globus ad E, & perforabit omnes illas cartas interpositas, nec comburet, ut experientia ostendit nec in distantia decempedali à sclopo, sed perforabit solum. nota ergo in singulis cartis, qua distantia à notato puncto axis, globus cartam perforet, & quia folia illa aequaliter distant; habebis praecise viam, quam signat globus delatus ab ore sclopi ad ultimum signum E." The associated diagram runs across p. 41. Note the frequent use of subjunctives in the passage, which I have ignored in my translation for ease of style. A culverin—*sclopus* in Latin—was a particularly large and long type of cannon.

this alone, "but I have assembled with me other observers, as I think ought to be done in physical matters."[12] The marginal gloss reads at this point: "How physical observations are made."[13]

The following year saw Torricelli's well-known discussion of ballistic matters in a correspondence with Giovanni Battista Renieri.[14] Torricelli suggested (without reference to Cabeo) the same procedure of paper screens, which Renieri proceeded to try. Renieri found nothing like a parabolic path; in this, Cabeo had been more fortunate, although also having had the good sense to cross-check with observations of the path of a jet of water.[15] But then, Cabeo knew "how physical observations are made."

Viviani, Galileo's disciple and biographer, is the literary source of the famous story concerning Galileo's dropping of balls from the Tower of Pisa. He wrote that Galileo showed bodies of different weights to fall at the same speed, "demonstrating this by repeated experiences performed from the height of the bell tower of Pisa in the presence of other lecturers and philosophers and all the student body."[16] The story, even if apocryphal, nonetheless shows the appropriate expectations attaching to this kind of empirical work in the 1650s, the date of Viviani's composition. A series of "repeated experiences" with witnesses, not a dramatic account of a specific test, constitutes scientific propriety. If Galileo was in many ways an anti-Aristotelian, it is now nonetheless possible to see his assumptions and those of his contemporaries as involving peculiarly Aristotelian convictions.[17]

II. Mersenne and Frequent Repetition

Marin Mersenne, a dogged promoter as well as critic of Galileo, emphasized the contingency of the created world in a way untypical of Galileo himself. The general framework within which he deployed and understood experience, however, was the same: it was scholastic-

---

12. Ibid., p. 42 col. 1: "Quod ego hic propono sum expertus non semel, neque bis, sed replicatis vicibus, neque solus, sed alios adhibui mecum observatores, ut sic puto faciendum in rebus physicis."
13. Ibid.: "Observationes physicae quomodo fiant."
14. On Torricelli's scheme see Segre, *In the Wake of Galileo* (1991), pp. 94–97; idem, "Torricelli's Correspondence on Ballistics" (1983); A. R. Hall, *Ballistics in the Seventeenth Century* (1952), pp. 91–100.
15. Cabeo, *Commentaria*, Lib. 3, pp. 43–44.
16. Galileo, *Opere* 19, p. 603; trans. Segre, *In the Wake of Galileo*, p. 35.
17. Cf. chapter 3, above, for examples of the rhetoric of indefinite or multiple repetition from Arriaga, Riccioli, and others.

Aristotelian.[18] But this is not to say that Mersenne passively exemplifies philosophical behavior quite alien to the development of the experimental science that emerged in the seventeenth century. The similarities and differences between himself and Galileo on the subject of falling bodies reveal important nuances.

In his "Preface au Lecteur" at the beginning of the *Harmonie universelle*, Mersenne advertises his subsequent discussions of Galileo's remarks in the *Dialogo* on falling bodies. In those discussions, he says, "I compare his experiences with mine, and I often confirm by very precise observations what he has proposed."[19] The remark easily reads like an account of experimental confirmation, or replication, in a modern sense. It does so even more when he goes on to ask the reader not to believe the experiences that he will produce until the reader has made them himself. At the time (the mid-1630s), this was still quite a novel thing to say. Galileo did not make a habit of telling his readers not to believe him (although that is itself, of course, a clear rhetorical attempt to gain credibility); but more to the point, scholastic writers, such as the many Jesuits and others who wrote on such subjects, always maintained an authoritative stance towards their own experiential assertions. Mersenne's antiauthoritative remark should not be overstressed, however, as his explanation of *why* the reader should confirm the experiences before believing them shows: the reader should do this, he says, "so as to have the pleasure of acting himself, and to wonder at the ignorance and carelessness of men" who believe erroneous things such as that bodies fall with speeds proportional to their weights.[20]

The scientific establishment of Mersenne's various propositions concerning natural motion requires the gradual extension to others of the experiences that justify them. But he is not saying (as Harvey, for example, had done) that this extension must occur through the creation of a community of experimenters each of whom has had exactly the same experiences and who therefore constitutes a scientific knower complete in himself.[21] Such would not, in fact, be a true community; it would be an atomized collection of individuals. Mersenne's desire that

---

18. Dear, *Mersenne* (1988); the best general discussion of Mersenne remains Lenoble, *Mersenne* (1943, 1971).

19. Mersenne, *Harmonie universelle* (1636–1637/1963), I ("Traitez de la nature des sons, et des mouvemens de toutes sortes de Corps"), "Preface au Lecteur," first page: "je compare ses expériences avec les miennes, & je confirme souvent ce qu'il a advancé, par des observations très-particulières."

20. Ibid.: "afin qu'il ayt le plaisir de se conduire soy-mesme, & d'admirer l'ignorance et le peu de soin des hommes," etc.

21. Cf. Wear, "William Harvey and the 'Way of the Anatomists'" (1983).

others should try his experiences seems to have something of a moral tone: he wishes that the reader should have the pleasure and *moral instruction* that would come of recognizing at first hand the errors of others, errors born of sloth.

Mersenne's experiences on motion in *Harmonie universelle* are not historical event experiments. Instead, in a manner similar to Galileo's, Mersenne tells the reader what happens in certain contrived circumstances, but not what happened on any specific occasion. He is chiefly concerned throughout to explore the implications and consequences of Galileo's claim about the increase of distance fallen in relation to time. But he is less than ready to accept Galileo's foundational assertion that the final speed acquired by a body falling along an inclined plane is solely a function of its initial vertical height, independent of slope. He says: "the experiences that we have made very precisely on this subject must be given here, so that one can follow what they yield."[22] Just like Galileo a year or so later in the *Two New Sciences*, Mersenne now describes his apparatus: "Having, therefore, chosen a height of five feet, and having had a plane grooved and polished, we have given it various inclinations, so as to roll a ball of lead, and of wood, very round, along the length of the plane." He then proceeds to generalize what he has learned from this apparatus, using exactly the form employed soon afterwards by Galileo—an imprecise statement of the sufficient frequency of trial followed by an account of what has been learned thereby. The rolling of balls, he says, has been done

> from several different positions following the different inclinations, while another ball of the same shape and weight fell from a height of five feet through the air; and we have found that while it falls perpendicularly from a height of five feet, it falls only one foot on a plane inclined by fifteen degrees, although it ought to fall sixteen inches.[23]

And so he continues, for differing inclinations.

Elsewhere, the same familiar commonplaces of presentation and conceptualization appear. Mersenne most frequently says simply "ex-

---

22. Ibid., "Traitez de la nature des sons," p. 111: "il faut icy mettre les expériences que nous avons faites très-exactement sur ce sujet, afin que l'on puisse suivre ce qu'elles donnent."

23. Ibid.: "Ayant donc choisi une hauteur de cinq pieds de Roy, & ayant fait creuser, & polir un plan, nous luy avons donné plusieurs sortes d'inclinations, afin de laisser rouler une boule de plomb, & de bois fort ronde tout au long du plan"; "de plusieurs endroits differens suivant les differentes inclinations, tandis qu'une autre boule de mesme figure, & pesanteur tomboit de cinq pieds de haut dans l'air; & nous avons trouvé que tandis qu'elle tombe perpendiculairement de cinq pieds de haut, elle tombe seulement d'un pied sur le plan incliné de quinze degrez, au lieu qu'elle devroit tomber seize poulces."

perience shows," whether referring to contrived tests or to the common experience of everyday life. In a more recondite case, akin to the example just considered, Mersenne echoes even more closely the expressions found in Galileo: discussing the matter of the ball falling to the foot of the mast of a moving ship, Mersenne says that this does indeed happen, "as all experiences show." But in computing the speeds involved, Mersenne also has occasion to refer to a ball of lead falling from a height of forty-eight feet: "now experience repeated more than a hundred times shows that it falls from this height in two seconds."[24] He had also stressed multiple repetition in a pamphlet of 1634 in which he had first publicly supported Galileo's odd-number rule for fall.[25]

In a specific case in *Harmonie universelle*, however, Mersenne warns against placing too much trust even in frequent repetition. He presents tables showing different mathematical progressions that might express distance fallen in successive equal periods of time, and observes that two of these progressions only diverge from each other gradually—the difference between them would, during the earlier part of the descent, be practically imperceptible, even though it subsequently becomes very large. "Whence one must conclude," he says, "that to make a principle it isn't enough that three or four experiences continually succeed, since the second, third, and fourth numbers of the third column [of the table], having approached so close to the truth, depart from it so greatly thereafter."[26] A similar warning appears at the end of the account of experiences made using inclined planes in which he stresses the exactitude with which everything had been done: he notes that, despite using these best of all possible conditions, the experienced behavior did not accord with rational expectation: "experience is not capable of engendering a science."[27]

III. Mersenne, Descartes, and Reasoned Experience

The similarities between Mersenne's presentational procedure and that of Galileo are striking, but they are not complete. At the end of the

---

24. Ibid., p. 154: "comme monstrent toutes les expériences"; "or l'expérience répétée plus de cent fois monstre qu'elle tombe de cette hauteur en 2 secondes." See also p. 87: "nos expériences répétées plus de 50 fois."
25. Mersenne, *Traité des mouvemens* (1634); see p. 35.
26. Mersenne, *Harmonie universelle*, I, "Traitez de la nature des sons," p. 126: "D'où il faut conclure qu'il ne suffit pas que 3 or 4 expériences reussissent continuellement pour en faire un principe, puis que le 2, 3, & 4 nombre de la 3 colomne ayant approché si prez de la verité, ils s'en éloigne si fort apres."
27. Ibid., p. 112: "l'expérience n'est pas capable d'engendrer une science."

material examining Galileo's claims, Mersenne appeals to the corroborating authority of witnesses: "Those who have seen our experiences, and who have helped in them, know that one cannot proceed in them with more exactness, whether for the plane, which is very straight, and very polished, and which constrains the moving body to descend in a straight path, or for the roundness and weight of the balls, and for the descents."[28] Mersenne is more forthcoming about the advantages of appealing to witnesses in his brief paraphrase, added in press to the *Harmonie universelle*, of Galileo's doctrines on fall found in the *Two New Sciences*. There Mersenne supports Galileo's first eight theorems on motion with "the witness of experience, which I have made from more than twenty-four fathoms high in the presence of learned persons, who have aided in it."[29]

Mersenne had invoked a similar image in a letter to the physician Jean Rey: "Truly, I am astonished at what you distrust of my experience of the equal speed of an iron ball and a wooden ball: for if several persons of quality who have seen and made the experience with me were simply to swear solemnly to you, they would witness it to you authentically."[30] Clearly, if there were witnesses who fulfilled the criterion of being learned or of being "persons of quality," then the experience itself should be believed by anyone who was not intent on disrupting the usual patterns of social behavior on which credibility rested. This is a feature of seventeenth-century experimental science that has been stressed by Steven Shapin for the case of England; we see it here strongly exhibited by Mersenne.[31] In addition, in all three of these examples, Mersenne mentions that the witnesses had *assisted* in the trials. His experience was their experience; they were not merely passive observers, people who saw an experience but did not have that experience themselves. The extension of experience was thereby

---

28. Ibid.: "Ceux qui ont veu nos expériences, & qui y ont aidé, sçavent que l'on n'y peut proceder avec plus de justesse, soit pour le plan qui est bien droit, & bien poli, & qui contraint le mobile de descendre droit, ou pour la rondeur, & la pesanteur des boulets, & pour les cheutes."

29. Ibid., first page following "Table des Matières": "le tesmoignage de l'expérience, laquelle j'ai faite de plus de 24 toises de haut en presence de personnes sçavantes, qui y ont aydé."

30. Mersenne, *Correspondance* (1932–1988), vol. 3, pp. 274–275: "Veritablement je m'estonne de ce que vous vous defiés de mon expérience de l'esgalle vistesse d'un boulet de fer et d'un boulet de bois: car s'ils ne tient qu'à vous faire signer solemnellement plusieurs personnes de qualité qui ont veu et fait l'expérience avec moi, ils vous le tesmoigneront authentiquement."

31. See above all Shapin, *A Social History of Truth* (1994); idem, "Who Was Robert Hooke?" (1989).

rendered as "authentic" (that is, as close to the original source) as possible, while at the same time that experience tended increasingly to take on the characteristics of an "experiment" as a discrete recorded event.

Nonetheless, Mersenne was more interested in the possibility of making calculations on the basis of theoretical models, such as Galileo's odd-number rule, than he was in the precise accuracy of those models when compared with experience. No doubt he would rather have had experience agree exactly with the mathematical model, but he was satisfied if the model was merely the best available—because that meant that it was thereby justified as the appropriate basis for such things as the gunner's tables in his *Cogitata physico-mathematica* of 1644. As he said in the *Cogitata*, the Galilean proportionality of speed to the square of the time was not contradicted by the senses.[32] Clearly, therefore, it was good enough for practical purposes. But Mersenne was not solely concerned with practical purposes. In admitting that the evidence of the senses was not sufficient to establish, rather than merely confirm, the Galilean acceleration of falling bodies, Mersenne pointed to the fact that he, like most of his contemporaries, sought demonstrative explanations that would be certain and hence would indeed serve to establish universal truths about nature. Descartes had dismissed Galileo's own work on falling bodies for just this reason—that Galileo "built without foundations."[33] Mersenne, too, had his own ideas about the proper relation of theoretical understanding to experience.

Mersenne argued that experience, because of its inevitable imprecision, should be corrected by reason. The clearest statement of this appears in 1634, in *Les preludes de l'harmonie universelle*. Mersenne compares the science of music to another classical mixed mathematical science, astronomy. Just as the astronomer's task is to "save the phenomena" by setting up hypotheses that accord with the appearances of the heavens, so the musician must establish a formal science of sounds that accords with the hearing.[34] Mersenne's point is that the senses must be disciplined by reason; otherwise there would be no science, merely descriptions of what the senses discern. Since the senses are imprecise, such descriptions would lack the solidity and constancy required of true scientific knowledge. Thus, the theoretical

32. Mersenne, *Cogitata physico-mathematica* (1644), "Ballistica"; see, e.g., pp. 45, 52.
33. See Descartes to Mersenne, 11 October 1638 (Descartes, *Œuvres* [1964–1976], vol. 2, pp. 380–402); cf. Shea, "Descartes as a Critic of Galileo" (1978).
34. Mersenne, *Les preludes de l'harmonie universelle* (1634), pp. 159–162. See Dear, *Mersenne*, pp. 147–149.

model, as with the astronomer's hypotheses, "saves the phenomena" by providing a consistent representation that does not contradict the senses, and that can even, if one has independent reason to trust its veracity, improve on the evidence of the senses by yielding greater precision than they are capable of attaining by themselves.

Specifically, Mersenne has in mind the Pythagorean association of simple ratios of whole numbers with musical consonances. He knows that the ear is not sensitive enough to discern the difference between a "just" consonance—that is, one produced by string lengths that accord with the Pythagorean ratios of small integers—and a musical interval produced by string lengths in a ratio very close to, but not exactly the same as, a Pythagorean ratio. Thus the ear could not distinguish between the interval produced by a length ratio of three to two—a Pythagorean fifth—and that produced by a ratio marginally differing from it. Indeed, Mersenne admits that even when a sensitive ear is able to distinguish between them, the imprecise, non-Pythagorean interval is not necessarily less agreeable, as the case of tempered scales often shows.[35] However, owing in part to his belief in the theory of the coincidence of air pulses as the physical basis of harmony, Mersenne is sure that the Pythagorean ratios are in a fundamental sense the true ones, and that the ear can therefore be "corrected" by reason when it hears a non-Pythagorean interval as if it were a consonance. Mersenne also adhered to a "rule" stating that, if identical strings are to sound notes an octave apart, their tensions should be in a ratio of four to one. His confidence in the "rule" was not shaken by the need to admit that, in practice, the ratio of tensions must be about four-and-a-quarter to one: his rule was guided by experience, but not governed by it.[36]

This view of the place of experience in the making of natural knowledge accounts for Mersenne's weakening adherence to Galileo's rules for falling bodies in the 1640s. Although he had, in the *Cogitata* of 1644, continued to support those rules as the best supported by experience of any yet suggested, he still looked forward to what Descartes might have to suggest on the matter. In 1646, writing to the young Christiaan Huygens, and then in 1647, in his book *Novarum observationum . . . tomus III*, he cast severe doubt on the Galilean rules precisely because they did not derive from a causal explanation of gravity. In the meantime, Descartes had put forward such an explanation that provided no basis for the Galilean rules, and Mersenne seems to have taken Des-

---

35. Mersenne, *Les preludes de l'harmonie universelle*, p. 164.
36. Mersenne, *Harmonie universelle*, III, "Des instrumens," pp. 123–126.

cartes's work as a reaffirmation of the fragility of experience in making a science.³⁷ As long as experience remained ungrounded in reason, it also remained indeterminate; reason might always change its verdicts. But even if experience were definitive, it could never, as Descartes saw, provide causal explanations: appearances and causal principles constituted logically distinct categories.

Mersenne's promotion of a new kind of scientific community—a promotion represented by his vast correspondence—involved a characteristically seventeenth-century attempt at making experience unproblematic, so that no disputes would arise concerning appearances even if they might arise concerning causal explanations for those appearances.³⁸ But, as his changing attitude towards Galileo's rules of free fall shows, he sometimes found the task difficult—even if, like Descartes, he believed that it was possible, at least sometimes, to establish universal knowledge-claims about the natural world on grounds other than experience alone.

Daniel Garber has pointed out that Descartes's method, as laid out in the *Rules for the Direction of the Mind*, only used experience as a way of establishing the details of what needs to be explained; the explanation itself proceeds, ideally, from intuitively known first principles. Thus when Descartes, in the *Météores*, corrected the angles for the primary and secondary rainbows that had been given by the sixteenth-century mathematician Maurolico, he did it on the basis not of new observations of rainbows, but of his own causal explanation of them. In words strongly reminiscent of Mersenne, Descartes observes that Maurolico's mistake "shows how little faith one ought to lend to observations that are not accompanied by the true reason."³⁹ That perspective echoes the positions in the 1640s of Honoré Fabri and J. B. Baliani on natural acceleration, considered below.⁴⁰ These were shared methodological commonplaces, to which Descartes could make routine appeal in weaving a plausible argument.

One of Descartes's best-known statements on the role of experience in the philosophy of nature appears in a short passage in part VI of the *Discourse on Method*. Its contextualization as part of a broader cul-

---

37. See, for a more detailed treatment, Dear, *Mersenne*, pp. 210–219.
38. See on this point Shapin and Schaffer, *Leviathan and the Air-Pump* (1985), esp. pp. 49–55.
39. Descartes, *Œuvres*, vol. 6, p. 340, quoted and discussed in Garber, "Descartes and Experiment" (1993), esp. p. 305; idem, "Descartes et la méthode en 1637" (1987); see also idem, *Descartes' Metaphysical Physics* (1992), chap. 2.
40. See Baliani's remark, making a very similar point to Descartes's, quoted in Moscovici, *L'expérience du mouvement* (1967), pp. 57–58.

tural setting serves to overcome our preoccupation with Descartes's use of the first person singular in the *Discourse* and the *Meditations* and his profession that all knowledge is grounded in the individual knower. Descartes wrote for a readership that held similar cultural values to his own and that was used to expressing them through similar practices. They were the people who knew how to read Descartes.

In this passage, Descartes says of "les expériences"

> that the further we advance in our knowledge, the more necessary they become. At the beginning, rather than seeking those which are more unusual and highly contrived, it is better to resort only to those which, presenting themselves spontaneously to our senses, cannot be unknown to us if we reflect even a little. The reason for this is that the more unusual observations are apt to mislead us when we do not yet know the causes of the more common ones.[41]

Despite the usual reading of these remarks as providing insights into a peculiarly Cartesian methodological conception, there is no reason to see them as anything other than fairly routine observations. Much the same views appear in a little book by Claude Mydorge. Mydorge was a mathematician with a special interest in optics; he was a colleague of Descartes's when the latter was living in Paris in the 1620s, and he and Descartes may well have collaborated on optical research.[42] In a work published in 1630, a good example of the contemporary genre of recreational mathematics, Mydorge makes the following observations as part of a discussion of spectral colors, especially as they appear in rainbows and fountains:

> In this subject of elevated speculation, as in all other phenomena of which we seek the causes, it is no small thing to have before us, and as it were in our hands, specific, familiar experiences and appearances, which we can compare with other, more recondite ones: for the more we find relations and common agreements, the more by the

---

41. Descartes, *Œuvres*, vol. 6, p. 63; trans. Robert Stoothoff in Descartes, *The Philosophical Works* (1985), p. 143. Clarke, *Descartes' Philosophy of Science* (1982), pp. 22–23, notes that Descartes explicitly distinguished between "common experience" and particular, contrived "experiments," each designated as *expériences*. See also Descartes, *Regles utiles et claires pour la direction de l'esprit* (1977), pp. 251–252, for Jean-Luc Marion's discussion of Descartes's reluctance to credit the experiental claims of others; cf. also Pérez-Ramos, *Francis Bacon's Idea of Science* (1988), p. 11.

42. Schuster, "Descartes and the Scientific Revolution" (1977), chap. 4; Smith, "Descartes's Theory of Light" (1987), pp. 27–28; Eastwood, "Descartes on Refraction" (1984). See also Shea, *The Magic of Numbers and Motion* (1991), chap. 7, esp. pp. 150–152; Shea rejects Schuster's account of the origin of the sine law as having been in collaborative work with Mydorge in the 1620s, however: see Shea, "Author's Response" (1993), pp. 27–28.

knowledge of those will we reach and approach knowledge of others: which is the surest means of philosophizing and reasoning on all subjects, even the most elevated.[43]

Mydorge's account of the primacy of everyday experience as a springboard for more recondite experiences makes it look like a methodological commonplace.[44]

IV. A Causal Science of Motion and the Limits of Experience

In a philosophical textbook of 1646, the French Jesuit theologian and natural philosopher Honoré Fabri had a number of things to say on these issues, and with specific reference to Galileo's work on falling bodies.[45] His overriding concern was to deflate the pretensions of Galileo's "science of motion" to proper (Aristotelian) scientific status.

Using the language of methodology as his weapon, Fabri focuses on Galileo's use of experience: Galileo used experience in his vindication of a formal science in a way that could not fulfill the demands put on it.[46] Fabri's word for "experience" is the Latin *experimentum* rather than the somewhat more common *experientia*, but his usage makes clear that he still means by it the Aristotelian concept of "experience." He defines his terms carefully:

---

43. Mydorge, *Examen du livre des recreations mathematiques* (1630), p. 92: "Or en ce sujet de haute speculation, comme en toutes autres apparences dont nous recherchons les causes, ce n'est pas peu d'avoir par devers nous, & comme en nos mains, des experiences & apparences particulieres & familieres, que nous puissions comparer aux autres plus eloignees: car plus nous trouvons de rapport & rencontres communs, & plus par la cognoissance des uns nous atteindrons, & approcherons à la cognoissance des autres: ce qui est le plus seur moyen de philosopher & ratiociner sur tous suiects, mesmes les plus relevez." Mydorge's work was a commentary of sorts written on a popular book of the 1620s credited to Jean Leurechon, *Recréations mathématiques*; see, for further details, Thorndike, *A History of Magic and Experimental Science*, vol. 7 (1958), p. 593; for questions on its attribution see Eamon, *Science and the Secrets of Nature* (1994), p. 420n.29.

44. This notwithstanding Mydorge's association with Descartes; the lack of formal argument to underpin Mydorge's assertions itself indicates his expectation that they would be seen as unproblematic by his readers. For another perspective on Descartes's historical narration, see Garber, "Descartes and Experiment."

45. The work appeared under the name of one of Fabri's students as being a version of Fabri's lectures at the Jesuit college at Lyons; it was apparently fully authorized and understood as such by contemporaries (see references in n. 55 below): Fabri, *Philosophiae tomus primus* (1646).

46. On the functions and characteristics of "method talk" in science, see Schuster, "Methodologies as Mythic Structures" (1984); also essays in Schuster and Yeo, *The Politics and Rhetoric of Scientific Method* (1986); Schuster, "Whatever Should We Do with Cartesian Method?" (1993).

A physical experiment *[experimentum]* is the behavior of some sensible thing, physically certain and evident—that is, such that it cannot fail this side of a miracle.[47] For example, at one time I see a stone move, at another I see it not move; I see the same thing with a sphere of lead and of wood; I feel the greater blow of a stone falling from a greater height, etc.[48]

He soon gives additional, somewhat more substantive examples of the sorts of things that "all experiences agree in"—things like projectile motion not lasting indefinitely, or the acceleration of the natural motion of freely falling bodies. These, he says, "are established from most certain experiences," meaning from reliable "common experience" of the Aristotelian kind.[49] Such experiences, of course, render immediately accessible the validity of the argument based on them; the reader is given no room to doubt them or to consider the trustworthiness of the author's assertion. These are just what everyone knows, and the acceptance of them therefore acts as a kind of condition of reading the text. This implied requirement had long been a strength of the scholastic commentary genre that works such as Fabri's were now tending to supplant.[50]

Fabri's expectations about what true scientific knowledge should accomplish were fundamentally the same as Galileo's, therefore; they were, in fact, commonplace. Fabri's terminology, however, is on occasion less commonplace: nonetheless, it relies on the same fundamental Aristotelian construal of the issues. He calls the basic experiential statements that act as premises in the syllogistic demonstrations of natural philosophy "physical hypotheses." A physical hypothesis derived from experience is therefore a universal statement such as "fall-

---

47. "Physical certainty" took its place between metaphysical certainty (the highest grade) and moral certainty (the lowest grade) in contemporary classifications; the terms find their echoes in Descartes. For a Jesuit exposition of the matter, see Arriaga, *Cursus philosophicus* (1632), p. 226 col. 1. Fabri lays out the division briefly in a set of *Controversiae logicae* (1646?), bound with separate pagination between Fabri's *Philosophiae tomus primus* and his *Tractatus physicus* (1646) in the Cambridge University Library copy, p. 77.

48. Fabri, *Philosophiae tomus primus*, p. 88, part of a chapter headed: "De Principiis & demonstrationibus Physicis": "Experimentum Physicum, est effetus aliquis sensibilis, certus & evidens Physicè, id est, ita ut citra miraculum fallere non possit, v.g. video lapidem modò moveri, modò non moveri; idem video in globo plumbeo, ligneo; sentio maiorem ictum lapidis ex maiori altitudine cadentis, &c."

49. Ibid., pp. 88–89: "omnia experimenta consentiant"; "quae ex certissimis experimentis constant."

50. See chapter 3, section I, above; P. Reif, "The Textbook Tradition in Natural Philosophy" (1969).

ing bodies accelerate."[51] Fabri's use of the word "hypothesis" is not intended to refer to a statement or group of statements that are conjectural, awaiting test through the empirical investigation of their consequences. Instead, he uses it to mean "fundamental statement," that is, a statement suitable to stand as a premise at the beginning of logical demonstration. A specifically *physical* hypothesis, furthermore, provides the relevant cause in a physical demonstration.[52] It is a physically certain principle founded on universal experience, the latter itself the product of many memories of the same thing.

In the scientific discourse in which Fabri (like Galileo) participated, "physical" means, among other things, "not mathematical."[53] When Fabri applies the label "physical hypothesis" to a science of motion, therefore, he signals that the science is properly to be seen as a part of physics. Galileo, by contrast, had clearly regarded his new science as mathematical. Fabri asserts that "no physical hypothesis"—such as Galileo's concerning uniform acceleration—"is to be sought from an experience that is not established with physical certitude."[54] Galileo is therefore guilty of solecism: he treats physics as mathematics.

Even had Galileo avoided that mistake, however, he would still have gone astray in resting his "hypothesis" on inadequate experiential foundations. Fabri argues against Galileo's odd-number rule for falling bodies on causal and analytical grounds in a separate treatise,[55] but

51. Fabri, *Philosophiae tomus primus*, p. 88. I have not noticed Fabri's category of *hypothesis Physica* elsewhere, but its meaning would appear to relate to the invocation of "physical certainty" (see n. 47, above) as well as to the role of such hypotheses in providing physical causes. See, on these questions in Fabri's work, Boehm, "L'aristotélisme d'Honoré Fabri" (1965), esp. pp. 335–338.

52. Fabri, *Philosophiae tomus primus*, esp. items VII,VIII.

53. For the institutional and intellectual importance to Jesuits of the disciplinary boundary represented (and maintained) by the distinction between "mathematical" and "natural philosophical" scientific genres, see above, chapter 2. "Mathematical" sciences, it should be remembered, included sciences of nature, such as geometrical optics, that utilized theorems drawn from arithmetic or geometry and which regarded only the quantitative properties of things; "physics," or "natural philosophy," addressed qualities. Although Aristotle's ideal of deductive scientific demonstration was modeled on Greek geometry, it was intended to apply to both mathematics and physics despite their distinct subject matters.

54. "Hinc nulla hypothesis Physica ab eo experimento petenda est, quod non est certum certitudine Physicâ": Fabri, *Philosophiae tomus primus*, p. 88. A brief account of this discussion appears in Lukens, "An Aristotelian Response to Galileo" (1979), pp. 115–118.

55. The *Tractatus physicus de motu locali* of 1646, also bearing the name of Mousnier and often bound together with the *Philosophiae tomus primus*. See, on its doctrines of fall, Lukens, "An Aristotelian Response to Galileo," chap. 4; Drake, "Free Fall from Albert of Saxony to Honoré Fabri" (1975); idem, "Impetus Theory and Quanta of Speed" (1974).

here he intends to show that claims to have established the rule from experience are false by their very nature: experience is incapable of demonstrating such a thing. Those people are simply wrong, he says, who assert the truth of the odd-number progression on the grounds that it agrees with "the most demonstrative experience that the space acquired in the second period of time, equal to the first, is triple that acquired in the first period," and his central point is that the senses cannot judge those distances and times precisely enough.[56] He had already explained that "the name *experimentum* ought to exclude all that which does not fall under the senses," giving as examples of improper usage its application to the equality of two times or the equality of two distances fallen. If one of those distances or times were one thousandth part greater than the other, he says, the senses typically would be incapable of discerning it. The essential problem with Galileo's odd-number rule was that it could not be based on experience, or "experiences," because sensory data could never provide sufficient precision to guarantee it.[57] Fabri, it should be noted, accepted the odd-number rule *phenomenologically;* as far as could be judged by the senses, it fitted the observed behavior of heavy bodies. The difficulty lay in the much more important matters of physical causation, strict demonstration, and their relation to experience.

Fabri's criticism of Galilean doctrine on falling bodies, then, amounts to the following. In Fabri's terms, Galileo's basic "hypothesis" is the odd-number rule—which Galileo claimed in effect to have evidenced with his inclined plane trials. But Fabri argues that this rule, by its very nature, cannot legitimately be established in that way, because sensory experience cannot vindicate mathematically precise ratios. Fabri asserts that only *doubtful* hypotheses can be drawn from doubtful experiences such as these, and doubtfulness is not science. As Mersenne had also argued, there are some kinds of natural knowledge that accounts of experience alone cannot create.

For discussion of reactions by contemporaries, with further references, see Dear, *Mersenne*, pp. 215–218. Fabri's position flows from his construal of the question as physical rather than mathematical and involves focus on causal explanation utilizing the idea of impetus, as does Baliani's: see below, as well as Boehm, "L'aristotélisme d'Honoré Fabri," p. 341–352.

56. Those who would maintain that the odd-number progression "probatissimo constet experimento; in secundo tempore aequali primo, acquiri spatium triplum acquisitum in primo tempore": Fabri, *Philosophiae tomus primus*, p. 88.

57. "Porrò experimenti nomine carere debet, id omne quod in sensum non cadit": ibid.

At around the same time, Cabeo too considered this question and reached similar conclusions. He remained unimpressed by Galileo's arguments because they relied on the shaky ground of apparent experience rather than being true demonstrations. "Galileo, in Day Two of his *Dialogue*, page 217, says that he has demonstrated that the proportion proceeds equally according to the odd numbers one, three, five, seven, nine . . . for which if it be demonstrated he will receive the greatest praise . . . but he has not demonstrated this principle, rather he has taken from experience that which I most eagerly awaited, and had wanted to be demonstrated." The best that can be said at the moment is that, disregarding some inconsistencies, falling bodies get progressively faster as they go.[58] Like Fabri, Cabeo holds Galileo to a strict standard of scientific demonstration.

Fabri's assertion that experience—*experimenta*—cannot support Galileo's odd-number rule was not meant to imply that statements about the ratios observed in natural acceleration can never be demonstrated scientifically (in the proper Aristotelian sense of the word). Galileo's talk about rolling balls down inclined planes suggested that those ratios could be established directly from experience, and that is the focus of Fabri's criticism; Fabri nonetheless has his own ideas on the matter that avoid the difficulties found in Galileo's. Fabri's own characterization of natural acceleration is justified according to a different methodological tactic: a statement about such a thing can be established, if not from direct experience, then either from another hypothesis that is itself validly derived from experience or from some other principle. In each case, however, the statement so established will not be a physical hypothesis, or, indeed, any hypothesis at all; instead, it will be a *theorem*. That is the status Fabri accords his own claim that the distances covered in successive equal periods of time by freely falling bodies follow the series of integers 1, 2, 3, 4. . . . [59] Fabri defines the difference between hypothesis and theorem in this way: a theorem both demonstrates *that* something is and *why* it is, causally. A hypothesis, by contrast, comes into play when the two are separated.[60]

58. Cabeo, *Commentaria*, Lib. 2, p. 76 cols. 1–2: "Galilaeus dialogo secundo pagina 217. dicit se demonstrasse, qua proportione procedat secundum numeros pariter impares unum, tres, quinque, septem, novem . . . quod si demonstretur suam habebit maximam laudem . . . sed hoc principium non demonstravit, immò assumpsit ex experientia, quod ego avidissimè expectabam, & voluissem demonstrari."

59. Lukens, "An Aristotelian Response to Galileo," chap. 4; Drake, "Free Fall from Albert of Saxony to Honoré Fabri"; idem, "Impetus Theory and Quanta of Speed."

60. Fabri, *Philosophiae tomus primus*, p. 89 (item VII). Fabri's expression to designate causal explanation is *propter quid sit*—recall the category of *demonstratio propter quid*, which provides the causal explanation for an effect.

The subtlety of the issues was made possible by a commonality of fundamental assumptions. Galileo had tried to derive his "new science of motion" formally from indubitable first principles, using geometrical demonstration and presenting theorems, the form championed by Fabri as well as by Cabeo. Galileo could not achieve the requisite effect of indubitability for his first principles, however, insofar as he could not be sure of their universal acceptance by his expected readership. He therefore threw in his inclined plane trials—"many memories of the same thing"—as a means of establishing one of his fundamental theorems on the alternative grounds of a proper Aristotelian "experience." Fabri, on the other hand (reacting not just to Galileo but also to other supporters of Galileo's science of motion), said that the odd-number rule cannot legitimately be derived in this way, even disregarding the issue of accessibility of results, because its nature as a mathematically precise statement necessarily goes beyond what sensory experience can endorse. Fabri thus forbade a mingling of two genres: mathematics, characterized by talk of quantity and ratios, and physics or natural philosophy, characterized by talk of qualities and "physical hypotheses."

Fabri's charge against Galileo had some validity. Another prominent writer on fall at much the same time, J. B. Baliani, attempted the same, somewhat illegitimate, mingling of categories: in Baliani's case, however, the methodological considerations were explicit. In his *De motu naturali* of 1646, Baliani portrayed the scientific character of his enterprise in a familiar way. Science is made through experience, which consists of often-repeated acts; thence arise the principles of the science, and hidden conclusions are disclosed through demonstrations based on them. "I have begun to search," Baliani declares; "others will decide whether I have discovered anything."[61] After having discussed the behavior of heavy bodies at considerable length, Baliani considers the character of the knowledge he has attempted to create, and wonders whether it is adequate:

> Hitherto, it seems to me that I have said as much as I can concerning the science of the natural motion of solid weights, as from certain properties familiar to the senses many things unknown have been deduced and disclosed. For according to Aristotle in this way alone is every science treated, as is seen in practice with Euclid, and others, who treat true and simple sciences: from whom [we see that] the geometer does not deal with the nature of quantity, nor the musician

---

61. Baliani, *De motu naturali* (1646), p. 8: "Rimari caepi; an deprehenderim aliorum erit iudicium."

with the nature of sound, nor the optician with the nature of light, nor the mechanic with the nature of weight. But truly my mind is not entirely satisfied if it does not grasp, or at least investigate, the prior causes from which these effects finally result.[62]

Baliani therefore gives himself leave to consider "not effects, but the natures of things," proceeding to present an explanatory account of natural acceleration.[63] He cautions the reader, however, that he does not affirm his account as certain, since this is, after all, physics.[64]

V. The Legitimate Sources of Scientific Experience

Although authoritative texts, whether containing the assertions of Aristotle or the stellar positions of Ptolemy, had long provided the most clear-cut repositories of scientific experience, in the early-modern period particular artisanal groups also came to represent sources of experience that could, through their proxy, be regarded as "common." Thus Galileo, to cite a famous example, appealed to the experience of gunners and shipbuilders in his work on mechanics. Such means of creating a commonality of belief acted as the social corollary of "self-evidence."[65] Another striking example of the generation of philosophical knowledge from artisanal knowledge comes from William Gilbert's *De magnete* of 1600: a large proportion of the phenomena that Gilbert discusses, to do with magnetic properties of the earth, magnetic variation and dip, and the apparatus that he describes as validating those phenomena, draw on craft knowledge and instruments that already formed part of the navigator's practical armamentarium. At one point Gilbert tells us of the mariner's compass: "There are in general use in Europe four different constructions and forms." He then describes

---

62. Ibid., pp. 97–98: "Hactenus mihi videor de scientia motus naturalis gravium solidorum satis pro viribus dixisse, dum ex quibusdam proprietatibus sensui notis, plures ignotae deductae, & patefactae sunt: in hoc enim solummodo ex Aristotele omnis scientia versatur; ut in praxi apud Euclidem, & alios, qui veras, & simplices scientias tractant, videre est: unde nec agit Geometra de natura quantitatis, nec Musicus de natura soni, nec perspectivus de natura luminis, nec mechanicus de natura ponderis. At vero meus intellectus non omnino acquiescit, ni causas priores, à quibus his effectus demum proveniunt, si non assequatur, saltem investiget."

63. Ibid., p. 98: "non effectus, sed rerum naturae"; pp. 98–102. See, for a discussion of Baliani's physical account of fall, Moscovici, *L'expérience du mouvement*, pp. 56–72. Baliani's account resembles Fabri's; see references in n. 55, above.

64. Baliani, *De motu naturali*, p. 102.

65. Shapin, *A Social History of Truth*, chap. 8, provides relevant discussion of the relation between artisanal and practical knowledge (and its possessors) on the one hand and "philosophical" knowledge on the other. See also Rossi, *Philosophy, Technology, and the Arts* (1970), on the rise in status of the practical arts.

their provenance, their construction and their use as well as the problems of inconsistency that result from the lack of standardization. Gilbert thereby ties his own work into the practical knowledge of navigators by taking it and turning it into a different kind of knowledge: natural philosophy.[66]

Such appropriation of experience from one context to another is a common feature of the period, but it often becomes invisible when historians treat "scientific practice" or "scientific method" as self-contained explanatory objects that transcend the specific occasions of their use. One of Alexandre Koyré's attempts to divorce Galileo from the image of an empiricist involved an account in Galileo's *Two New Sciences* of an apparent experiment with water and wine. Koyré intended to destroy its plausibility so as to show that Galileo had never really tried it.[67] Galileo has been asserting a natural conflict between water and air as an explanation of what were later called surface-tension phenomena. He compares the behavior of water when put in contact with air to its behavior when put in contact with wine:

> If I fill with water a glass ball that has a small hole, about the size of a straw, and I turn it thus filled mouth downward, then, though water is quite heavy and prone to descend in air, and air is likewise disposed to rise through water, being light, they will not agree the one to fall by coming out through the hole, and the other to rise by entering it, but both remain obstinate and contrary. But if I present to that hole a glass of red wine, which is almost imperceptibly less heavy than water, we promptly see it slowly ascending in rosy streaks through the water, while water with equal slowness descends through the wine, without their mixing, until finally the ball will be filled entirely with wine, and the water will drop quite to the bottom of the glass below.[68]

According to Koyré, Galileo's claim is clearly impossible because wine dissolves in water; the separation that Galileo describes is impossible,

---

66. Gilbert, *De magnete* (1600), pp. 165–166, quote p. 165: "Vulgò igitur in Europâ 4 sunt diversae compositiones & formae"; cf. Gilbert, *On the Loadstone* (1952), p. 83. See Zilsel, "The Origins of William Gilbert's Scientific Method" (1941). On the "mathematical practitioners" among whom Gilbert is properly numbered, important ingredients of mathematical science in the seventeenth century, see Bennett, "The Mechanics' Philosophy and the Mechanical Philosophy" (1986); idem, "The Challenge of Practical Mathematics" (1991). See also, for a balanced view of the role of Gresham College in London, with whom several English "mathematical practitioners" were associated, and as a corrective to older literature, Feingold, *The Mathematician's Apprenticeship* (1984), chap. 5.
67. Koyré, "Galileo's Treatise *De motu gravium*" (1968), on pp. 82–84.
68. Galileo, *Opere* 8, pp. 115–116; trans. Drake in Galileo, *Two New Sciences*, pp. 74–5.

so Galileo must not have done the experiment.[69] However, in 1973 James MacLachlan reported the success of his own attempt to conduct the experiment, concluding that Galileo probably had carried it out as he described.[70]

The disagreement between Koyré and MacLachlan centers on the characterization of this manipulative procedure as an *experiment*. It takes on a quite different complexion, however, when considered in light of the fact that the contested experiment was a standard part of the natural magic repertoire. It appears, for instance, in Della Porta's sixteenth-century compendium *Natural Magick*. Della Porta describes the trick as a way of "making sport of those that sit at table with us." It requires a specially made cup that consists of an inverted cone, with a narrow hole at the apex, set into a hollow glass ball. The perpetrator of the trick pours water into the cup so as to fill the ball but not the cone. He then gradually adds wine, which, says Della Porta, will not mix with the water because the hole is narrow and water is heavier than wine. Everyone at the table can see this procedure, but diluting wine with water was a perfectly standard practice and hence would arouse no suspicion. The perpetrator drinks first from the cup and gets all the wine, and the victim, who drinks second, gets only water. But if the victim tries to challenge, and insists that the procedure be conducted in the opposite order, with the wine being put in first, he can still be tricked. Della Porta advises as follows:

> stay awhile, and hold him in discourse; for the water will sink down by the narrow mouth, and the wine by degrees will ascend as much, and you shall see the wine come up through the middle of the water, and the water descend through the middle of the wine, and sink to the bottom; so they change their places: when you know that the water is gone down, and the wine come up, then drink, for you shall drink the wine, and your friend shall drink the water.[71]

Another version of the device appeared in 1630 in a recreational mathematics collection by Claude Mydorge, the Parisian mathematician and friend of Descartes. It takes the form of a way of making foul-smelling wine potable. Mydorge describes an arrangement whereby a container with a narrow neck is upturned so that the neck slots into the wider neck of a lower container, the whole making a closed system.

69. Koyré, Galileo's Treatise *De motu gravium*," pp. 82–84.
70. MacLachlan, "A Test of an 'Imaginary' Experiment of Galileo's" (1973).
71. Della Porta, *Natural Magick* (1658/1957), p. 383.

Initially, the top flask is filled with water and the bottom flask with wine, but, as Mydorge says, because the wine is lighter, it ascends in place of the water. After a time, therefore, the wine and water will have changed places: "And in this penetration the wine will lose its vapors and fumes."[72]

These three presentations are sufficiently similar, as well as sufficiently different in their details, to make it clear that they represent a fixture of literature on natural magic or mechanical wonders—on a par with tricks, found in Hero of Alexandria, using siphons or expanding air. Galileo did not propose something unknown, even if many of his readers might not have been familiar with it. Instead, he appropriated it for philosophical purposes, in order to obtain natural knowledge. Thus the question of whether Galileo actually performed his "experiment" misses the point. It matters little whether Galileo himself tried it, because the phenomenon was not novel and was apparently uncontroversial. MacLachlan showed that Koyré's denial of the phenomenon was illegitimate, but Koyré's fundamental claim still holds: Galileo's assertion did not rest on his having carried out an experimental manipulation. The contrivance to which Galileo appealed was not an experiment; it was a curiosity, even when it was dressed up as an illustration of the principles of buoyancy. Galileo turned it into an argument that would support a philosophical conclusion in a way similar to William Gilbert's transformation of the navigator's practical experience. It was no more a piece of experimental science than were Galileo's appeals to the experience of gunners and shipbuilders. Instead, it was a piece of experience available from a common, public stock, just as scholastic natural philosophers drew on empirical materials in the authoritative texts on which they commented.

Another common stock of experience that sometimes came to the aid of philosophers resided in the games of children. Thus Aguilonius, in his optical treatise, illustrated an account of the necessity of using both eyes in order to judge distance through reference to a children's game. After having given a geometrical demonstration of the matter, he observes that it is almost impossible to judge distance with one eye closed (see figure 4). He then adduces a "kind of playful experience"

---

72. Mydorge, *La seconde partie des recreations mathematiques* (1630), pp. 58–9. Gaspar Schott also gave a version of the trick: Schott, *Ioco-seriorum naturae et artis* (1666), p. 178, in turn referring back to his own *Magia universalis* (1657–1659), part 3, lib. 5, as his proximate source and to Della Porta as the ultimate source. Cf. Findlen, "Jokes of Nature" (1990), which concentrates especially on that aspect of this genre, and Eamon, *Science and the Secrets of Nature*, chap. 6.

Figure 4. On the difficulty of judging distance with one eye, from Franciscus Aguilonius, *Opticorum libri sex* (1613), p. 151. Courtesy of Division of Rare and Manuscript Collections, Carl A. Kroch Library, Cornell University.

that he learned from children—"but we have judged [it] fit for a philosopher, inasmuch as through consideration of it we arrive at knowledge of this remarkable property."[73] Jacques Pierius, a contributor to the controversies surrounding Torricelli's purported vacuum in the 1640s, discussed the expansive properties of air by reference to

> an experience taken from a children's game. They carefully seek out and dry thoroughly a pig's bladder; into this is admitted air until it is slightly more than half full, and still appears flaccid: it is closed in a way known to the children, it is exposed to very light ashes, the

---

73. Aguilonius, *Opticorum libri sex* (1613), p. 154: "Simile & illud est ludicri experimenti genus, quod à pueris olim didicimus, sed philosopho dignum sumus arbitrati, utpote cuius consideratione in eximiae huius proprietatis cognitionem devenimus."

whole thing will be filled and on account of the fullness it will appear hard.[74]

Pierre Guiffart, considering the same issues the year before, made appeal to the same source of common experience:

> Children themselves show it to us, when in a barrel they close up air between two plugs: for when they press the second of them, the air trapped between the two compresses, and then it occupies less space than previously; and if they press it still further, the forced air pushes and makes the first [plug] shoot out with impetuosity.[75]

The advantages of stressing the puerile provenance of such examples lay in their obviously commonplace character. These were things so much a part of common experience that they were well known to children, the commonest class of all humanity.

Steven Shapin's studies of credibility in seventeenth-century English experimental philosophy examine the social criteria by which truth-telling about events was accredited to particular sources.[76] Such an analysis is appropriate for a kind of natural philosophy that rested on discrete, singular experiences presented historically—the hallmark of the new English philosophy of the period. For most of the century and in most places, however, reliance on isolated testimony was irrelevant because prepackaged philosophical universality was the norm. From various repositories of experience, phenomena could be deployed like commonplaces in rhetoric, ready to be used in the composition of an argument.[77] These repositories included, in addition to accredited literary sources, recognized social groups such as navigators, gunners, or even children: it would be a valuable, though vast, project of social topography to determine which social groups, and what kinds of social groups, counted as legitimate collective authorities, and where educated Europeans drew the lines between reliable expertise and tall

---

74. Pierius, *Ad experientiam nuperam* (1648), p. 6: "Adde experientiam à lusu puerorum desumptam. Curiosè quaerunt & exsiccant visicam [sic] porcinam; in hanc immitatur aër donec sit paulò plusquam semi plena, & flaxida adhuc appareat: claudatur ea ut norunt pueri, exponatur cineribus levissime calidis, tota implebitur & prae plenitudine dura apparebit."

75. Guiffart, *Discours du vuide* (1647), p. 65: "Les enfans nous le monstrent eux mesmes, quand dans une canonniere ils enferment l'air entre deux tampons: car lors qu'ils en pressent le dernier, l'air prins entre deux se resserre, & lors il occupe moins d'espace qu'auparavant; et s'ils le pressent encores d'avantages, l'air forcé pousse & fait sortir le premier avec impetuosité."

76. See esp. Shapin, *A Social History of Truth*.

77. Cf. Slawinski, "Rhetoric and Science" (1991), section 4.

tales in the making of natural knowledge. Shapin's lines for experimental philosophy are drawn according to the seventeenth-century English social hierarchy; prolix gentlemen make good experimental philosophers. Other lines may be drawn for older and competing forms of philosophy, whereby special, shared expertise belonged to particular occupational groups (although members of those groups might not themselves be able to act as philosophers).

One of these latter lines of legitimacy certainly encompassed academic and court astronomers, whose own status as philosophers, as we have seen before, was sometimes contested along scholastic disciplinary lines.[78] The disciplinary boundary between mathematics and natural philosophy that characterized school learning throughout this period played a particularly delicate role for Jesuit mathematicians, owing to their especial concern with the scientific status of their work. But the acceptability of astronomical knowledge in its guise as the common stock of the astronomical craft stood independent of the key issue of scientificity that so concerned Jesuit mathematicians; the natural philosopher could quite happily accept empirical truths about the heavens from his mathematical colleague even if he denied the scientific status of the astronomer's knowledge. Scientific knowledge involved causal explanation, but empirical statements themselves remained legitimate as data—even highly refined "data" of the kind that Blancanus called "observations"—in the absence of causal status. The formal division between mathematics and natural philosophy was not, however, solely maintained by practitioners of the latter discipline at the expense of the former. In the earlier part of the seventeenth century, mathematicians themselves found that the demarcation could do positive work for their own pretensions to scientificity, and set the stage for a co-option of natural philosophy itself—the emergence of "physicomathematics."

78. See chapter 2, section I, above.

Six ART, NATURE, METAPHOR: THE GROWTH OF PHYSICO-MATHEMATICS

I. Mechanical Nature

Aristotle's distinction between the artificial and the natural was a commonplace of natural philosophy in the early seventeenth century. Nonetheless, one of the century's most famous innovations, the metaphor of the world as a machine, is an apparent conflation of the two. The widespread adoption of various forms of "mechanical philosophy" went along with a drastic weakening of the art/nature distinction in philosophical thought and provided ontological vindication of the primacy of the mathematical sciences. This process cannot be understood, however, without a recognition that such categories as "art," "nature," and "machine" are mutually interactive: their meanings change as their relationships are reconfigured.[1]

The statement "the world is a machine" asserts meaningful similarities between the world and (to take a typical example) a cathedral clock: it is an attempt actually to make the world machinelike.[2] More than that, however, it also calls into question the meaning of the word "machine" itself. Mary Hesse, in common with a number of other philosophers, has maintained that a metaphor is not merely descriptive of one

---

1. Henry, "Occult Qualities and the Experimental Philosophy" (1986), issues important caveats regarding the "mechanical philosophy," its character and acceptance in the seventeenth century, especially in England. On metaphor, the best starting place is still Ortony, *Metaphor and Thought* (1979). The principal positions are discussed in Ricoeur, *The Rule of Metaphor* (1975). I leave aside work concerned principally with the aesthetic and literary dimensions of metaphor.

2. Cf. Mayr, *Authority, Liberty and Automatic Machinery* (1986), chap. 3, on "The Clockwork Universe."

concept in terms of another, but becomes constitutive of the meaning of both.³ Hesse applies her well-known "network model" of language and reference to understanding how the meanings of words depend on those of all the others making up the network.⁴ Her idea may be compared in its broad outlines to the arguments of Kuhn and Feyerabend regarding incommensurability: a change in the meaning of a single term itself changes, if only very subtly, the meaning of every other term in the language, because meaning is determined by use and words can only be used in relation to other words.⁵ Thus the statement "the world is a machine" inevitably affects the meaning of "machine" as well as the meaning of "world"—the interconnections of the network make that unavoidable. Gisela Loeck has claimed that the machine metaphor in its modern sense is a true invention of the seventeenth century, even though comparisons between the heavens and various kinds of clockwork mechanism can be found in medieval and ancient writers such as Cicero. Loeck argues that when those authors described the cosmos as a machine, they understood machines themselves in a way different from that purportedly introduced by Descartes. Whether or not Loeck's argument is historically accurate, it draws attention to the importance of perceiving the flexibility of all terms in a metaphorical identification. Her point may be glossed in this way: creating a particular machinelikeness of the world necessarily at the same time created a particular worldlikeness of machines—the metaphor was one of the ways by which to understand the nature of a "machine," not merely a way to understand the "world."⁶

The metaphor's realization required that it become accepted and entrenched within a language community—leading to the possibility that people might cease to view it as a metaphor at all: the world would simply *be* a machine. Geoffrey Cantor has discussed the eighteenth-century metaphor of light as projected corpuscles, noting the way in which it came to be constitutive of what light *was* for many

---

3. Most recently in Hesse and Arbib, *The Construction of Reality* (1986), chap. 8. This view is known as the "interaction" theory of metaphor; one of its principal exponents was Max Black: Black, *Models and Metaphors* (1962). See in general Ortony, *Metaphor and Thought*; also Bloor, "The Dialectics of Metaphor" (1971).

4. The "network model" was first put forward in Hesse, *The Structure of Scientific Inference* (1974). It was given its sociological reading by Barnes, "On the Conventional Character of Knowledge and Cognition" (1983).

5. Kuhn, *The Structure of Scientific Revolutions* (2d ed. 1970); Feyerabend, "Explanation, Reduction, and Empiricism" (1962).

6. Loeck, *Der cartesische Materialismus* (1986), pp. 219–234; Funkenstein, *Theology and the Scientific Imagination* (1986), esp. pp. 317–319, discusses similar material. Gabbey, "The Mechanical Philosophy and its Problems" (1985), confronts the imprecision in notions

British natural philosophers—so that they could even attempt to determine the masses of these corpuscles.[7] Here again, metaphor became practical identity.

Robert Boyle provided an excellent illustration of the cultural relativity of clockwork in his *Hydrostatical Discourses:*

> [If] I had been with those Jesuits, that are said to have presented the first watch to the king of *China,* who took it to be a living creature, I should have thought I had fairly accounted for it, if, by the shape, size, motion &c. of the spring-wheels, balance, and other parts of the watch I had shewn, that an engine of such a structure would necessarily mark the hours, though I could not have brought an argument to convince the Chinese monarch, that it was not endowed with life.[8]

There could be no way of forcing a change in the emperor's perception of the nature of the watch (the prototypical machine for this European metaphor), because his construal of its meaning was entrenched in a different network of concepts: the emperor of China was not a European.

The art/nature distinction impinged on the use of artificial contrivance in the making of natural knowledge—that is, it compromised the legitimacy of using in natural philosophy the sorts of procedures used by mathematicians. The full integration of artificial experience into natural philosophy in the seventeenth century therefore involved a reconfiguration of the relationship between art and nature, one represented by the new metaphor of the world as a machine.

II. Nature and Art

The distinction between the natural and the artificial made by the learned world of early seventeenth-century Europe was, like the

---

of "mechanical explanation" in the seventeenth century; see also idem, "Cudworth, More, and the Mechanical Analogy" (1992); idem, "Henry More and the Limits of Mechanism" (1990) pp. 19–35. Gabbey has also looked at the crucial question of what "mechanics" actually meant in the seventeenth century: see chapter 8, section III, below, for references and further discussion.

7. Cantor, "Light and Enlightenment" (1985).

8. Boyle, *Hydrostatical Discourses,* quoted for different purposes in Shapin and Schaffer, *Leviathan and the Air-Pump* (1985), p. 216. Cf. the story in the Port Royal *Logique* about the Chinese attribution of a "sonorific virtue" to a clock to explain its ticking: Arnauld and Nicole, *La logique ou l'art de penser* (1662/1965), p. 264 (i.e., part III, chap. 17, section III). This story is mentioned in Funkenstein, *Theology and the Scientific Imagination,* p. 324, with what appears to be an inaccurate citation.

universality of scientific experience, an aspect of the scholastic-Aristotelian view of nature as an object of scientific knowledge.⁹ Nature was the principle of motion, or change, and change was teleologically conceptualized as the effect of process, or the striving towards a goal. Thus the flux of generation and corruption received its rational ordering through final causes. To give a scientific explanation was to give an account of a thing's particular nature or form, and a thing's nature was only revealed by discovering the end towards which it strove. The paradigm originally employed by Aristotle seems to have been biological: one learns the nature of an acorn, for example, by observing its development into an oak tree.¹⁰

This view of nature implied a particular character for scientific experience. Natural processes do not always achieve their ends—acorns do not always develop into oak trees. Processes might fail to fulfill themselves for a variety of reasons: a concatenation of various accidental causes, each in itself "natural," could pervert the course of a given process; most radically, for Christianized Aristotelianism, God could always exercise his *potentia absoluta* to bring about an irregular occurrence.¹¹ This was a source of difficulty in building a science of the natural world, because scientific demonstrations had to express necessary connections between cause and effect; if natural processes lacked absolute uniformity, that necessity would seem to be lacking. Albertus Magnus and Thomas Aquinas developed a way around the problem: the technique known as reasoning *ex suppositione*.¹² The process to be explained, such as the development of acorns into oak trees, is assumed to have actually occurred without impediment. The essential properties of an acorn explain why it grows into an oak tree, given that it actually does so; if it does not, accidental impediments must have prevented it. Thus the demonstration *ex suppositione* itself remains properly scientific.¹³ The rationale behind this technique serves to expose the significance of the universality of experiential statements: sin-

---

9. See, for a useful exposition, Weisheipl, "Aristotle's Concept of Nature" (1982).

10. See exposition and references in Dear, "Miracles, Experiments, and the Ordinary Course of Nature" (1990), as well as chapters 1 and 2, above. Aristotle, *Physics* II.8, is a basic reference. On Aristotle and the biological paradigm, see Grene, *A Portrait of Aristotle* (1963), for an attempt to represent Aristotle's thought as having developed from early biological work.

11. Weisheipl, "Aristotle's Concept of Nature," esp. p. 152; Funkenstein, *Theology and the Scientific Imagination*, pp. 124–152.

12. See Wallace, "Albertus Magnus on Suppositional Necessity in the Natural Sciences" (1980); idem, "Galileo and Reasoning *ex suppositione*" (1981); see also idem, *Galileo's Logic of Discovery and Proof* (1992), pp. 144–146.

13. William A. Wallace has claimed an important role for this kind of *ex suppositione* reasoning in Galileo's thought: see especially Wallace, "Galileo and Reasoning *ex suppos-*

gular instances might not represent the ordinary course of nature and so could not establish the nature of a particular process. An acorn failing to develop into an oak tree would shed no light on the nature of acorns. Only experience formed from the instances supplied habitually by the senses could have scientific relevance; there could be no "crucial experiments."[14]

The contrast with "art" rested on just this view of nature. The natural course of a process could be subverted by man-made, artificial causes, because art replaced nature's purposes with human purposes. An aqueduct, for example, is not a natural watercourse; it reveals the intention of its human producer, which thwarts that of nature. The characterization of each "intention" in terms of the other was the root metaphor (no doubt unrecognized as such) in Aristotelian natural philosophy. The breakdown of the distinction between nature and art that occurred in the seventeenth century required its complete restructuring.

A common way of overcoming the absolute separation of the two categories seems to have been the one adopted by Francis Bacon. Bacon argued that art was only a matter of setting up situations in which nature will produce a desired result—so that art is the human exploitation of nature rather than an activity outside of nature. Thus, contrived situations were no longer fundamentally different from natural ones; except for the accident of how they came into being, each behaved according to the same principles. At the root of Bacon's position was a different view of the aim of natural knowledge. The Aristotelian distinction between art and nature depended on seeing human purposes as separate from natural ones and hence irrelevant to the creation of a true natural philosophy. For Bacon, by contrast, human purposes were paramount: his natural philosophy aimed at creating knowledge of how to achieve *human* ends. An independently existing realm of natural purposes thus became a strictly irrelevant category.[15]

---

itione"; idem, "Aristotle and Galileo: The Uses of *Hupothesis* (*Suppositio*) in Scientific Reasoning" (1981). The argument has been challenged, especially regarding uses of the term in Galileo's *Dialogo*, by Wisan, "On Argument *ex suppositione falsa*" (1984), and is much toned down in Wallace, *Galileo and His Sources* (1984), pp. 340–343, or in idem, *Galileo's Logic of Discovery and Proof*, e.g., p. 224.

14. To the extent that there was an abandonment of the art/nature distinction in the seventeenth century, it was a rejection of the kind of teleological explanation that focused on process. It was not generally a rejection of immanent teleology; thus one finds natural theology, the argument from design, alive and well at the end of the century. These turned out to be two very readily separable kinds of teleology, which was quite contrary to Aristotle's original thrust: compare Aristotle, *Physics* II.8.

15. Bacon, *Novum organum* I, aph. LXVI, concludes a discussion of whether matter is indefinitely divisible or atomic by describing these matters as "things which, even if

156   Chapter Six

Even though Bacon thereby rejected the validity of the art/nature distinction, the same basic perspective informed the discussions of many of those who retained it as part of the common apparatus of philosophy. A properly Aristotelian subdistinction between artifice as a way of achieving particular ends and artifice as something productive of new, artificial forms allowed any appearance of inconsistency to be avoided.[16] Artificial forms were always categorically distinct from natural forms (Aristotle had used the example of a bed made from wood—if the wood sprouted it would produce a tree, not a new bed); the achieving of ends, however, bypassed that issue altogether, and its increasing prominence resonates strongly with Bacon's utilitarian ideas. Patricia Reif has noted that two prominent philosophical textbooks of the early seventeenth century, by Eustace of Saint Paul and Rafael Aversa, speak of artists utilizing the natural properties of natural materials, while Aversa examines the paradigmatic case of the clock: a clock's motion is caused through nature, in the form of the descent of its driving weights, whereas the determination of that motion is caused by art.[17]

A typical example of this view appears in a little book, famous for its accounts of Pascal's early work on the Torricellian vacuum, published in Rouen in 1647. Called *Discours du vuide*, it was written by

---

true, can do but little for the welfare of mankind": trans. Spedding in Bacon, *Works* (1860–1864), vol. 8, p. 97, from Latin of vol. 1, p. 272. The art/nature distinction lost its meaning partly to the extent that the worth of practical knowledge, especially that of engineers and architects, came to be reevaluated and redefined: see especially Rossi, *Philosophy, Technology, and the Arts* (1970). This aspect of the matter, associated with the rather diffuse social changes that can easily be related to the rise of capitalism, parallels the more concrete emergence of new intellectual roles outside the universities that helped to undermine the existing hierarchy between mathematical sciences and natural philosophy.

16. The subdistinction forms a part of Arriaga's discussion of the issue, for example: see, e.g., Arriaga, *Cursus philosophicus* (1632), p. 319 col. 1, on artificial bodies. He explicitly exempts from his discussion arts that achieve their ends through the application of natural agents (as with mixing and distilling).

17. M. R. Reif, "Natural Philosophy in Some Early Seventeenth Century Scholastic Textbooks" (1962), pp. 166–169; pp. 228–241, esp. p. 234 on Eustace and Aversa; also Wallace, *Galileo and His Sources*, pp. 209–211; the central source is Aristotle, *Physics* II.2–3. Aversa discusses the matter in Aversa, *Philosophia metaphysicam physicamque complectens* (1625), pp. 266–269; see esp. p. 266 col. 2 on clocks and drugs as examples of artificial things working through natural processes (and note the use of the term *determinatio* here, calling to mind Descartes). This appears to have remained the standard school distinction: Micraelius, *Lexicon philosophicum* (2d. ed., 1662/1966), col. 172: "artificialiae non habent internum motus principium, ut naturalia, sed externum simul cum interno, ut videre est in automatis seu horologiis." See also Hooykaas, "The Discrimination between 'Natural' and 'Artificial' Substances" (1948).

Pierre Guiffart, a physician and a teacher at the local college.[18] Guiffart fully accepts a distinction between art and nature, but he explains it in such a way as to undermine its relevance to natural philosophy:

> There is a very notable difference between art and nature: art cannot produce anything without nature; it not only needs nature to furnish the material, but it also needs nature's natural inclinations to go along with it, so that thereby it supplements nature's rules and produces its own works, and as soon as nature is lacking to it, those works are promptly destroyed, and without its help they can never be recovered. But nature has no need of art; it is always similar to itself, from the moment of its birth until its end.[19]

Nature never changes, being, as it was to be for Newton, "always similar to itself."[20] Art, however, is subject to the vagaries of human purpose and cannot guarantee that nature will always cooperate.

A subtle redefinition of natural knowledge—and thus of nature itself—is apparent in Guiffart's discussion. Guiffart talks about nature's "rules" (*ses regles*), by which nature pursues its "intentions"; in effect, human art makes use of these rules, when it can, to achieve its own ends. Knowledge of nature, rather than being about identifying purposes, is now, insensibly, becoming about characterizing "rules" (or what in the seventeenth century were increasingly called "laws") of nature.[21] The rules governing nature's behavior take the place of the purposes for the sake of which that behavior occurs, much as one

---

18. The most comprehensive discussion of this culture is Brockliss, *French Higher Education in the Seventeenth and Eighteenth Centuries* (1987). See also chapter 7, below.

19. Guiffart, *Discours du vuide* (1647), pp. 48–49: "Il y a une tres notable difference entre l'Art et la Nature; l'art ne peut rien produire sans elle, il y a besoin non seulement qu'elle luy fournisse de matiere, mais aussi qu'elle l'accompagne de ses inclinations naturelles, afin que sur cela il adiuste ses regles & produise ses ouvrages, & si tost qu'elle luy manque; ils sont incontinent destruits, & sans son secours ils ne se retrouvent jamais. Mais la Nature n'a pas besoin de l'art, elle est toujours semblable à soy, & depuis le moment de sa naissance jusques à sa fin."

20. Famously, Newton, in the *Opticks* (4th ed., 1730/1952), p. 397, remarks on nature being "very conformable to her self." Guiffart's remarks could just as well have been translated, in keeping with Newton's remark, with a feminine "nature," a very familiar image from that period: see, e.g., Thorndike, *History of Magic and Experimental Science*, vol. 8 (1958), pp. 243–244, on nature as a beneficent mother, and p. 593, for more of Newton's use of the feminine image. *Scientia* was also routinely female: see Schiebinger, *The Mind Has No Sex?* (1989), chap. 5.

21. Milton, "The Origin and Development of the Concept of the 'Laws of Nature'" (1981); idem, "Laws of Nature" (forthcoming); Oakley, *Omnipotence, Covenant, and Order* (1984), pp. 77–92; Zilsel, "The Genesis of the Concept of Physical Law" (1942); also Funkenstein, *Theology and the Scientific Imagination*, pp. 192–193. Ruby, "The Origins of Scientific 'Law'" (1986), stresses a more medieval development.

might want to know that traffic is obliged by law to drive on the right-hand side of the road rather than to know its destination. This is a change from nature seen on the Aristotelian model of the purposeful activity of an individual to nature being conceived in terms of a social group governed by nonteleological rules of communal behavior. People and society, in keeping with what we have seen of metaphorical identifications, are no doubt themselves reconstituted also.[22]

Experimental contrivance was permissible when the goal was operational knowledge rather than teleological knowledge of nature. In Aristotelian terms, this is a kind of artifice the human goal of which is to make operational, or at least phenomenological, knowledge; this cannot, by definition, be a natural goal. It was exemplified most clearly in the mathematical sciences: artificial contrivances allowed the investigator to discover what nature would allow him to do.

The very concept of nature implied universality:[23] experimental contrivance implied that the behavior of nature, which is always similar to itself, will now become known by identifying it with the particularity of a contrived situation. To say that what happens in such a situation is what happens in nature is therefore a metaphorical identification.[24] But such a metaphor was not a mere matter of playing with words: a whole concatenation of actions and meanings went to make up an experimental event. It is also possible, however, to draw attention to the distinction between the *semiotic,* focusing on signs and representation, and the *mimetic,* focusing on imitation.[25] Its establishment in the philosophical practice of the seventeenth century is an important feature of the metaphorical functioning of experiment.

Guiffart again provides an example. In discussing Pascal's Torricel-

---

22. That nature and society are mutually constituted is the chief burden of Latour, "Postmodern? No, Simply Amodern" (1990); idem, *We Have Never Been Modern* (1993); broadly related points may be found in Elias, *The Civilizing Process* (1994; originally published in German 1939), esp. pp. 477–478, 496–497.

23. Cf. Collingwood, *The Idea of Nature* (1945); Lenoble, *Esquisse d'une histoire de l'idée de nature* (1969).

24. Perhaps more properly one should say "metonymic." I only require the loosest and most permissive of the accepted senses of the word "metaphor," however. And cf. Slawinski, "Rhetoric and Science" (1991), esp. pp. 95–96, for a similar idea.

25. Galison and Assmus, "Artificial Clouds, Real Particles" (1989), have identified a particular "mimetic tradition" in nineteenth-century science, which they use in discussing C. T. R. Wilson and the invention of the cloud chamber, a case where the object was in effect to make atmospheric clouds on a laboratory scale. Such a "mimetic" strain in natural philosophy can also be found much earlier than the nineteenth century, as Gooding, Pinch, and Schaffer, *The Uses of Experiment* (1989), p. 9, observe.

lian liquid-filled glass tubes, which supposedly demonstrate the possibility of a vacuum, he says: "In them one sees a little miniature of the world, in which, holding the enclosed elements in our hands, and at our disposition, they make known what they are and what they can do."[26] (He means chiefly to draw attention to the effects of air on other kinds of matter.) The fact that one of Pascal's glass tubes puts constituents of nature "at our disposition" is, of course, one of the primary characteristics of an experimental apparatus. That by this means "they make known what they are and what they can do" is the other primary characteristic: the experimental manipulation generates knowledge about nature, and especially knowledge of an operational kind.

This is mimetic, not semiotic. The glass tube does not here *signify* an aspect of the world according to a particular set of local conventions; Guiffart takes it to be actually *like* that aspect in some essential way—there is a movement towards identity. A metaphor, according to Aristotle in the *Poetics*, is the application to a thing of a name that properly belongs to something else. Guiffart portrays himself as applying to Pascal's glass tubes and their manipulation a characterization that properly belonged to the world. Metaphors, as we have seen, tend to turn into practical identities when people start treating them that way, which is why Lakoff and Johnson have written of "metaphors we live by."[27] Guiffart has started to live by a new metaphor whereby playing with glass tubes becomes identical to learning about the atmosphere.

One of the most notable examples of mimesis in seventeenth-century natural philosophy is William Gilbert's work on the magnet in 1600. His central piece of artifice, which is manipulated to generate knowledge about nature, is a spherical magnet made by turning a piece of lodestone on a lathe. Gilbert calls it a *terrella*, a little earth, but this does not imply a simple analogical relationship between the carved lodestone and the earth. The little earth and the great earth are interactively related so as to set up a kind of identity: the *terrella* does not simply imitate the earth, so as to elucidate properties of the latter, and the earth does not represent a gigantic *terrella*, so as to elucidate properties of magnets. Instead, the earth *is* a magnet, and the *terrella*, possessing the proper shape for a magnet, *is* a little earth.[28]

Gilbert accomplishes this feat by first of all appealing to a concep-

26. Guiffart, *Discours du vuide*, pp. 56–7: "En elles on voit un petit racourci du monde, dans lequel tenans les elemens enfermez entre nos mains & à nostre disposition, ils donnent à connoistre ce qu'ils sont & ce qu'ils peuvent faire."
27. Lakoff and Johnson, *Metaphors We Live By* (1981).
28. Gilbert, *De magnete* (1600), book I, chap. 3.

tion of the earth already well entrenched in astronomy—the earth as a sphere with poles. He then remarks that magnets also have poles. The argument proceeds:

> Inasmuch as the spherical form—which, too, is the most perfect—agrees best with the earth, which is a globe, and also is the form best suited for experimental [that is, manipulative] uses, therefore we propose to give our principal demonstrations with the aid of a globe-shaped loadstone, as being the best and the most fitting.[29]

He then instructs the reader to make a ball by turning a flawless piece of lodestone on a lathe.

> The stone thus prepared is a true homogeneous offspring of the earth and is of the same shape, having got from art the orbicular form that nature in the beginning gave to the earth, the common mother.[30]

For Gilbert, the *terrella* is the same as the earth in every way that counts. He goes so far as to discount the significance of the many "impurities"—nonlodestone materials—found in the earth by referring to similar impurities found in lodestone itself as it is dug from the ground.[31] The chief evidence that the interior of the earth consists almost entirely of lodestone is the fact that the *terrella* moves spontaneously to align itself with the astronomical poles. Gilbert sees the alignment of the poles of a magnet with those of the earth as a matter of the *terrella* aligning itself in the universe in the same way as does the earth; the phenomenon is not to be understood as a matter of the earth's poles drawing the magnet's poles. The earth affects the motion to the extent that a larger and more powerful lodestone always rules a smaller, weaker one, but the identity of the underlying behavior remains uncompromised. The *terrella*, says Gilbert, retains the primordial form of the earth from which it was made.[32] The *terrella* thus shows that the earth is a lodestone, but it is only because a lodestone has a tendency to align its poles in the same direction as those of the earth

---

29. Ibid., p. 12: "Sed quoniam forma sphaerica, quae & perfectissima, cum terra globosa maximè consentit, & ad usus & experimenta maximè idonea sit, praecipuas igitur nostras per lapidem demonstrationes, globoso magnete fieri volumus, tanquam magis perfecto & accommodato"; trans. P. Fleury Mottelay in Gilbert, *On the Loadstone* (1952), p. 9.

30. Gilbert, *De magnete*, p. 12: "Hic ita praeparatus lapis, vera est, homogenea, eiusdemque figurae, telluris soboles: formam arte orbicularem nacta, quam communi matri telluri à primordijs natura concessit"; trans. P. Fleury Mottelay, pp. 9–10.

31. Gilbert, *De magnete*, book, I, chap. 17, esp. p. 42.

32. Ibid., book I, chap. 17, esp. pp. 41–42.

that Gilbert can also argue that a spherical lodestone is an earth—a *terrella*.

Guiffart's lip service to the art/nature distinction notwithstanding, the barrier to experimentation in the making of natural knowledge was eminently breachable. The notion of manipulation acted as the Trojan horse: once admitted as a legitimate part of natural inquiry, the use of contrivance might overwhelm the art/nature distinction that warranted the separation of natural philosophy from mathematics. If artificial contrivances could allow the investigator to discover what nature would allow him to do—and could even, as with the *terrella*, stand for other things not directly manipulable—the potential extent of new philosophical experience was unlimited.

III. Demonstrative Mathematics versus Causal Physics

An increasing rejection in practice of the art/nature distinction could have implied a wholesale rejection of Aristotelian natural philosophy, but that move would have been unattractive to those such as the Jesuits who were a part of the academic establishment within which Aristotelianism had its life. For them, the art/nature distinction was not something to be rejected; it was a legitimate categorical division expressing the way things were done in the study of nature. One of its most important functions was to keep natural philosophy separate from, and superior to, mathematics, but the distinction also served to protect the use of artificial contrivance in the mathematical sciences from the criticism that, being unnatural, it could not contribute to knowledge of nature.

Blancanus's concern, for example, to include telescopic observations in his *Sphaera mundi* underlines his awareness of how many pieces of astronomical knowledge—"observations"—were instrumentally constructed. Similarly, in book III, part I of his optical treatise Scheiner treated the telescope (the *tubus oculus*) along with lenses and pinhole images.[33] It is probable that when he referred to things that "do not come into being" without the "industry of special empirics," he meant especially "experiences" constructed with lenses. Aguilonius described how, in his optical work, "I consider things intentionally al-

---

33. Scheiner, *Oculus* (1619); he appears to be greatly indebted to Kepler. Blancanus, *Sphaera mundi* (1620), in an appendix called "Apparatus ad mathematicas addiscendas, et promovendas," p. 413, refers to Scheiner's book and to the great potential value of the telescope and its relation to optics. He also mentions the many "experiments" in reflection and refraction brought forward by Scheiner. Later (p. 445), Blancanus refers to Scheiner's "abstrusis experimentis."

tered by me," as well as those provided by occasion.³⁴ Such experiences were, by definition, artificial; they did not, therefore, represent the "ordinary course of nature"—Clavius had regarded Galileo's telescopic discoveries as "monstrosities." The sharp disciplinary boundary between mathematics and natural philosophy prevented constructed experiences in the mathematical sciences from invading physics.

That disciplinary boundary was warranted by Aristotelian arguments concerning the differences between the subject matters of mathematics and natural philosophy. But Jesuit mathematicians in the early seventeenth century, concerned as they were about methodological challenges from the natural philosophers, were able to use such arguments to their own advantage. Although the mixed mathematical disciplines concerned physical phenomena, they only considered the quantitative aspects: physical causes fell inside the domain of physics, and outside that of mathematics. Because mathematics had to do only with quantity, not with process and teleology, the Aristotelian dichotomy of natural and artificial did not apply. The mathematician did not seek to reveal essential natures, and could therefore interfere with natural processes quite legitimately; to do so, however, required a constant policing of the disciplinary boundary between mathematics and natural philosophy. Jesuit mathematicians maintained the integrity of the methodological structure of the mathematical sciences by emphasizing whenever necessary that certain questions were beyond the purview of the mathematician. For example, Blancanus's *Sphaera mundi* was a treatise of astronomy and cosmography (mathematics), not of cosmology (physics); when Blancanus discusses the earth's motion he stresses that his subject is not *terraemotu*, the motion of earth as an element, but *motu terrae*, the motion of the earth itself, "for the former has nothing astronomical, and accordingly is to be left entirely to the physicists."³⁵ Disciplines such as geometrical optics focused on the

34. See chapter 2, section III, above.
35. Blancanus, *Sphaera mundi*, p. 74: "Non de terraemotu, sed de motu terrae hic agendum est: ille enim nihil habet Astronomicum, ac proinde totus physicis reliquendo est." Contrast this with Paul Guldin's designation of the same issue as "physicomathematical": see below, text to n. 75. See similar demarcations in Otto Cattenius's mathematical lectures of 1610–1611, printed in Krayer, *Mathematik im Studienplan der Jesuiten* (1991), pp. 181–360, concerning the corporeal nature of the heavens, e.g., p. 289: "id curae physicis, non mathematicis, est. Illorum nam est, ut iam diximus, leges praescribere astronomis, sed ab iisdem suas in caelesti materia accipere." The entire issue of the demarcation between physics and mathematics, and the importance to the Jesuits of maintaining it, is dealt with from a different perspective by Feldhay, "Knowledge and Salvation in Jesuit Culture" (1987).

quantitative properties of things and their mathematical implications; an optical apparatus was a contrivance the properties of which were those of an artificial object. The separation of mathematics from natural philosophy preserved the methodological propriety of experimental procedures by denying that they pretended to the creation of natural philosophical knowledge. Mathematical sciences that applied to the physical world were not taken to be in conflict with qualitative Aristotelian natural philosophy, but were typically seen as being about different things.

Because the mathematical sciences did not employ natural causes, the admitted tendency of the latter to be obscure highlighted the contrasting clarity of mathematical demonstrations, thereby supporting Clavius's arguments for mathematics' superior status vis-à-vis other disciplines. The prestige of mathematics for its Jesuit practitioners therefore relied to a considerable extent on preserving the sharp differentiation from physics, a philosophical distinction mirroring the disciplinary and curricular one. And since the eschewal of natural causes permitted the use of empirical principles that were established artificially, the enforcement of the disciplinary boundary between physics and the mathematical sciences promoted the methodological health of the latter.

Concern to enforce the boundary appears especially clearly on those occasions when the mathematician oversteps it and makes a point of apologizing. Blancanus's *Aristotelis loca mathematica* contains a fairly full paraphrase and commentary on the pseudo-Aristotelian *Mechanical Questions*, a text that hovers on the border of applying physical principles to mechanics.[36] At the end, where the peripatetic author touches on vortices and their tendency to push heavy objects inwards towards the center of rotation, Blancanus excuses himself from treating the matter on the grounds that it is physical, not mathematical. Having provided that *caveat*, he then feels free to criticize it anyway (Aristotle was wrong because bodies get thrown outwards, not inwards, "for experience teaches [it]").[37] Again, in *Sphaera mundi*, Blancanus gives an account of the new stars seen in the late sixteenth and early seventeenth centuries (up to 1604).[38] What they might be, and whether they

---

36. Blancanus, *Aristotelis loca mathematica*, bound with his *De mathematicarum natura dissertatio* (1615), "In mechanicas quaestiones," pp. 148–195. See, on other treatments of the work that Blancanus discusses, Rose and Drake, "The Pseudo-Aristotelian *Questions in Mechanics*" (1971), which does not, however, do more than list Blancanus's discussion.

37. Ibid., p. 195: "experientia enim docet."

38. Blancanus, *Sphaera mundi*, pp. 344–351.

constitute generation in the heavens, are matters for physicists, he says, and not appropriate for discussion here.[39] But, he continues, perhaps they are like comets, approaching and receding so that they are not always visible.[40] Blancanus signaled his brief foray across disciplinary boundaries precisely because those boundaries were too important to be blurred.

The same concern to maintain the autonomy of the mathematical sciences even when briefly considering physical questions can be found in Orazio Grassi's *Libra astronomica* of 1619, part of the exchange with Galileo over comets.[41] The chief issues were optical and astronomical, but owing to the wide-ranging character of Galileo's arguments Grassi, the mathematician, found himself obliged to venture outside his proper territory. Thus, at one point Grassi addresses Galileo's suggestion that comets and sunspots are composed of the same matter: accepting the suggestion, he uses the concept of natural motion to cast doubt on Galileo's claims for cometary paths—because sunspots travel in circles around the sun. Grassi concludes the digression abruptly with the words: "But these things are physical rather than mathematical." The next paragraph begins: "Now I come to the optical reasons by which it is proved far more effectively that the comet never was a vain apparition, nor did it ever wander as a specter among the nocturnal shadows, but that it displayed itself to the view of all in one place and that it was always one and the same in appearance."[42] For Grassi, mathematical arguments were to be strictly segregated from physical; in addition, they were far more effective and conclusive, as Clavius had argued.

The disciplinary boundary also acted, therefore, as a way of preserving the certainty of demonstrations in the mixed mathematical sciences. Blancanus's *Sphaera mundi* contains a chapter headed "On the parts of the world, and first on the elementary part." It begins by carefully delimiting its territory: "This inferior part of the world, which is composed from the elements (which whether they be three or four we leave to the disputation of the physicists)...."[43] The implication that he will discuss firm knowledge, not the conjectures of the physicists,

---

39. Ibid., p. 350.

40. Ibid., p. 351.

41. Grassi, *Libra astronomica* (1619), trans. by O'Malley in Drake and O'Malley, *The Controversy on the Comets* (1960); see chapter 3, section V, above.

42. Grassi, *Libra astronomica*, pp. 140–141; O'Malley, pp. 90–91 (quoting from p. 141/p. 91).

43. Blancanus, *Sphaera mundi*, p. 67: "Haec inferior mundi pars, quae ex Elementis componitur (quae tria ne, an quatuor sint, Physiologis disputandum relinquimus)...."

is clear. The same attitude appears in Scheiner's *Rosa ursina* of 1630, a large work on astronomy that concentrated on his own sunspot observations. It defines the subject matter of astronomy as follows:

> The astronomer, according to Ptolemy, *Almagest* book I, chapter 1, considers the quantity of celestial bodies, and indeed Aristotle himself declares in the *Categories*, and teaches in the *Physics*, that quantity is considered as much by the physicist as by the mathematician, but in different ways.[44]

The differences were important. As Scheiner said further on,

> In physical matters many things are unknown about what concerns the heavens; few things are known for certain; there are many doubtful things, many false things are asserted; many true things denied.[45]

The mathematical approach, Scheiner seems to suggest, is surer than the physical.

The disciplinary division sometimes served to protect the demonstrative claims of mathematics from taint. Blancanus employed it to remove what he appears to have seen as a pollution of astronomy by astrology. Clavius had divided astronomy into theoretical and practical branches, the first comprising all the apparatus of positional astronomy, including use of instruments and the drawing up of tables, and the second, "which others call judicial, or prognostic, or divinatory," consisting of astrology. Clavius sanctioned only natural astrology, including the effect of celestial bodies on temperament, and condemned judicial astrology proper as superstitious and theologically suspect (it was to be officially denounced once more by Urban VIII in 1631).[46] In his own writings, however, Clavius neglected treatment of astrology entirely, restricting himself to the "theoretical" branch of astronomy. Blancanus, too, failed to discuss astrology in the *Sphaera mundi*, but unlike Clavius he gave reasons, reasons denying the propriety of Clavi-

---

44. Scheiner, *Rosa ursina* (1630), p. 602 col. 2: "Astronomus teste Ptolemaeo, lib. I. Alm. cap. I. considerat Corporum caelestium Quantitatem, & quidem illam quam in Praedicamento declarat ipse Aristotiles [sic], & in physicis docet tam à physico quam à Mathematico considerari sed diversimodè."

45. Ibid., p. 606 col. 2: "In rebus Physiologicis circa ea, quae caelitus contingunt, plurima ignorantur; pauca certo sciuntur; dubia sunt plerique, asseruntur multa falsa; negantur plurima vera."

46. Clavius, "In sphaeram Ioannis de Sacro Bosco commentarius," in *Opera mathematica* (1611–1612), vol. 3, p. 3: "quam alij Iudiciariam, seu Prognosticam, id est, Divinatricem dicunt." On the condemnation see Walker, *Spiritual and Demonic Magic* (1958), pp. 205–206; also Ernst, "Astrology, Religion and Politics in Counter-Reformation Rome" (1991).

us's division of astronomy. The "most noble" of all the mathematical sciences on account of its subject matter, astronomy is subordinate both to geometry and to arithmetic; astrology, Blancanus continues, does not really deserve its common assignment to astronomy. This is because, even if astrology were a valid science, celestial influences are natural causes, and demonstration through natural causes belongs to physics, not mathematics.[47] In his subsequent listing of the divisions of practical mathematics, Blancanus includes among them the compilation of astronomical tables.[48] Here too he differs from Clavius, who had reserved the category of practical astronomy for astrology alone: Blancanus's attitude suggests that excluding the latter from the mathematical science of astronomy preserved astronomy's integrity and "preeminence" stemming from the sureness of its demonstrations.

But although Jesuit mathematicians, at least during the earlier decades of the century, were content to leave natural causes to the physicists, they did not defer to natural philosophy as if to the higher discipline. Clavius's largely successful attempts to give mathematics a significant place in the Jesuit educational curriculum had created teachers of mathematics in the colleges whose academic position in principle equaled that of the natural philosophers. Clavius, and above all Blancanus, aimed at achieving a genuine equality for the mathematical disciplines by claiming them as genuine sciences worthy of high respect.[49] Since they wished to portray their subject as a partner of natural philosophy, therefore, the Jesuit mathematicians had no interest in maintaining the disciplinary boundary between them solely by disqualifying themselves as competent to treat physical questions. On the contrary, the physicists could learn things from the mathematical disciplines; indeed, mathematics was indispensable to physics, as Clavius always insisted. Aguilonius's *Opticorum libri sex* bears the advertisement that it is "useful as much to philosophers as to mathematicians." In the preface Aguilonius praises Archimedes over Plato for judging the union of mathematics with matter "in no way to derogate anything from mathematics, but indeed even to adorn and perfect it." He goes on to say: "We, trusting in his authority, have undertaken the cultivation of this part of mathematics, which comprehends both genera, namely the philosophical and the mathematical."[50] Mathematics

47. Blancanus, *Sphaera mundi*, p. 390.
48. Ibid.
49. See chapter 2, above.
50. Aguilonius, *Opticorum libri sex* (1613), title page: "Philosophis iuxtà ac Mathematicis utiles"; preface "Lectoris," 10th p.: Archimedes "non modò nihil Mathesi derogare, sed verò exornare eam etiam ac perficere existimavit. Cuius nos auctoritate freti

was relevant to natural philosophy insofar as it dealt with the physical world: anyone who has seriously considered the question, says Aguilonius, "will approve very greatly the various ways of explaining [of mathematical demonstrations] to be employed for the diversity of things."[51]

Scheiner was particularly concerned to make this point in his *Rosa ursina*. Following the passage quoted earlier, in which he set out the difference between the treatments of quantity proper to physics and to astronomy, Scheiner lists some of the things that astronomy, as a mathematical science, was competent to treat. All kinds of change in the heavens—local motion, duration, augmentation, diminution, and so forth—came under the auspices of astronomy. In the case of the *Rosa ursina*'s special subject matter, therefore, Scheiner could present results such as these: sunspots are on the surface of the sun; the sun (from the observed motion of sunspots) is a globe; the sun rotates around its own center.[52] A subsequent section of the work is headed: "Many truths are disclosed to celestial physics from the appearance of the sun."[53] The advent of the telescope, Scheiner says, now allows access to things in the heavens hitherto inaccessible, things having relevance for physics; in particular, the newfound rotation of the sun has important physical implications.[54] Scheiner thus points out the indebtedness of physics to the new "mathematical" discoveries and demonstrations. His marginal gloss avoids confusion, however, by observing that the physical questions emerging from the new knowledge of solar appearances are to be settled by physicists.[55]

The disciplinary division allowed empirical principles drawn from artificial contrivance to be used in the study of the physical world, as long as this was done by the mathematical sciences rather than by physics. While physics was understood as a search for physical, or natural, causes, in which artificial contrivance had no proper place, mathematics sought demonstrations based only on the formal properties of magnitudes. The two domains related to radically different kinds of questions about the world. Nonetheless, by the middle of the

---

hanc Matheseos partem, quae utrumque genus, Philosophicum scilicet & Mathematicum comprehendit, suscepimus excolendam."

51. Ibid.: "pro rerum diversitate varios usurpari explanandi modos maximopere comprobabit."

52. Scheiner, *Rosa ursina*, pp. 602 col. 2 through 604 col. 1.

53. Ibid., p. 606 col. 2: "Ad physiologiam caelestem plurimae Veritates panduntur è Phaenomeno solari."

54. Ibid., p. 607 col. 1–2.

55. Ibid., gloss to p. 607 col. 2.

seventeenth century a new term had become established in the writings of both Jesuit and non-Jesuit writers on mixed mathematics. "Physico-mathematics" simultaneously exploited and overrode the standard scholastic disciplinary division between physics and mathematics: it advocated mathematics as a tool for the creation of genuine physical knowledge, but did so by means of the Aristotelian characterizations of their subject matters. The route of its emergence and acceptance may be traced clearly in the increasingly ambitious claims of mathematicians in the first few decades of the century. Physico-mathematics was a bid for disciplinary authority over knowledge of nature.

IV. Physico-Mathematics

It will be recalled that the conventional description of the mathematical disciplines at the beginning of the century drew on Aristotle's account of subordinate sciences.[56] Thus Blancanus had simply referred to the *Posterior Analytics* as providing the universally accepted view.[57] Aristotle's basic approach is to put each mixed mathematical science under an appropriate branch of pure mathematics: optics under geometry, mechanics under solid geometry, music under arithmetic. The subordinated science makes use of the conclusions of demonstrations from the superior, subordinating science. Blancanus had made the most of the scheme, arguing for the causal status of demonstrations in these sciences. Astronomy, for example, demonstrates such things as the cause of lunar eclipses or the slower motion of the sun through the zodiac in summer than in winter, and optics provides the final cause for the sphericity of the eye.[58] Small but significant alterations to the strict Aristotelian position are common, however, in both Jesuit and non-Jesuit sources. Chief among them is a growing tendency to expand the subordination of the mixed mathematical sciences to include not just the relevant branch of pure mathematics (or, especially for astronomy, both branches), but also physics itself.

Aguilonius, in 1613, had already described optics, with its close concern for the behavior of light and the anatomy of the eye in addition to geometrical inferences, as comprehending "each genus, that is, phil-

---

56. See chapter 2, above.
57. Blancanus, *De mathematicarum natura dissertatio*, p. 29. See esp. Aristotle, *Posterior Analytics* I, chaps. 7, 9, 13. Clavius also gives an untroubled conventional account i. Clavius, "In disciplinas mathematicas prolegomena," *Opera mathematica*, vol. 1, pp. 3–4. See chapter 2, section I, above.
58. Blancanus, *De mathematicarum natura dissertatio*, p. 30.

osophical and mathematical."⁵⁹ But a more unequivocal view may be found in a pamphlet of 1622, containing "selected propositions" concerning the chief branches of mathematics, from the Jesuit college of Pont-à-Mousson—an unremarkable source likely to be revealing of emerging pedagogical commonplaces in mathematics. Its section on mechanics provides the following characterization: "Mechanics, or the science of machines, is subordinated to geometry and physics, and [comes] from their principles."⁶⁰ Optics is described as treating visual rays and light rays; however, its particular concern with geometrical objects (lines, angles, cones) means that it is "subordinated not so much to physics, but especially to geometry"—a concession careless of Aristotle's failure to subordinate it to physics at all.⁶¹ Arriaga's 1632 *Cursus philosophicus* provides a similar casual implication of physics in the structure of a mathematical science in its remarks on music. Discussing the familiar tripartite disciplinary structure of metaphysics, physics, and mathematics, Arriaga says that some people include music under the latter heading: "however, I do not see in what way music, considering sounds, abstracts from sensible matter, any more than one might say that a science concerning colors abstracts from sensible matter. It ought better to be reduced, therefore, to a mixture of physics, on account of sensible sound, and mathematics on account of number and harmony."⁶²

A more formal set of such characterizations appears in Hugo Sempilius's textbook *De mathematicis disciplinis libri duodecim* (1635). Optics is "subordinated to geometry and physics"; mechanics "or the science of machines" is "subordinated to geometry and physics"; music is "subordinated to arithmetic and physics."⁶³ These formalisms go along with protestations, resembling those of Clavius and Blancanus, concerning the indispensability of geometry and arithmetic for physics itself. "For what can a physicist, without the work of geometry, discuss

---

59. Aguilonius, *Opticorum libri sex*, 10th p. of preface: "utrumque genus, Philosophicum scilicet & Mathematicum comprehendit."

60. *Selectae propositiones* (1622), p. 10: "Mechanica sive Machinatrix scientia, subalternatur Geometriae ac Physicae, & ex illarum principijs."

61. Ibid., p. 30: ". . . non tantùm Physicae sed praesertìm Geometriae subalternatur."

62. Arriaga, *Cursus philosophicus*, p. 44 col. 1: "non video tamen quomodo Musica, considerans sonos, abstrahat à sensibili materiâ, perinde ac si diceret, scientiam agentem de coloribus abstrahere à sensibili materiâ. Meliùs ergo ad mixtam ex Physicâ propter sonum sensibilem, & ex Mathematicâ propter numerum & concentum deberet reduci."

63. Sempilius, *De mathematicis disciplinis libri* (1635), pp. 61, 15, 107: "Tertium Mathesis membrum est Optica, Geometriae & Physicae subordinata"; "Mechanica, seu machinatrix scientia, Geometriae & Physicae subalternata"; "Musica Arithmeticae & Physicae subalternatur." There seems to be no equivalent statement in this text for astronomy.

reliably about points, lines, surfaces, indivisibles (whether they be positive or negative, real or imaginary)? What of rarefaction or condensation, the occupying of a greater or lesser space, without addition or subtraction of parts?" A physicist without geometry is blindfolded.[64] Similarly, physics needs arithmetic to cope with such things as degrees of qualities.[65]

The gradual emergence of the term "physico-mathematics" should occasion no surprise in light of this push towards the practical inversion of the mathematics-physics relationship. The idea that mathematics, and the mixed mathematical disciplines in particular, could yield genuinely causal scientific knowledge of natural bodies and phenomena became a virtual commonplace, especially in the mathematical writings of the Jesuits, during the early decades of the seventeenth century.[66] "Physico-mathematics" served to elevate the status of mathematical sciences to a par with physics without formally violating the longstanding, and well-entrenched, disciplinary division between mathematics and physics. True physical causes could still be portrayed as differing from mathematical causes, but such a distinction now worked to the detriment of the former.

The origin of the expression seems quite obscure even though its historical meaning is clear. An entry from late 1618 in the journal of the Dutch mechanical philosopher and schoolmaster Isaac Beeckman gives some indications of its novelty, however. Beeckman had recently met and worked with the young René Descartes, who had been deeply impressed by Beeckman's ideas. These amounted to an attempt to extend the scope of the mixed mathematical sciences further into the territory of physics, and involved micromechanical corpuscular explanations.[67] In his journal, Beeckman refers to an observation of Des-

64. Ibid., p. 54: "Quid enim Physicus sine Geometriae ope disputare solidè potest de punctis, lineis, superficiebus, indivisibilibus, an sint positiva vel negativa, an realia aut imaginaria? Quid de rarefactione aut condensatione, modò maiorem modò minorem locum occupante, sine additione aut detractione partium?"

65. Ibid., p. 57. Cabeo, *In quatuor libros Meteorologicorum Aristotelis commentaria* (1646), vol. II, p. 186, observes that explanations of the rainbow are not just geometry, or just physics, but part of each. It is significant that Cabeo was as much a *physicist* as a mathematician.

66. Baldini, *Legem impone subactis* (1992), chap. 1, esp. pp. 32–36, 41–45, discusses conceptual aspects of the relationship between physics and mathematics, and the role of mixed mathematics as mediator, among Jesuit mathematicians in the early decades of the seventeenth century. For discussion of some of the pressures that promoted the revaluing of mathematical knowledge see Biagioli, "The Social Status of Italian Mathematicians" (1989).

67. Schuster, "Descartes and the Scientific Revolution" (1977), chap. 2; idem, "Descartes' *Mathesis universalis*" (1980), pp. 48–49; Van Berkel, "Beeckman, Descartes et 'la

cartes's regarding the relationship between the physical properties of a string and the musical note it emits; he is pleased by its accordance with his own "hypothesis" about the corpuscular basis of sound production.[68] He then continues proudly:

> This Descartes has been educated with many Jesuits and other studious people and learned men. He says however that he has never come across anyone anywhere, apart from me, who uses accurately this way of studying that I delight in, with mathematics connected to physics. Also neither have I told anyone apart from him of this kind of study.[69]

Beeckman's marginal gloss summarizing this paragraph says simply: "Very few physico-mathematicians."[70]

The next three decades or so witnessed the appearance of many more. Descartes himself engaged in a style of natural philosophy very similar to Beeckman's; the Parisian mathematician Claude Mydorge was a friend and collaborator of Descartes's a few years later and in 1630 was comfortable enough with the new category to mention it casually in his commentary on problems in recreational mathematics. After having discussed a "problem" about a way to make water appear to boil by rubbing a wetted finger around the rim of its glass container, Mydorge remarks:

> These things reduced to the truth of appearances, we leave for the present to the more curious to discover their true causes. And we set aside the matter for now so as to bring to light some day, with God's help and by means of greater leisure, what we have examined and resolved in these matters in our physicomathematical disquisitions.[71]

---

philosophie physico-mathématique'" (1983); idem, *Isaac Beeckman* (1983). (It should be noted that the section of material in Descartes, *Œuvres*, vol. 10, labeled "Physico-mathematica," is so called by its modern editors, whereby Beeckman's term is borrowed for writings produced by Descartes for Beeckman.)

68. See on this aspect of Beeckman's work Van Berkel, "Isaac Beeckman"; Buzon, "Descartes, Beeckman et l'acoustique" (1981); idem, "Science de la nature et théorie musicale" (1985).

69. Beeckman, *Journal* (1939–1953), vol. 1, p. 244. "Hic Picto cum multis Jesuitis alijsque studiosis virisque doctis versatus est. Dicit tamen se nunquam neminem reperisse, praeter me, qui hoc modo, quo ego gaudeo, studendi utatur accuratèque cum Mathematicâ Physicam jungat. Neque etiam ego, praeter illum, nemini locutus sum hujusmodi studij." "Picto" is Descartes; cf. ibid., p. 237.

70. Ibid., "Physico-mathematici paucissimi." Beeckman generally preferred the term *mathematico-physicâ*; see, e.g., ibid., vol. 4, pp. 196, 200, a letter of 1630 to Descartes where Beeckman is again somewhat exercised about his priority. See Shea, *The Magic of Numbers and Motion* (1991) pp. 77–86, on their stormy relationship.

71. Mydorge, *Examen du livre des recreations mathematiques* (1630), p. 75: "Ces choses reduictes à la verité de l'apparence nous laissons quant à present aux plus curieux à en rechercher les vrayes causes. Et nous reservons à faire voir quelque iour avec l'aide de

By clear implication, Mydorge (like Beeckman and Descartes) saw physico-mathematics as a causal science of nature transcending the limitations conventionally assigned in the schools to mixed mathematics. For Beeckman in 1627, the "mathematical-physical philosophy" amounted to "the true philosophy."[72]

The new category made it easier for mathematical scientists to make philosophical claims that had previously been fiercely contested. Galileo's dispute over floating bodies in 1612 had taken the form of an assertion of the rights of mathematics over those of physics.[73] A process of disciplinary imperialism, whereby subject matter usually regarded as part of physics was taken over by mathematics, operated to upgrade the status and explanatory power of the mathematical sciences. The label "physico-mathematics" made the move explicit to all. One of its earliest applications appears in the work of Scheiner's collaborator on sunspot observations, the Jesuit mathematician Paul Guldin.[74] Guldin published his *Dissertatio-phisico-mathematica de motu terrae* in 1622.[75] The cosmological-mathematical double significance of the work's subject is at once evident, this being the arena within which Galileo attempted his own mathematical takeover of natural philosophy.

In 1631, the Jesuit polymath Athanasius Kircher, at Würzburg, wrote a dissertation for a student's defense (a standard procedure in this period) that also employed the new term. Its title is *Ars magnesia, hoc est, disquisitio bipartita-empeirica seu experimentalis, physico-mathematica de natura, viribus, et prodigiosis effectibus magnetis.*[76] Thus the work aims at revealing the nature, as well as the properties, of magnets. The little book's contents are further described on the title page in such a way as to clarify the meaning of its physico-mathematical characterization: this disquisition is "proposed in both theorem and problem form, and propounded by a new, apodictic or demonstrative method, confirmed by various applications and by daily experience."[77] The work, which

---

Dieu, & moyennant plus de loisir, ce que nous en avons examiné & resolu dans nos disquisitions physicomathematiques."

72. See Hooykaas, "Isaac Beeckman" (1970), on p. 567. Hooykaas also notes the posthumous work *D. Isaaci Beeckmanni medici et rectoris apud Dordracenos mathematico-physicarum meditationum, quaestionum, solutionum centuria* (Utrecht, 1644).

73. This is discussed at length in Biagioli, *Galileo, Courtier* (1993), chap. 4.

74. See chapter 4, section III, above.

75. See Baldini, *Legem impone subactis*, p. 62 n. 27, for brief discussion.

76. Kircher, *Ars magnesia* (1631).

77. Ibid: "Cùm theoreticè, tùm problematicè propositam, novâque methodo ac apodicticâ seu demonstrativâ traditam, variisque usibus ac diuturnâ experientiâ comprobatam."

concerns magnets, is thus to have a mathematical structure, much as Cabeo's treatise on this subject had appealed to the same model two years before.[78] As advertised, the first part is made up of theorems, the second of problemata (that is, constructional problems). Each of the ten theorems (with some corollaries added at the end of the part) consists of a statement of that which is to be proved, followed by an *experimentum* that bears out the proposition. Thus the first *theorema* proposes that "a magnet draws iron." This universal proposition is supported by an *experimentum* that begins by requiring the following: "You allow a steel needle fastened to a little cork boat to move about freely on the top of water gathered in a bowl." This having been done, a magnet brought nearby will cause the needle to move.[79] This is the form of all ten of the theorems, the *experimenta* appearing as instructions or recipes for the production of an effect confidently asserted to result.

An important feature of the new term is that its ideological function as a token of the new pretensions of the mathematical sciences overshadowed the greater specificities of its usage. Guldin's book had concerned the motion of the earth; Beeckman associated it with corpuscular explanations; Kircher used it to designate a mathematically structured treatise on the "nature" as well as properties of magnets. From the 1630s onwards, the term appears frequently: in 1631 a *Dissertatio physico-mathematica de camera obscura* appeared in Sweden, while in 1636 it was applied to a collection of marvels in natural magic written, apart from the title, in German.[80] A further example of the rapid diffusion of the new category, also from 1636, appears in another disputation written for the examination of a student. Produced by a professor of mathematics at Königsberg, it contains, as its title says, "physico-mathematical controversies." These concern such things as gravity and whether it is a tendency towards the center of the universe (Galileo's discovery of Jupiter's satellites and the evidence that sunspots circle the sun supply reasons for doubting the proposition); whether the earth is the center of the universe (the sun is asserted to be much more probable as the center—"multò verò probabilius"); the complex-

---

78. See chapter 3, section I, above.

79. Kircher, *Ars magnesia*, p. 3: "Infixo subereae raticulae chalybeo obelo, eum super aquam pelui exceptam fluctuare libere permittas; quo facto admoto eminus magnete videbis ilico nutare."

80. Vallerius, *Dissertatio physico-mathematica de camera obscura* (1631), evidently a work in the Keplerian optical mold; Schwenter, *Deliciae physico-mathematicae* (1636), discussed in Thorndike, *A History of Magic and Experimental Science*, vol. 7 (1958), pp. 594–595. The latter resembles Della Porta's *Natural Magick* (as well as some of Mersenne's collections of mathematical curiosities) in its miscellany of contents, including how to stand an egg on one end or how to see stars in daylight.

ity of stellar motion; whether daily motion is due to the rotation of the earth or the heavens. This last point was presumably designed to invite consideration of the Tychonic system, but, interestingly, it finishes with the observation that the question cannot be settled by astronomical hypotheses, but requires physics.[81]

That "physico-mathematics" should appear both in popular vernacular texts (recall also Mydorge's French work) and in workaday academic settings demonstrates its newly acquired currency. In the 1640s Marin Mersenne invoked the category in one of his mathematical compilations. Mersenne had corresponded with Beeckman, visiting him in 1630, and was Descartes's contact with the learned world of Paris after Descartes moved to the Netherlands; he also knew Mydorge. His intellectual sympathies made "physico-mathematics" a perfect description and legitimation of his scientific work. Thus, in 1644, he published *Cogitata physico-mathematica*, which discusses such subjects as harmonics, pneumatics, ballistics, hydraulics, and the mechanics of moving bodies as recently developed by Galileo and others; he treated similar matters in the "Reflectiones physico-mathematicae" forming part of his *Novarum observationum . . . tomus III*.[82] Mersenne took the mixed mathematical sciences as his models for understanding nature, and part of their attraction for him lay precisely in the fact that they were not physics. The pretensions of physics, or natural philosophy, to qualitative causal explanation based on knowledge of the essential natures of things seemed to Mersenne both unrealistic and dangerous. Sceptical arguments stood ready to indict them as unrealistic, and the apparent impossibility of achieving consensus on many properly physical questions signaled the dangers of dissension.[83] If there was one thing upon which both physicists and mathematicians always agreed, it was that mathematical demonstrations were especially certain. The Jesuit-educated Mersenne was familiar with the texts in which Clavius had paraded this property as a mark of the superiority of mathematics over physics;[84] the appeal to Mersenne of mixed mathematics as

---

81. Linemannus, *Disputatio ordinaria continens controversias physico-mathematicas* (1636). The quoted phrase is from the fourth page of the unpaginated pamphlet (British Library).

82. Mersenne, *Cogitata physico-mathematica* (1644); idem, *Novarum observationum . . . tomus III* (1647).

83. Lenoble, *Mersenne* (1943, 1971); Dear, *Mersenne* (1988); on scepticism see Popkin, *The History of Scepticism* (1979), esp. chaps. 5–7. For just such an arena of dissension, see below, chapter 7.

84. Dear, *Mersenne*, pp. 37–39.

an exemplar of a new philosophy of nature was of a piece with the widespread, and growing, acceptance of the idea of a true physico-mathematics that combined mathematical demonstration with physical subject matter. The loss of the Aristotelian physicists' fundamentally teleological causes had come to matter little.

By the middle of the century, the term "physico-mathematics" appears routinely. Apart from Mersenne's use of it in 1644, Honoré Fabri labeled the first appendix to his 1646 *Tractatus physicus de motu locali* "Appendix prima physicomathematica, de centro percussione." Fabri's fourth appendix, furthermore, indicates particularly clearly the ambition and scope of this category by outlining a kind of mathematical approach to physics that would have had no place in school natural philosophy a few decades earlier: "On the physical principle of the duplicate physical ratio." This includes such phenomena as falling bodies (the time-squared law), the action of siphons, the propagation of light, the action of pendulums, and the relation of musical string tension to pitch.[85] Fabri's Jesuit brother Riccioli defined the central subject of his *Almagestum novum* in 1651 as follows: "Astronomy is a physico-mathematical science concerning the terminated quantity of celestial bodies and their terminating sensible accidents." It is, he says, subordinated to both physics and mathematics, although chiefly to the latter.[86] Later, after having criticized the claims of Cabeo and Arriaga on falling bodies (see chapter 3, above), Riccioli introduces his own work on the subject by asserting that "we will proceed from the following experiments, not by way of likely conjectures but according to infallible physico-mathematical science, to certain conclusions."[87] It is important to note Riccioli's characterization of physico-mathematics as scientific, infallible, and certain: ever since Clavius, Jesuit mathematicians had striven to invest their subject with those properties, and always with an eye towards the prestige of physics.

An especially noteworthy example, again by a Jesuit, of the encroachment of mathematics into physics under the guise of physico-

---

85. Fabri, *Tractatus physicus* (1646), pp. 420, 443, "De principio physico rationis duplicatae Physicae." Fabri's clear ambition was to turn mathematical science into physics by introducing properly physical causal explanations, as evidenced by his insistence on invoking the word "physical" (see chapter 5, section III, above).

86. Riccioli, *Almagestum novum*, pt. 1, p. 2 col. I: "Astronomia est *Scientia Physico-Mathematica de Coelestium corporum quantitate terminata, & eorum accidentia sensibilia terminante.*"

87. Ibid., pt. 2, p. 383 col. I: "non ex verisimilibus coniecturis, sed ex infallibili scientia Physico-Mathematica ab experimentis sequentibus ad conclusiones certas procedemus."

mathematics appeared in 1658 with title of *Microcosmi physicomathematici*.[88] Francesco Eschinardus's volume presents a form of disciplinary interaction wherein the subject matter of physics is, so to speak, parceled out to the various mathematical sciences. Thus the structure of the work has the mathematical disciplines (including geography) placed under headings that one would ordinarily expect to see in a physical treatise, as the subtitle of the work proclaims: *In quo clarè, & brèviter tractantur praecipuae mundi partes, caelum, aer, aqua, terra; eorumque praecipua accidentia*. The work as a whole is advertised as having five parts, of which the first four correspond to the elements while the fifth is "De numeris." "For thus we have divided the entire object of mathematics, quantity, of course into continuous and discrete; and we have indeed divided the continuous into four chief parts, the heavens, air, water, and earth."[89] The book is thus a kind of cosmographic survey fleshed out with material generally of an instrumental mathematical kind.[90] Its generic resemblance to a nineteenth- or early twentieth-century school text of the "heat, light, and sound" variety, now bearing the old label "physics" with a quite different meaning from that of the seventeenth century, is strikingly instructive; we have here the two termini of a pedagogical tradition three centuries long. The construction of Eschinardus's treatise, not around subject-matter but around the principal parts of the universe, places it beyond mixed mathematics and, more ambitiously, into physico-mathematics.

Some years later, in 1665, Riccioli's collaborator in the work on falling bodies, Francisco Maria Grimaldi, published a work announcing the diffraction of light. It is entitled *Physico-mathesis de lumine, coloribus, et iride*,[91] and adopts the geometrical style of presentation that was

88. Eschinardus, *Microcosmi physicomathematici* (1658). This work, billed as "tomus primus," was, in fact, the only volume of the work ever to appear: Sommervogel, *Bibliothèque de la Compagnie de Jésus* (1890–1932/1960), vol. 3, p. 432 col. 2. On Eschinardus see Feldhay and Heyd, "The Discourse of Pious Science" (1989), esp. pp. 111–123, discussing a later work by Eschinardus, the *Cursus physico-mathematicus* of 1689. Eschinardus is also discussed in Middleton, "Science in Rome" (1975).

89. Eschinardus, *Microcosmi physicomathematici*, "Index tractatum" (unpaginated): "Sic enim divisimus totum obiectum Mathematicae, nempè quantitatem in continuam, & discretam; & continuam quidem in quatuor praecipuas partes divisimus Coelum, Aerem, Aquam, Terram."

90. Eschinardus provides discussions of such things as hydraulic clocks (in the section "De coelo" as well as in a separate appendix), devices for measuring specific weights, siphons, and other contrivances in the tradition of Hero's *Pneumatica* (in "De aqua," which also discusses wine-and-water mixing tricks of the kind found in Della Porta), in addition to descriptive geographical accounts of the main winds in the Mediterranean and European region, of tides, and of the behavior of rain and snow.

91. Grimaldi, *Physico-mathesis de lumine* (1665).

standard for optical treatises. Even in the case of its central claim—an empirical novelty of great significance for Grimaldi himself as well as for later students of light—Grimaldi avoids any weakening of its scientific status that might result from making it appear contingent on his own reports of particular experiments. The book's first proposition asserts that "light is propagated or diffused not only directly, refractively, and reflectively, but also by another certain fourth way, DIFFRACTIVELY."[92] Its supporting experiments are all of a universalized or recipe form, with appropriate use of the passive voice, employing such formulations as "we assert," "we devise," and, for the outcomes, "it is observed."[93] Grimaldi also makes the usual invocation of unnumbered frequency as a means of removing particularity: "It truly has been manifestly established by very frequently repeated experiment. . . ."[94]

The remarkable aspect of Grimaldi's treatise is its use of "physico-mathesis" to argue for properly *physical* conclusions—this is not simply the mixed-mathematical science of optics under a new label. Grimaldi means to sustain the Aristotelian thesis of the accidental, as opposed to substantial, or material, character of light. He does so in good scholastic fashion, first presenting arguments (utilizing his "new experiments") for the substantiality of light, and then refuting them, all for the purpose of showing that the peripatetic doctrine can still be sustained. But Grimaldi's final position, in contrast to Riccioli's boasts for the "certainty" and "infallibility" of his own physico-mathematical conclusions, does not involve assertions of mathematical certainty. Instead, Grimaldi describes his favored position as being "probably" (*probabiliter*) sustainable; that is, sustainable with probability rather than with demonstrative certainty—a typical qualification for any scholastic physical thesis,[95] but one explicitly disavowing the pretensions to certainty that had been the mathematician's chief weapon.

The mingling of physics with mathematics in the new physico-mathematical genre, then, ran the risk of weakening mathematical claims to certainty. Such a view, at any rate, emerges from a consider-

---

92. Ibid., p. 1: "Lumen propagatur seu diffunditur non solùm Directè, Refractè, ac Reflexè, sed etiam alio quodam Quarto modo, DIFFRACTÈ."

93. Ibid., p. 1 col. I and pp. 1–11 passim: "ponamus," "fingamus," "observetur." A typical recipe-like instruction for success commences: "So that the experiment succeeds correctly, the light of the sun is required. . ." ("Ut experimentum rectè succedat, requiritur Lumen solis. . .": p. 10 col. I).

94. Ibid., p. 9 col. II: "Id verò constitit manifestè repetito saepius experimento. . . ."

95. The title of Grimaldi's treatise continues thus: ". . . aliisque adnexis libri duo, in quorum Primo afferuntur Nova Experimenta, & Rationes ab ijs deductae pro Substantialitate Luminis. In Secundo autem dissolvuntur Argumenta in Primo adducta, & probabiliter sustineri posse docetur Sententia Peripatetica de Accidentalitate Luminis."

ation of mechanics by the Jesuit disciple of Athanasius Kircher, Gaspar Schott. Schott's *Magia universalis* was published in four volumes between 1657 and 1659, and part III includes the identification of two distinct approaches to mechanics. One is mathematical, in keeping with the traditional categorization of mechanics as a mixed mathematical science subordinate to geometry. The other, however, is called "physico-mechanical," and is distinguished from the first by its search for physical causes. Among those taking the former approach, says Schott, are Guidobaldo del Monte, Simon Stevin, Galileo, and Mersenne, while the latter is taken by Aristotle (in the *Mechanical Questions*) and certain followers of his approach, including the French Jesuit Honoré Fabri, all of whom used the idea of differential potential velocities to explain equilibrium.[96] Schott concludes his consideration of these alternative approaches by observing that there must be some physical cause of the increase of force in simple machines beyond the mere disposition of weights and forces—he means the law of the lever and its kin—because disposition itself is not *activa physice*. However, he says, we do not know what that cause is. Accordingly, Schott proceeds to abandon consideration of the "physico-mechanical" approach, and continues with a conventional treatment of statics, the simple machines, and hydrostatics falling under his "mathematical" category.[97] Unlike Grimaldi, Schott was unprepared to adopt the probabilistic stance of the physicist.

Isaac Barrow provides a useful picture of the meaning of "physicomathematics" in the 1660s, by which time it had long become established as a widely used category. In discussing mathematical terms and categories as part of the introductory material to his *Mathematical Lectures*, Barrow makes the usual distinction between "pure" and "mixed" mathematics. He notes that the latter deals with physical accidents rather than the nature of quantity in itself and that some people are wont to call its divisions "Physico-Mathematicas."[98] Barrow insists

96. On the Pseudo-Aristotelian work, see especially Rose and Drake, "The Pseudo-Aristotelian *Questions of Mechanics*"; Laird, "The Scope of Renaissance Mechanics" (1986); De Gandt, "Les *Mécaniques* attribuées à Aristote" (1986). The last two items in particular discuss the role of the work in creating a new, prominent role for mechanics as a mathematical science.

97. Schott, *Magia universalis* (1657–1659), part III, pp. 211–225. In his *Cursus mathematicus* (1661), Schott uses the term "physico-mathematical" in connection with mechanics, listing for book XV on contents page "Prooemium totius operis" (unpaginated): "Accedunt in fine Theoreses Mechanicae Physico-Mathematicae novae, à docto Mathematico conscriptae."

98. Barrow, *Lectiones mathematicae* (1683), reprinted in Barrow, *The Mathematical Works* (1860/1973), p. 31 (lect. 1); see also p. 89 (lect. 5).

throughout the lectures that physics and mathematics are, in fact, strictly inseparable: all physics implicates quantity and hence mathematics.[99] Barrow's position resembles closely Isaac Newton's idea of "mathematical principles of natural philosophy"—a terrible category mistake by earlier standards.[100] It sums up neatly, however, the direction that arguments concerning the potential of the mixed mathematical sciences had taken during preceding decades.

99. Ibid., p. 41 (lect. 2).
100. See chapter 8, below.

Seven   PASCAL'S VOID,
NATURAL
PHILOSOPHERS, AND
MATHEMATICAL
EXPERIENCE

I. Pascal's New Experiences

The two most famous event experiments in the seventeenth century are surely Isaac Newton's exploit with a prism, reported in his 1672 letter to the Royal Society, and the ascent of the Puy-de-Dôme by Blaise Pascal's brother-in-law Périer. But despite this apparent natural affinity, the contemporary meanings of these two episodes differed radically. Newton, as we shall see in chapter 8, recast a piece of mixed mathematical science into the form of a true event experiment so as to make it a piece of "experimental philosophy." Pascal, by contrast, used the Puy-de-Dôme adventure as an ingredient of a mathematical argument, and to that end tried to shed it of all particularity. He was even less tolerant of particularity than the mathematical Jesuits whose influential work we have already considered.

During 1646 and 1647 Pascal had performed in Rouen, both alone and with Pierre Petit (who brought knowledge of it from Mersenne in Paris), various practical trials of a new and sensational phenomenon from Italy which has since become known as the Torricellian experiment.[1] A glass tube open at one end is entirely filled with mercury; the operator then seals the opening with a finger. The tube is upended and made to stand vertically in a basin that already contains a couple of inches' depth of mercury. Given a long enough tube (about three feet in the present case), upon removal of the finger the column of mercury descends partway down the tube, its excess flowing into the basin, and then hangs, as if suspended. A mysterious space now occupies the

---

1. See Taton, "L'annonce de l'expérience barométrique" (1963).

tube's vacated portion.[2] Pascal also performed the trick using water and wine in place of mercury and needed considerably longer tubes in proportion to the lesser specific gravities of these liquids.[3]

In October 1647 Pascal announced the results of his investigations in a pamphlet entitled *Expériences nouvelles touchant le vide*—an outline, he says, of a projected longer treatise.[4] Following an introductory section "Au lecteur," the purported abridgement of the first part of the unknown treatise begins with eight itemized "experiences."[5] These are not, however, historical reports of individual trials; instead, they are given as statements of what happens under particular contrived conditions. They are universal statements about how things happen in the world, suitable for Pascal's subsequent use of them in deriving other universal statements about nature. Thus the first begins: "A glass syringe with a good, tight piston, submerged entirely in water, and the opening of which one closes with a finger so that it touches the bottom of the piston, putting for this purpose the hand and arm in the water; one needs only a mediocre force to withdraw it, and make it separate from the finger. . . ."[6] The third, which can be related to a specific occurrence known independently, commences as follows:

> A glass tube forty-six feet long of which one end is open and the other sealed hermetically, being filled with water, or preferably with

---

2. The classic form of the experiment also has a layer of water over the mercury in the basin. See De Waard, *L'expérience barométrique* (1936), for detail especially on the early Italian work; Segre, *In the Wake of Galileo* (1991), pp. 80–88; Middleton, *The History of the Barometer* (1964), chap. 2 (early Italian), chap. 3 (Pascal). There is a good brief discussion in Dijksterhuis, *The Mechanization of the World Picture* (1986), p. 444–455.

3. See the account in Pascal, *Œuvres* (1904–1923), vol. 2, p. 8.

4. Pascal, *Œuvres* (1963), p. 195 (title).

5. Ibid., pp. 196–197. I translate *expérience* as "experience" throughout. For an illuminating discussion of these experiences (involving barometer tubes, siphons, and syringes with water, wine, and mercury), see Harrington, *Pascal philosophe* (1982), pp. 48–51. Harrington shows how each experience builds on suppositions derived from its predecessors, as, for example, in moving from small transparent pieces of apparatus to larger opaque ones: the behavior of liquids inside the latter is taken as evident, even though not visible, because of implicit analogies with the earlier observable arrangements. See also the treatment by Akagi, "Pascal et le problème du vide" (3 parts, 1967, '68, '69), in (1967), pp. 196–199.

6. Pascal, *Œuvres* (1963), p. 196 col. 2. In 1646 Pascal had put on a public show in Rouen employing such tubes, one containing water and the other wine, so as to demonstrate theatrically that the space at the top of the tube is not, as some claimed, filled with the fumes or vapor of the liquid: he encouraged such critics to predict whether the column of wine or the column of water would stand higher. Because wine is more volatile than water, it was apparently easy to get them to predict that the wine would stand lower, its fumes supposedly capable of filling a larger space above the column. Pascal

very red wine, to be more visible, then plugged up and lifted in this condition, and brought perpendicularly to the horizon, the opening blocked at the bottom, in a vessel full of water, and submerged about a foot; if one unplugs the opening, the wine in the tube descends to a certain height, which is about thirty-two feet above the surface of the water in the vessel. . . .[7]

As the form of this example illustrates, the first part of these experiences typically describes the setup, while the second describes the behavior of the apparatus when employed in various ways.

Seven "maxims" follow the eight experiences. These are the universal statements about nature that Pascal's experiences are taken to justify. He introduces them by gesturing at an unexplicated store of further experiences that will be contained in the full treatise (which never appeared and may never have been written): "From which experiences [the eight just given] and from many others reported in the complete book, where appear tubes of all lengths, thicknesses, and shapes, charged with different liquids, immersed diversely in different liquids . . . [etc.], the following maxims are manifestly deduced."[8] The reader is, of course, in no position to argue. The first maxim maintains "that all bodies have repugnance to separating themselves one from the other and admitting this apparent void in their interval; that is, that nature abhors this apparent void."[9] The maxims become increasingly experientially specific, the seventh starting out by asserting "that a force greater, by as small an amount as one likes, than that with which water of thirty-one feet high tends to flow downwards, suffices to produce this apparent void."[10]

The term "maxim" had a double reference. It was sometimes used in the seventeenth century as a synonym for "axiom" in the Aristotelian or mathematical sense; that is, a self-evidently true statement such as "the whole is greater than its proper part."[11] But it also commonly referred to general philosophical rules that acted in scholastic argu-

---

then enacted the demonstration to show the opposite. The falsification of the "vapor" explanation had greater logical (and, under the circumstances, dialectical) force than any direct positive inference to the presence of a vacuum. See Pascal, *Œuvres* (1904–1923), vol. 2, pp. 7–8. On the logic of falsification see Pascal, letter to Noël, in Pascal, *Œuvres* (1963), p. 202 col. 2; Popkin, "Scepticism, Theology and the Scientific Revolution" (1968), p. 14.

7. Pascal, *Œuvres* (1963), p. 196 col. 2.
8. Ibid., p. 197 col. 2.
9. Ibid.
10. Ibid., p. 198 col. 1.
11. See, for example, Locke, *Essay Concerning Human Understanding* (1690), IV.12.9, 15, for "maxim" used as a synonym for "axiom" in a specifically mathematical context.

ments as received truths: a famous instance is Descartes's use in his *Meditations* of a standard scholastic maxim holding that a cause must always be at least as great as its effect.[12] Patricia Reif has compiled a considerable list of such maxims, which formed a significant component of scholastic philosophical textbooks during this period.[13] Pascal's maxims in the present case, however, serve neither as self-evident axioms, insofar as they are justified through reference to experience, nor as general tenets of philosophical reasoning. Pascal regards them as universal truths about nature that experience, in the form of many and diverse contrived "experiences," reveals.[14] Collectively, they are a particularly clear instance of the kind of mathematical science discussed in chapter 2; they could serve as the premises of the syllogistic demonstrations that constituted the methodological core of an Aristotelian science. Although they are not *self*-evident, they are *experientially* evident truths and beyond question to those who have acquired the requisite experience. After all, lifelong cave dwellers would not know that the sun rises in the east, yet that truth counted as an unimpeachable scientific premise.[15]

In the succeeding parts of his "abridgement" Pascal flaunts a methodological prissiness regarding his central concern, the question of the void. Following the second part of the treatise, which will advocate and demonstrate certain "propositions" concerning the absence of any sensible matter from the space at the top of the tube, Pascal's conclusion will establish new maxims about the void that differ from the

---

12. This is something of an oversimplification; Descartes's argument involves the relationship between "formal reality" and "objective reality": *Meditations* III. A simpler version appears in *Discourse on Method* IV; the purpose is to prove the existence of God, the cause of my idea of perfections in various qualities, through reference to a being that actually possesses those perfections. Descartes of course justifies his argument by appealing to "clarity and distinctness," but its content relies on a scholastic maxim. See Descartes, *Discours de la méthode* (1930), pp. 324–325.

13. P. Reif, "The Textbook Tradition in Natural Philosophy" (1969). Maxims also found their place in early-modern English legal thought: for a general discussion of the significance of the maxim as a token of particular culturally embedded ways of doing things, see Shapin, *A Social History of Truth* (1994), pp. 239–240.

14. Indeed, to the extent that they approximate to a mathematical category, they are less axioms than postulates in the preferred sense of Proclus or Clavius (see chapter 8, section II, below). The Port Royal logicians, Arnauld and Nicole, underlined the acceptability of Pascal's approach by using the commonplace stress on the multiplicity of repetitions underlying an experiential assertion: Pascal *(un grand esprit)* has proved the role of the weight of the air in "fear of the void" phenomena "par *mille* experiences tresingenieuses" (my emphasis): Arnauld and Nicole, *La logique ou l'art de penser* (1662/1965), p. 267.

15. Cf. the astronomical issues involved in chapter 2, above.

earlier set in only one, systematic way. The original maxims had invariably used the term *vide apparent* when needing to refer to the putative void. These new maxims, on the strength of the preceding propositions (taken as proven), state exactly the same generalizations, word for word, except that the *vide apparent* has been systematically substituted by *vide*.[16] Pascal wants to make true statements about the nature of the world from experiences relating only to the appearances of the world. This requires a methodological trick reminiscent of Newton's "Rules of Philosophizing": Pascal establishes a favorable "burden of proof" condition.[17]

Introducing the group of ontologically upgraded maxims, Pascal says that, having shown that no sensible matter occupies the space in the tube, "my sentiment will be, until the existence of some matter that fills it be shown to me, that it is truly void, and destitute of all matter. That is why I will say of a true void [*vide véritable*] what I have shown of an apparent void [*vide apparent*], and I will hold as true of an absolute void the maxims posed and stated above concerning the apparent [void], to wit in this way."[18] The list follows. Pascal's later response to the Jesuit Noël, who was to criticize Pascal's conclusion that a true void existed in the space above the column of mercury, reiterates his "burden of proof" argument for the apparent void being a real void, although stressing in another letter that this is not tantamount to an assertion that it is really so.[19]

The main thrust of Pascal's letter to Noël is that Pascal has stuck close to the principles of logical demonstration, following a "universal rule" which is the chief part of the "way in which the sciences are treated in the schools, and the one which is in use among people who seek out that which is truly solid and which fills and satisfies the mind completely." Clearly, Noël is being told that arguing with Pascal will amount to arguing with received school logic. The rule holds that a decisive judgment concerning a "hypothesis" may be made only if it either appears so "clearly and distinctly" that the mind cannot possibly doubt its certitude—"and it's this that we call *principles* or *axioms*, such as, for example, *if to equal things one adds equal things, the totals will*

---

16. Pascal, *Œuvres* (1963), p. 198 col. 2.
17. Cf. Feyerabend, "Classical Empiricism" (1970). William Whewell noticed that Newton's rules really serve to justify his particular claims in the *Principia*: see Blake, "Isaac Newton and the Hypothetico-Deductive Method" (1960), on p. 143 n. 114 (note on p. 324). See chapter 8, section V, below.
18. Pascal, *Œuvres* (1963), p. 198 col. 2. I here depart slightly from a literal translation because of the original's clumsy construction.
19. Ibid., p. 202 col. 2; p. 209 col. 2.

be equal"—or if it is demonstrated necessarily from such principles.[20] The property of clarity and distinctness possessed by the axioms of a scientific demonstration may appear "to the senses or to reason, according as it is subject to the one or the other." Thus Pascal implicitly jettisons the strict Aristotelian requirement that a demonstrative axiom must be not only evident but also hold of necessity, such that it could not be otherwise. Empirical principles had long functioned that way in practice, but only in the mathematical sciences, where essential definitions were not available to provide the grounding of claims to necessity.[21]

In the *Expériences nouvelles*, then, Pascal is concerned with turning accounts of particular contrived events (events being by their nature of a private, or at least a spatiotemporally restricted, character) into publicly available, and in that sense evident, experiences. His accounts needed, that is, to function *as if* they were evident, to enable them to undergird demonstrations. Pascal's readers had to accept his experiential assertions on faith, but in order to win that faith Pascal had to make those assertions rhetorically effective.

Pascal wrote at the beginning of his pamphlet that he had decided to publish this outline of a fuller work in part because of the following consideration:

> ... having made experiences with much expense, care, and time, I feared that another, who would not have employed the time, the money, or the care, anticipating me, would give to the public things that he had not seen, and which in consequence he would not have been able to report with the exactitude and order necessary to connect [*déduire*] them properly: there being no one who has had tubes and siphons of the length of mine; and few who would wish to give themselves the trouble necessary to get them.[22]

The experiences on which Pascal based his arguments were, therefore, in large part ones that only he had properly had; they were, furthermore, experiences that very few people would be likely to go to the trouble of acquiring. Pascal's accounts of those experiences, rather than the reader's own acquisition of them, will act as the basis of proper inferences about nature, which is why it was important to ensure that the accounts would be capable of yielding correct inferences: other people could not be relied upon to report the experiences reliably.

Pascal's assumption that his own accounts were unproblematic was

20. Ibid., p.201 col. 1.
21. See chapter 2, above.
22. Pascal, *Œuvres* (1963), p. 196 col. 1.

borne out by the criticisms of Étienne Noël that followed hard upon the publication of Pascal's pamphlet: the Jesuit's objections all concerned the validity of Pascal's inferences, not the truthfulness of the described experiences.[23] Pascal's experiences—or rather, the accounts that, for all practical purposes, constituted them—were presented in such a way as to avoid contradicting anything that was already generally accepted as true. Trouble arose only when they were used to argue for the possibility of a void in nature—that is, when the experiences gave way to the maxims. Pascal's experiences remained secure to the extent that they were accepted as philosophically neutral appearances.[24]

Nonetheless, as the earlier quotation indicates, Pascal was also concerned to establish his own priority in the invention of these experiences. The final paragraph of his opening address "Au lecteur" treats this motive as a quite proper one: "And as gentlemen *[les honnêtes gens]* join in the general inclination that all men have to maintain themselves in their rightful possessions, [namely] that [inclination] to refuse the honor that is not due to them, you will doubtless approve that I defend myself equally, and from those who would wish to take from me some of the experiences that I give you here."[25] This is because these are "of my invention"; Pascal sharply distinguishes them from the original experience from Italy, "having the intention to give only those which are proper to me and of my own creativity *[génie]*."[26] Experiences are property, the possessions of a gentleman like Pascal.[27]

The notion of intellectual property was a new and contested one in this period.[28] It was surely especially problematic when applied to something the function of which was to serve as a foundation for a scientific demonstration. How could one *discover* something that is *evi-*

---

23. For another example, see Pierius, *An detur vacuum* [1646?]. There is no title page, but there is no doubt of its author, who is named at the end (p. 14). For speculation on its exact date of publication see Mesnard's notes in Pascal, *Œuvres complètes II* (1970), p. 360. Pierius praises Pascal for his "accuratissimo experimento" even though he disagrees with the conclusion of a vacuum (ibid., p. 361).

24. Cf. Shapin and Schaffer, *Leviathan and the Air-Pump* (1985), esp. chap. 2, on the importance to Boyle of the "matter of fact"; and Schaffer, "Making Certain" (1984). The social/epistemological function of Pascal's "matters of fact" seems somewhat different, however. For a discussion of Pascal's "experiences" in this text and their idealized nature, see Koyré, "Pascal savant" (1968).

25. Pascal, *Œuvres* (1963), p. 196 col. 1.

26. Ibid.

27. On being a French "gentleman," see Motley, *Becoming a French Aristocrat* (1990). This is an issue central, for the English setting, to Shapin, *A Social History of Truth*.

28. See on this Iliffe, "'In the Warehouse'" (1992); also references on early-modern patents given in Eamon, *Science and the Secrets of Nature* (1994), p. 383 n. 213.

dent? The answer seems to require a further examination of the distinction that Pascal makes between an experience and a maxim. A maxim was, as we have seen, a universal experiential principle of the familiar Aristotelian kind; an experience was a similarly universal statement of a specific natural behavior, but this time one that served as part of the justification (or source of evidentness) for a maxim.[29] Pascal's claim to proprietorship rests not on the invention of a maxim (which would be like owning a law of nature), but on the invention of an experience, which is tantamount to the invention of the *technique* that produces the experience.[30]

II. Credit, Controversy, and the Empirical Establishment of Truth

Shortly before Pascal's pamphlet appeared, a claim to independent discovery of the Torricellian phenomenon was published by an Italian Capuchin monk resident in Poland, Valeriano Magni. Magni's original claim was one of actual invention, and when subsequently challenged he continued to deny that he had heard of it from its earlier source. Magni's first text on the subject, with an initial date of July 1647, had been published in Warsaw and appeared again in Paris shortly thereafter together with an earlier letter from Pierre Petit to Pierre Chanut.[31] The whole is prefaced by an unsigned introduction "Au lecteur," which is concerned with establishing credit. Magni, it says, has not invented this thing; it comes from Italy, and is owing especially to Galileo rather than to Torricelli.[32] The credit for being the first to succeed with it in France, however, at a time when many people were trying it and having trouble, belongs to Petit.

Petit's letter is quite short and commences with a historical narrative of what he had done, at Mersenne's suggestion, in the company of Pascal at Rouen. The narration is followed by a discussion of possible

29. There is a general similarity here with Scheiner's practical distinction, in the *Oculus,* between *experimentum* and *experientia:* see chapter 2.

30. And cf. Bachelard, *La formation de l'esprit scientifique* (1938), on the notion of *phénoméno-technique;* also Gaukroger, "Bachelard and the Problem of Epistemological Analysis" (1976).

31. Petit, *Observation touchant le vuide* (1647). Cf. Pascal, *Œuvres* (1904–1923), vol. 2, pp. 15–20, 503–509; Middleton, *The History of the Barometer,* pp. 42–43. The question of the exact date is tricky because Magni uses two different dates (apart from the "Approbatio") for the work's two parts: one in July and one in September. For a discussion of Magni's arguments and their conceptual presuppositions see Fanton d'Andon, *L'horreur du vide* (1978), pp. 52–63. Chanut was the man who later persuaded Descartes to make his ill-fated trip to Sweden as Queen Christina's instructor in philosophy.

32. On the question of the preface's author, see the short discussion in Middleton, *The History of the Barometer,* p. 40 and n. 31.

explanations for the mysterious space, directed towards the establishment of it as a genuine void.[33] Magni's text follows, proclaiming itself a *Demonstratio ocularis*. Magni sets up his argument by reviewing the standard Aristotelian arguments against the possibility of a void in nature.[34] Unlike Pascal in the *Expériences nouvelles*, but like Petit in his letter, Magni then provides an historical narrative, in a section headed "Facti historia." This title means, roughly, "account of what was done," and such is its form. We are introduced to a glass tube of over two cubits in length, closed hermetically at one end, and told: "With quicksilver I filled this tube, the open mouth of which I stopped up with a finger," and so on.[35]

This is quite different from Pascal's form of presentation, and different from the forms we have seen repeatedly in such presentations by others during and before this time. Magni is intent not so much upon establishing a knowledge-claim as upon establishing priority in the invention of a new artificial phenomenon. He specifies that this work was done in the presence of the king and queen of Poland, more illustrious witnesses (together with their courtiers) than Pascal had been able to assemble from among the local worthies in Rouen who had witnessed some of his demonstrations.[36]

The historical structure is maintained throughout Magni's modest forty-five page treatise. There is one notable deviation from the pattern towards the end, however. "I have indeed often experienced that no air from the standing mercury ascends to the empty part of the tube: for not even the smallest portion of air is driven upwards through the cylinder of mercury, as far as can be seen."[37] Magni's concern for credit almost subordinates this claim to philosophical expertise, so reminiscent of Riccioli's rhetorical procedure in the matter of falling bodies. Magni reports on some things that he did and saw, and then allows the reports to stand for what he knows. But on the question of whether the emptiness of the space above the column of mercury (an emptiness

33. Petit, *Observation touchant le vuide*, pp. 1–22.
34. Magni, *Demonstratio ocularis* (1647), pp. 31–34. The pagination is continuous with Petit's letter.
35. Ibid., pp. 34–35: "Hanc fistulam implevi argento vivo, cuius orificio patens obturavi digito. . . ."
36. Ibid., p. 35, and see the report on Magni's show written by Des Noyers to Mersenne, dated 24 July 1647, in Pascal, *Œuvres* (1904–1923), vol. 2, pp. 15–18. Pascal's spectators included Guiffart and Pierius, on whom see text at nn. 43–55, below.
37. Magni, *Demonstratio ocularis*, p. 66: "Ego verò saepiùs expertus sum, nil aëris, stante mercurio, salijisse ad partem fistulae vacuam: non enim vel minima portio aëris sursum agitur per cylindrum mercurij, non conspicua intuenti: in quo casu nil est, quod aut rarescat, aut densetur, prout saepiùs expertus sum."

that Magni maintains against the peripatetics) is compromised by the intrusion of air through the mercury, no specific experimental trial determines the matter; instead, he tells us what he has "indeed often experienced."

These early investigations of the Torricellian device were chiefly concerned with deciding on the nature of the created space at the top of the tube. Why the mercury remained suspended above its level in the bowl beneath was at first of secondary concern, and usually attributed either to a limited *horror vacui* or to the balancing weight of the air, as Torricelli had supposed. Pascal orchestrated the Puy-de-Dôme exploit as a means of demonstrating the latter cause, although there were other means by which to do it as well as other interpretations suited to the actual result. W. E. Knowles Middleton credits Torricelli with the "invention" of the barometer precisely because the latter was the first to suggest using the device to track variations in air pressure under changing meteorological conditions.[38] But of course the nature of the Torricellian device remained mutable until the eventual formation of an authoritative consensus.

For most investigators in the 1640s it remained a technique for creating the mysterious space; Magni called it "an art for exhibiting a vacuum."[39] The properties of the space might be examined, as when Athanasius Kircher, like others, tried ringing a bell enclosed within it,[40] but the device itself played no role beyond acting as the means for its artificial production. When, by contrast, the device was used in such a way as to examine possible reasons for the elevation of the mercury column, it became a piece of apparatus that allowed, if all went well, perception of an underlying cause. This is a form of what Latour and Woolgar called "splitting and inversion," whereby the laboratory results on the basis of which an entity is postulated change status to become effects of that entity once the latter is accepted as real.[41] The device created a preternaturally elevated column that could serve as evidence that an appropriate cause for the elevation existed; once established, the cause (the pressure of the air) is evidenced by that elevation. What was once a technique has become an apparatus that devolves into an instrument.[42]

One of Pascal's neighbors in Rouen stoutly maintained the impossi-

38. Middleton, *The History of the Barometer*, p. 29.
39. Magni wrote "De inventione artis exhibendi vacuum": see Fanton d'Andon, *l'horreur du vide*, p. 60.
40. Kircher, *Musurgia universalis* (1650/1970), pp. 11–13.
41. Latour and Woolgar, *Laboratory Life* (2d ed., 1986), pp. 176–178.
42. Cf. Schaffer, "Glass Works" (1989), esp. pp. 70–71.

bility of a true void and denied Pascal's inferences. Jacques Pierius, a local scholastic professor of philosophy, published a pamphlet towards the end of 1646 that mentioned Pascal's name and denied his work's value as evidence for a true vacuum in nature.[43] Although Pierius was responding to very recent events, he uses only a universalized recipe form to describe the experiences; there is no question of their authenticity, only of their proper interpretation. Soon afterwards, in 1647, another neighbor, a physician, took up Pascal's defense: Pierre Guiffart published a short work called *Discours du vuide, sur les experiences de Monsieur Paschal, et le traicté de Mr. Pierius*. Its intended role in the affair is evidenced by the several pages of poems (following a dedication to the president of the *parlement* of Normandy), in Latin, French, and even Spanish, in which various local worthies praise Guiffart and his book. The collective repudiation of Pierius is quite striking, and gives an interesting picture of the provincial salon culture of the period.[44]

Guiffart's rhetorical tactics are noteworthy for their carefulness; his first chapter resembles some of Francis Bacon's writings in its discussions of the past and the care one must take in dealing with the legacy of antiquity. The central message is that respect for antiquity, or aversion to novelty, should not be allowed to get in the way of disinterested investigation.

> That is why one must proceed to the examination of the opinion that there can be no void in nature, which is so universally received and so obstinately defended by those who follow the most famous philosophers that they would rather put up with it in ruins, and endure the return of the original chaos, than to admit the least part [of the doctrine of a void] that one could imagine; since we see that several new experiences seem to witness the contrary, resting on very considerable reasons, which I mean to represent here.[45]

---

43. Pierius, *An detur vacuum*. Guiffart quotes from this work in translation at some length.

44. Guiffart, *Discours du vuide* (1647). Compare the *conférences* of Théophraste Renaudot: Solomon, *Public Welfare, Science and Propaganda* (1972); or a provincial philosophical society of a few years later: Lux, *Patronage and Royal Science in Seventeenth-Century France* (1989). See also Sutton, "A Science for a Polite Society" (1981).

45. Guiffart, *Discours du vuide*, pp. 6–7: "C'est ainsi qu'il faut proceder à l'examen de l'opinion qu'il n'y peut avoir de Vuide en la Nature, laquelle est si universellement recevë & si opiniastrement deffenduë par ceux qui suivent le plus fameux des Philosophes; qu'ils en souffriroient plustot le debris & le retour du premier cahos [sic] que d'en admettre le moins qu'on se puisse imaginer; puis que nous voyons que quelques experiences nouvelles semblent tesmoigner le contraire, appuyées de raisons assez considerables, lesquelles j'ay dessein de representer icy."

That, then, will be the main task ahead: "But it's appropriate before getting down to business to speak of the subject that has obliged me to undertake this work."[46] Some time ago, M. Pascal performed several experiences in the town "in the presence of all the most learned men of his acquaintance," Guiffart himself being invited to the last two. Pascal, in wishing to show that a void is possible in nature, made it evident, says Guiffart, that no void existed in his mind. Guiffart had been reluctant to say anything about this for a while, because it was controversial, but Pierius's publication was unsatisfactory and has prompted him to respond.[47]

Among a variety of topics, to do with art and nature, the scholastics' arguments against a void, and various pieces of experiential evidence on the matter, Pascal's demonstrations form the linking theme. Even the initial focus, the authority of antiquity, reenters once Pascal's work is addressed directly. "Although the experiences of M. Pascal look new to us, they have some appearance of having been formerly practiced, and several ancients having taken from them the basis for maintaining that there could be a void in nature, or even that it was a principle of nature." Democritus, Leucippus, Diodorus, Epicurus, and Lucretius are noted in the margin.[48] Those who have subsequently denied the void have based their position on reason alone, not on the certitude of the senses. However, Guiffart continues, even if these experiences are in fact new, this would be so much the better, since "the first flowers of spring are the most agreeable, and the first fruits were formerly the sacred portion that God reserved to Himself."[49]

The ambiguity that Guiffart is happy to invoke regarding the novelty or antiquity of Pascal's experiences underlines clearly the requirement of their universality. In describing some of them in a degree of detail later in the book, Guiffart uses the familiar universalized form: "when one allows air or water to rise in the quicksilver"; "if you turn the tube upside down"; "you will feel it draw with force"; "to the degree that one allows air to enter the tube in which the quicksilver is contained,

---

46. Ibid., p. 7: "Mais il est à propos avant que d'entrer en matiere de dire le sujet qui m'a obligé d'entreprendre cet ouvrage."
47. Ibid., pp. 7–8.
48. Ibid., p. 54: "Quoy que les experiences de Mr. Paschal, nous paroissent nouvelles, il y a de l'apparence qu'elles ont esté autresfois pratiquées, & que plusieurs anciens ont prins de là sujet de maintenir qu'il y pouvoit avoir du Vuide en la Nature, voire mesme qu'il en estoit un principe."
49. Ibid., p. 55: "les premieres fleurs du Printemps sont les plus agreables, & les premices des fruicts estoient autresfois le partage sacré que Dieu se reservoit."

it empties itself of quicksilver to that degree at the bottom into the vessel."[50] These are throughout typical of Guiffart's descriptions of the experiences; he never gives historical accounts of Pascal's performances. Thus, whether they were known to the ancients or were new, the experiences were in themselves already universals. "These reasons," says Guiffart after describing the basic Torricellian procedures and results, "are so strong and so convincing for bringing about acknowledgment of the void, that they can pass for demonstrations."[51] This, of course, is precisely what Pascal aimed for: demonstrative experiences of the kind found in mixed mathematics.

In 1648 Pierius published a lengthier text on the same subject, still taking the experiences themselves as unproblematic universals ("experience shows" is a characteristic expression).[52] Pierius does allow room for a historical account, however, covering the experiences of Petit and Pascal, then Magni, and finally Roberval's recent public exhibition.[53] The account is a way to dramatize the recent flurry of interest in the matter and to stress the lack of demonstration of a void provided by the experiences. Roberval's work, as the most recent and best-known in Paris, especially rouses Pierius's ire:

> Lately Mr. Roberval, most meritorious professor of the mathematical disciplines, has exhibited this experience and many others related to it to public view, by which it seemed clearly to furnish a way to the light and to everyone an absolute answer; I, however, have been able to derive nothing of any precision from his explications. Clearly it reveals that that superior part of the tube does not remain a vacuum, which is of great moment in the mathematician who had taught the contrary opinion to the point of doggedness. In recent memory there is nothing that I might come up with which he has proclaimed so effusively: I would propose, however, with his indulgence, things seen by me that are difficult to understand according to his explication and opinion.[54]

50. Ibid., pp. 129 (first), 130 (rest): "quand on permet à l'air ou à l'eau de monter dans le vif argent"; "si vous renversez la sarbatane"; "vous le sentirez tirer avec force"; "autant qu'on donne entrée à l'air dans la sarbatane en laquelle est le vif argent, Il se vuide autant de vif argent en bas dans le vaisseau."
51. Ibid., p. 133: "Ces raisons sont si fortes & si convainquantes pour faire advoüer le Vuide, qu'elles peuvent passer pour demonstrations."
52. Pierius, *Ad experientiam nuperam circa vacuum . . . responsio* (1648); "patet experientiâ," p. 13.
53. Ibid., pp. 13–15.
54. Pierius, *Ad experientiam nuperam circa vacuum . . . responsio*, pp. 14–15: "His diebus Dominus de Roberval Mathematicarum disciplinarum professor meritissimus hanc experientiam publicè videndam exhibuit, aliasque multas huic affines, quibus videbatur sibi prae parare aditum ad luculentam & omnibus numeris absolutissimam responsio-

Pierius is, as we have come to expect, not in the least interested in arguing over the putative experiences provided by the apparatus. All he cares about are the conclusions that may or may not legitimately be drawn from them. Roberval is criticized for his opinions, not for his inept experimentation or for the improper singularity of the experiences he has produced: indeed, in mentioning the trials of Petit and Pascal, Pierius had earlier used the standard formula "repeated many times."[55] Such is his approach throughout.

The most celebrated of Roberval's demonstrations in 1648 was that of the "void in the void." This trial, which perhaps had already been performed by Pascal with a different setup, argued for the role of the weight of the air in supporting the column of mercury (see figure 5).[56] In effect, one Torricellian apparatus is put inside another, and the space evacuated by the mercury in the latter proves unable to support any mercury in the former. The introduction of air into that space, however, immediately serves to elevate the mercury in the enclosed tube. An account of the experience was published soon afterwards by another opponent of the void, the Jesuit Étienne Noël of the Collège de Clermont in Paris, whose correspondence with Pascal the year before had prompted the latter's first major methodological pronouncements.[57] Roberval was a mathematician whose epistemological views squared with those of many other mathematicians by this period: physics was inferior to mathematics because physical causes too often remained hidden, thereby compromising the quality of the statements made regarding them.[58]

> Mathematics has all the fine prerogatives of physics as regards being true, immutable, and invincible, but it is not so hidden from men: it

---

nem; nullam tamen nisi valde indefinitam ex eius explicationibus potui colligere. Aperte fatetur partem illam tubi superiorem vacuam non remanere, quod magni est momenti in Mathematico qui contrariam sententiam huc usque mordicus docuerat. In tam recenti memoria nihil est quod recenseam quae fusissimè enarravit: proponam tamen, cum bona eius venia, quae visa sunt mihi difficilia intellectu in eius explicatione & sententia." On Roberval, see Auger, *Un savant méconnu* (1962), esp. pp. 117–133 on the "question du vide."

55. Pierius, *Ad experientiam nuperam circa vacuum . . . responsio*, p. 13: ". . . quam etiam multis alijs experientijs & multoties repetitis illustraverat Dominus Paschal." Part of this material is reprinted in Pascal, *Œuvres complètes II*, pp. 641–642; I have a transcription different from Mesnard's, p. 642 line 2: in place of "experientijs" (i.e., "experientiis") in this quotation, he has "experimentiis."

56. Cf. Pascal, *Œuvres complètes II*, p. 634; see, for a thorough discussion of the chronology here, Akagi, "Pascal et le problème du vide" (1968), esp. pp. 171–181.

57. Biographical material in Pascal, *Œuvres* (1904–1923), vol. 2, pp. 79–81.

58. See esp. Gabbey, "Mariotte et Roberval" (1986).

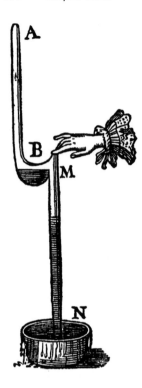

Figure 5. Pascal's void-in-the-void apparatus, reproduced in Blaise Pascal, *Œuvres* (1904–1923), vol. 3, p. 237.

loves evidence [i.e., evidentness], and it makes it appear clearly and distinctly in its proper object (which is size or number) . . . and not in the composition of material things.[59]

Roberval goes on to describe mixed mathematics, which, unlike pure mathematics, does regard the composition of material things. The principles here are, he says, the same as for physics; but these are too hidden from men. Accordingly, mixed mathematics (not so called in this passage)

> takes for the foundations of its reasoning facts [*des faits*] which are averred by an unchanging, constant experience [*une experience constante de tout temps*], and on these foundations it establishes mechan-

---

59. Pascal, *Œuvres* (1904–1923), vol. 2, p. 50. Roberval's use here of the apparently Cartesian expression "clearly and distinctly" supports Gabbey's supposition that his employment of it in *L'optique et la catoptrique du Reverend Pere Mersenne Minime* (Paris, 1651) is not, contrary to Robert Lenoble's opinion, meant sarcastically. Gabbey, "Mariotte et Roberval," p. 225 n. 45.

ics, optics, astronomy, music, and the other particular sciences combined from geometry, arithmetic, and physics.[60]

Thus Roberval's science is to be based, not on singular experimental events, but on those unchanging universals of experience that had always underpinned mixed mathematics.

Noël's published account of Roberval's void-in-the-void demonstration (also known from other contemporary evidence) uses it in a way not all that alien to Roberval's own position. Noël was concerned to disallow the inference of a true void in nature, but Roberval, as the above-quoted material indicates, was himself enough of a philosophical sceptic to doubt the demonstration of a physical conclusion on the basis of a "mathematical" trial. This is much like the contemporary tendency by some mathematical writers on optics, which we have seen before, to present theorems on the behavior of light independent of inferences about the true physical nature of light. Indeed, along with observations of the presence of small bubbles in some trials, Roberval's contemporaneous displays of the expansion of a carp's swim bladder inside the super-mercurial space led him to suppose that perhaps this was indicative of rarefaction (as Aristotelians would have it) rather than a true void—else how could the bladder's inflation be explained?[61]

Noël's account is instructive, not for its unusual form, but for the routine way in which it provides the historical placement of a universal experience. This is quite usual in these early accounts of the Torricellian device and its descendents. He describes a "new experiment" in some detail: "let there be a glass tube three feet long, open at each end. Let there be a small flask of glass, shaped like a human heart...."[62] After a lengthy account of the apparatus given in this way, Noël informs the reader: "Let the lower opening of the large tube be uncovered; the mercury descends, and ether is drawn to the top of the small tube on the right in place of the descending mercury."[63] (Note that the rising ether is part of Noël's perception of the behavior of the device.) "This recent experiment the illustrious Mr. Roberval, regius professor of the mathematical disciplines, with his kindness and singular indus-

---

60. Pascal, Œuvres (1904–1923), vol. 2, p. 51. Note the inclusion of physics alongside arithmetic and geometry; cf. chapter 6, section IV, above.

61. Ibid., p. 295; cf. Auger, Un savant méconnu, pp. 128–131; also discussion in Akagi, "Pascal et le problème du vide" (1968), pp. 172–173. Cf. Pierius's remarks about Roberval, text to n. 54, above.

62. Beginning of passage transcribed in Pascal, Œuvres complètes II, pp. 637–639, from Noël, Stephani Natalis ... gravitas comparata (1648), pp. 77–80.

63. Pascal, Œuvres complètes II, pp. 638–639.

try, has shown to us."[64] In Noël's version of the matter, the "recent experiment" is nonetheless timeless; there is no question of reporting *what Roberval did*.

Noël had by this time been won over to the opinion that the column of mercury was indeed supported by the weight of the air, pressing down on the dish of mercury at the base of the tube. Only the question of the space above the mercury column—whether it was a void or not—separated him from Pascal, his recent correspondent.[65] For Pascal, however, things had by this time moved along. Having decided that he had established the existence of a void in nature as far as any reasonable methodology could warrant it, Pascal was now interested in giving a demonstration that the column of mercury was supported by the air outside. This was precisely the point that, by mid-1648, many others in Paris were prepared to accept on the basis of the void-in-the-void trial.[66] But Pascal did not, apparently, regard this as sufficient.

III. A Scientific Event and Its Erasure

The most striking feature of Pascal's brother-in-law Florin Périer's trip up the Puy-de-Dôme is that Pascal presents it as a detailed historical record. Pascal's pamphlet of 1648, the *Récit de la grande expérience de l'équilibre des liqueurs*, is designed to show that the column of mercury in the Torricellian tube is supported by the weight of the air. He had not claimed this in his earlier pamphlet, which aimed only at establishing that the space at the top of the tube was actually, as far as could be judged, a void; he simply took it that nature's abhorrence of a vacuum had its limits. Pascal explains how in the earlier work "I had employed the maxim of the 'fear of a void,' because it was universally accepted, and I hadn't yet any convincing proofs at all to the contrary."[67] Contradicting common opinion was not something to be done lightly. As Pascal says in the included letter to Périer, "I do not consider that we are permitted to depart lightly from maxims that we hold from antiquity, if we are not obliged to do so by indubitable and invincible proofs."[68] The sentiment closely echoes one by Pascal's Rouen promoter, Guiffart, who had written in the previous year regarding the great men of antiquity that "it is reasonable that we defer to their precepts: and that when

64. Ibid., p. 639.
65. For the correspondence between Noël and others see Pascal, *Œuvres* (1963), pp. 199–221.
66. Akagi, "Pascal et le problème du vide" (1968), p. 175; Roberval, as Akagi notes, thought the Puy-de-Dôme result unnecessary for demonstrative purposes.
67. Pascal, *Œuvres* (1963), p. 221.
68. Ibid., p. 222.

the thread of our reason proves too short to guide us in the labyrinth of difficulties, their authorities should be the oracles that end our doubts, and the sovereign decrees that decide our differences."[69] We should nonetheless, he stresses, not follow them blindly.

Pascal's letter to Périer, asking him to make the ascent, is dated November 1647. That is almost a year before the trip took place; the letter serves as Pascal's main claim to priority in conceiving the idea (a subject that clearly bothered him).[70] Doubts have been raised as to the date's authenticity, and Mersenne and Descartes touted as alternative originators of the idea. Even with Pascal's dating left intact, Descartes still has strong claims to the idea—if one disregards his quite different interpretation of its meaning.[71] The fact remains, however, that Pascal, with his publication of Périer's report towards the end of 1648, was eager to see the thing done even after the void-in-the-void results had become known. He notes that talk of it had become widespread without anyone having actually tried it, and asserts in the letter that the void in the void, while persuasive of the role of the air, could still be interpreted in terms of a *horror vacui;* only climbing a mountain could be "decisive." Why the same objection could not also apply to the latter is not explained.[72]

The present report is, of course, Pascal's, not Périer's, insofar as it was Pascal who made it the centerpiece of his own publication. This centerpiece is a detailed historical narrative, in Périer's first-person authorial voice, of events that had taken place on Saturday, 19 September 1648; it includes the names of various other people who had assisted or witnessed the proceedings.[73] Pascal's usual appeal to universalized Aristotelian "experiences," of the kind found in the *Expériences nouvelles,* makes the Puy-de-Dôme account seem incongruous; it looks as

69. Guiffart, *Discours du vuide,* p. 3.
70. Cf. material in "Nouvelles expériences," above, and 1651 letter to De Ribeyre, esp. Pascal, *Œuvres* (1963), p. 229 col. 1.
71. For a survey of discussions casting doubt on the authenticity of Pascal's letter, see Middleton, *The History of the Barometer,* pp. 45–48; also Rochot, "Comment Gassendi interprétait l'expérience du Puy de Dôme" (1964). On Descartes's role, see Garber, *Descartes' Metaphysical Physics* (1992), pp. 136–143; see also Pascal, *Œuvres* (1904–1923), vol. 2, p. 46 n. 4; and Akagi, "Pascal et le problème du vide" (1968), pp. 182–184, for a different view.
72. Dijksterhuis, *Mechanization of the World Picture,* p. 452, seconds Pascal's judgment without explaining its logic. Akagi, "Pascal et le problème du vide" (1968), pp. 179–180, 184, suggests that Pascal may have conceived the idea of the Puy-de-Dôme trial even before his own version of the void in the void, in late 1647, rather than afterwards.
73. For Gassendi's independent, secondhand account, with contextual discussion and French translation, see Rochot, "Comment Gassendi interprétait l'expérience du Puy de Dôme."

if he is using a set-piece event as the foundation of a philosophical argument. But Pascal took care to *avoid* characterizing his account of the events on the Puy-de-Dôme as a foundational singular experiment, and endowed the "decisive experience" with a philosophical meaning derived from the mathematical sciences.

The account portrayed what Pascal described as making "the ordinary experience of the void several times in the same day," using the same apparatus, at the bottom and the top of a high mountain.[74] Périer's report describes making the "ordinary experience" a number of times at different locations on the mountain and in the garden of the Minim convent at the mountain's foot—without detailing each trial separately. Périer also describes a combined technique of calibration and control, whereby a tube like the one he carried up the mountain was left in the care of a monk in the convent garden, being set up in a vessel of mercury in the usual way. The tube Périer took with him was initially set up next to it, to check that the two columns of mercury were of exactly the same height, and they were compared again when he returned. In the meantime the monk kept watch over the tube in the garden; he reported that the height of the mercury had remained constant while the others had been gone. Meteorological variations were thereby ruled out.[75]

Périer's report tells, in some circumstantial detail, of the various measurements that he and his companions made of the mercury's height at different places on the mountain, both during the ascent and during the descent. The results were entirely consistent with Pascal's expectations: the higher up the mountain the apparatus was deployed, the lower in the tube the column of mercury rested. But Pascal, in his publication, did not rest his claims on the story of the trip up the Puy-de-Dôme. At the conclusion of Périer's letter, Pascal tells his readers that, taking Périer's reported measurements as calibration standards, one can infer the generalization that a difference in altitude of six or seven fathoms makes a difference in height of mercury of about half a line (a line being a twelfth of an inch). He notes the ease of trying

---

74. Pascal, *Œuvres* (1963), p. 222.
75. This is a rather complicated matter: meteorological variability would only be an issue liable to cast doubt on the meaning of changes in the height of the mercury during the ascent if it were assumed that such changes were indeed due to variation in the weight of the air pressing on the mercury (and that the weight variation might also correlate with changes in the weather). But that assumption was the central point at issue. All that the control really did, therefore, was to support the proposition that the changes in the height of the mercury observed during the trip up the mountain were due to the decreasing weight of the superincumbent air with increasing altitude, given that the mercury column's height depended on the weight of the air.

this using tall buildings in Paris and remarks on his own consequent attempts, which had obtained results corresponding to Périer's. He gives very little detail, with no dates, and only one specific location. Pascal then concludes: "All the curious can test it themselves whenever they like."[76] A remarkable and eagerly awaited historical exploit has thus turned, in the space of a paragraph, into an easily acquired, routine philosophical *experience*. The ascent of the Puy-de-Dôme had served its purpose.[77]

A comparison between Pascal's use of Périer's ascent of the Puy-de-Dôme and his use in the *Traité de la pesanteur de la masse de l'air* of an apparently similar event underlines the true character of the former. In 1649 or 1650, Pascal had himself ascended the Puy-de-Dôme carrying a partially inflated bladder which he found to expand and distend as he went up the mountain and become flaccid again as he descended.[78] If the original Puy-de-Dôme experiment of Périer had acted as the model, a historical account of this ascent of the Puy-de-Dôme would have appeared in the treatise. But instead, that independently known event seems to have only occasional relation to the following passage: "If one takes a balloon half-full of air, flaccid and limp, and one carries it at the end of a string on a mountain five hundred fathoms high, it will come about that, in proportion to one's ascent, it will swell of its own accord,"[79] and so forth. This is so universalized an experience that even the Puy-de-Dôme itself has vanished, to be replaced by generic and infinitely multiplicable "mountains five hundred fathoms high." Pascal ensures that the experience's significance is fully understood: "This experience proves everything that I have said about the mass of the air with a totally convincing force: so it was necessary to establish it well, because it is the foundation of this entire discourse."[80] The establishment of the "experience" is not the presentation of a credible report of a specific historical event, Pascal strolling up the Puy-de-Dôme with a bladder; it is a presentation of a universal experience, as if a collective familiarity could be unproblematically brought to mind concerning this kind of behavior of bladders on high mountains. It is, in other words, a classic example of a mixed mathematical postulate.[81]

The demonstrative knowledge that Pascal was concerned to make

---

76. Pascal, *Œuvres* (1963), p. 225.
77. Cf. Périer's description of Pascal's Parisian versions of his Puy-de-Dôme exploit in his introduction to the two treatises published in 1663: Pascal, *Œuvres* (1963), p. 236.
78. Pascal, *Œuvres* (1904–1923), vol. 3, p. 200 n.
79. Pascal, *Œuvres* (1963), p. 245.
80. Ibid.
81. Cf. chapter 2, section II, above, and chapter 8, section II, below.

in his work on the void was, then, a kind modeled on the mathematical sciences and on the place of experience within them. There was no real controversy over the proper role of experience in such sciences, and Pascal did not court it; he simply employed experience in a normal way, and it was a useful weapon against natural philosophers like Noël. The inappropriateness of singular experiences in the creation of such a science (although the same would apply equally to qualitative physics) is underlined in an even more striking fashion by Pascal's and Périer's conception of the calibration apparatus left in the Minim garden during Périer's ascent of the Puy-de-Dôme: the tube in the vessel of mercury had been left "en expérience continuelle."[82] This terminology was also used by Pascal, Descartes, and others to describe mercury tubes left to exhibit changes in height over time as a presumed function of the weather.[83] The category of *expérience continuelle* (later to be applied to experimental controls) enabled the establishment of the perpetual behavior of nature in deliberately contrived situations.

In material from an unfinished work appended to his two posthumously published treatises of 1663, Pascal discusses the variation of barometric readings according to the weather. He includes measurements taken at Dieppe, and this is how he presents them:

> One will see that in Dieppe, when the weather is the most oppressive [*chargé*], the mercury will be at the height of 28 inches 4 lines, measured from the mercury in the curved end. And when the weather will unload itself, one will see the mercury lower, perhaps by 4 lines. The next day, one will see it perhaps lowered by 10 lines; sometimes one hour later it will reascend by 10 lines; some time afterwards one will see it either higher or lower, depending on whether the weather will be laden or unladen [with rain].[84]

The clumsy translation here is intended to emphasize the use of tenses: Pascal gives an account of *what one will find*, not *what he found*. He continues:

> And from the one to the other of these periods, one will find 18 lines of difference, that's to say, that it will sometimes be at a height of 28 inches 4 lines, and sometimes at a height of 26 inches 10 lines.
>
> This experience is called the *expérience continuelle*, because one observes it, if one likes, continually, and one finds the mercury at almost

---

82. Pascal, *Œuvres* (1963), p. 223.
83. On this point, see Humbert, *Cet effrayant génie* (1947), p. 129.
84. Pascal, *Œuvres* (1963), p. 259.

as many diverse points [on the scale] as there are different weathers where one observes it.[85]

This is universal experience, not experimental data in the modern sense. The numbers given are not evidence for a general claim; they are simply illustrations of it.

Pascal's subsequent discussions in this text of the weather's relation to variation in the height of barometric mercury lays out broader generalizations of a similar kind, culminating in four predictive claims. These are not probabilistic assertions but are as confident as any of his previous statements. The final one, quite typical, says "when one sees together the air oppressive and the mercury high, one can be assured that the bad weather will continue, because assuredly the air is very oppressive." Pascal's confidence stems from his belief that he is dealing with demonstrable causal connections, not probabilistically interpreted conjunctions. Having told the reader, for each of the four predictive tips, that "on peut s'assurer," he then observes that "it isn't that a supervening wind could not frustrate these conjectures, but ordinarily they succeed," because the height of the mercury is a direct effect of the oppressiveness of the air.[86] This combination of demonstrative argument and possible accidental violation of its conclusions is a classic case of the *ex suppositione* form of argument well known in scholastic philosophy of nature.[87]

Périer's ascent of the Puy-de-Dôme to see what would happen to the height of the mercury had been eagerly awaited, regarded by many in the philosophical circles of Paris and beyond as an important indicator of the cause of its suspension. But some, such as Roberval, disdained it, and Descartes had affected to be so sure of the outcome as to be insouciant of the whole affair.[88] Pascal, having got the result he wanted, promptly established the requisite philosophical foundations not on a historical report, but on universal experience. The difference, between a narration and a piece of science, was well understood by all concerned.

IV. Mathematical Tables

The Torricellian device was one thing as a means of producing a mysterious space which might or might not be a true void and another as

---

85. Ibid.
86. Ibid., p. 261.
87. See chapter 1, section II, above.
88. Indeed, Descartes did not receive a report on the outcome until much later, if ever: Garber, *Descartes' Metaphysical Physics,* pp. 139, 142.

a contrivance to manifest the weight of the air. As the former it was simply productive of an artificial phenomenon; in the latter it provided empirical evidence for a philosophical conclusion. In the minds first of Torricelli, then of Pascal and others, that conclusion quickly became established; the device then devolved into a mathematical instrument (in the standard seventeenth-century sense) for measuring atmospheric variation—a barometer.[89]

Jean Pecquet's 1651 *Experimenta nova anatomica* discusses the Torricellian device and its philosophical implications in a way that treats the weight of the air itself as a trivial point scarcely worth mentioning: "I would treat of the Air's ponderosity, yea in its proper (as they say) place, except it were an Argument known to all. For who doth not see the Air of its own accord to descend into the Chinks and Ditches, yea even into the lowest Center of the Earth, if you delve so deep?"[90] Pecquet is already prepared to present a question at the center of philosophical speculation only a handful of years before as if it were a part of common experience. "Who doth not know," he continues, "that a little Bladder, the more turgid it is, is so much more heavy than it self being flaccid? Who, if he weigh a Gun burthened with condensed Air (they call it a Windgun) will not observe that its weight then is heavier than 'tis when it is discharged?"[91] Pecquet's words call to mind Simplicius's questioning in late antiquity of Aristotle's claims for the weight or levity of the air, done by reference to the weighing of a bladder; the quarrels of a few years earlier are forgotten, and the implication is that well-known phenomena have always established the point.[92] Pecquet subsequently adduces evidence for the "elatery" (expansivity), not the weight, of the air, under the telling label "Experimenta physico-mathematica de vacuo."[93] He describes historically not only one of Roberval's famous Parisian experiments, but also the Puy-de-Dôme trial. The account of the latter concludes on this note: "Neither wonder thou that this Experiment doth not agree with those that

---

89. On Torricelli's dubbing the device an "instrument" see Middleton, *The History of the Barometer*, pp. 29, 32.

90. Pecquet, *Experimenta nova anatomica* (1651), p. 48; trans. from Pecquet, *New Anatomical Experiments* (1653), extracts reprinted in M. B. Hall, *Nature and Nature's Laws* (1970), p. 187.

91. Ibid. Windguns were described in Marin Mersenne's well-known *Cogitata physico-mathematica* (1644), among other places.

92. On Simplicius see Sambursky, *The Physical World of Late Antiquity* (1962), pp. 80–81. Simplicius claimed, contra Aristotle, that there should be no difference in weight between an empty and an inflated bladder, while allowing that moist air expelled from the lungs might account for the claimed increase.

93. Pecquet, *Experimenta nova anatomica*, p. 50.

I tried in Paris by falling of Quick-silver: For both the difference of the foot of measure in Avernia from ours, which exceeds in some lines (which I exactly observed) and the divers distance of Places may be from the Center of the World (even thou being Judge) may not keep equality in these experiments."[94] Again, as in Pascal's own account, the singularities of the singular event are carefully removed so as to leave only the underlying universal truth; the device is already a barometer.

Pascal's most fully accomplished treatises on these matters, composed in the early 1650s but published posthumously in 1663 by Périer, themselves fall squarely into the genre of the mixed mathematical sciences.[95] They discuss hydrostatics and the weight of the air, the latter being little more, in Pascal's treatment, than a subsumption of the behavior of air to the behavior of liquids; Pascal's hydrostatics is, generically speaking, an extension of the classical mathematical science of mechanics. The treatises present accounts of contrived experiences in a universalized form appropriate to the establishment of experiential premises within formal scientific argument: in the usual manner of the mixed mathematical sciences, the experiences appear as sets of instructions or conditionals providing procedural details, and incorporate confident statements of what happens when the detailed steps are accomplished.[96] Thus the *Traité de l'équilibre des liqueurs*, referring to appropriate naturalistic diagrams of apparatus (see figure 6), begins: "If a number of vessels are attached against a wall, one like that of the first figure; another leaning . . . ; another very large . . . ; another narrow . . . ; another that is only a small tube . . . ; and one fills them all with water to the same height, and makes all the openings equal at the bottom, which one plugs to retain the water; experience shows *[l'expérience fait voir]* that the same force is required to prevent all the plugs from coming out."[97] This is the format within which Pascal's own trip up the Puy-de-Dôme with a bladder was incorporated and within which it vanished.[98]

94. Ibid., pp. 50–55, quote on p. 55, trans. in M. B. Hall, *Nature and Nature's Laws*, pp. 189–193, quote on p. 193. Apart from these two celebrated historical events, Pecquet's experiments take a generic instructional form. On variations in local measurement scales, see Rochot, "Comment Gassendi interprétait l'expérience du Puy de Dôme," on p. 280, where the differences between the figures given in Périer's report and those given by Claude Mosnier, one of Périer's party on the ascent, are explained by reference to the difference between a Parisian foot and a *pied de Mâcon*.

95. *Traités de l'équilibre des liqueurs et de la pesanteur de la masse de l'air* (Paris, 1663) which includes some additional material discussed below; Pascal, *Œuvres* (1963), pp. 233–263.

96. See chapter 2, above, on this routine form.

97. Pascal, *Œuvres* (1963), p. 236.

98. See above, section III.

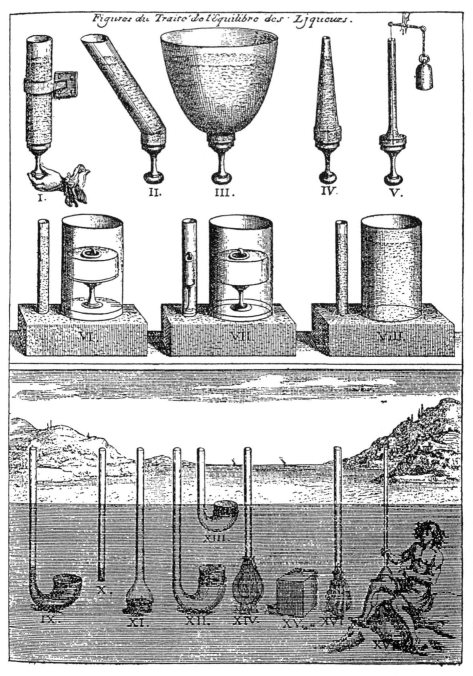

Figure 6. Pascal's hydrostatical apparatus, reproduced in Blaise Pascal, *Œuvres* (1904–1923), vol. 2, p. 157.

The publication of 1663 concludes with a fragment immediately following the meteorological material discussed above. It consists largely of tables that at first glance appear to present figures concerning barometric readings at Paris and Clermont, the latter derived from Périer's Puy-de-Dôme readings. In fact, however, these tables are not compilations of data, but numbers calculated, in part from explicitly inaccurate simplifying assumptions, to show the breadth of mathematical results determinable from Pascal's barometric rules relating altitude to the height of a mercury column. Four locations—Paris, Clermont, the top of the Puy-de-Dôme, and a place called La Font de l'Arbre, lying partway up the mountain at which Périer had taken measurements—are designated along the vertical axis of these tables, with various heights or weights alloted to them along the horizontal. La Font de l'Arbre is treated as if it were halfway up the mountain, although Pascal observes that it is in reality much nearer the mountain's foot. The seven extant problems (not here explicitly so called) for which these tables serve as solutions are of this kind: "To assign a cylinder of lead, of which the weight be equal to the resistance of two polished bodies applied one against the other, when they are separated." The numerical solutions are introduced by the following reasoning: "This resistance is equal to the weight of a cylinder of lead, having for its base the common face, and for its height, *when the air is laden,*" the numbers listed in the table immediately following. Appropriate figures are provided for each of the four places.[99] Another typical problem attended by a table reads: "To assign the height at which water raises itself and remains suspended in the ordinary experience."[100] The "ordinary experience," a term used also for the case of mercury, means simply the basic Torricellian setup. The clear implication is that results for the resistance to separation applicable to such things as columns of lead are determinable from empirical figures for the height of liquids suspended by the weight of the air, the resistance itself resulting from the same cause.

The significance of the tables lies in their mathematical style and distance from raw empirical reportage. Périer might have walked up the Puy-de-Dôme on a specified date and recorded various measurements, but when that event is fully digested, the result is a mixed mathematical science that relies neither on singular events nor on their constant replication. The science develops from its roots; they do not grow with it. An empirical report need not necessarily stand for "what happens" in nature under specified circumstances. It might function

99. Pascal, *Œuvres* (1963), pp. 261–263, quotes p. 261.
100. Ibid., p. 263.

only to stand for "what happened" on a particular occasion. In order for the singular to stand for the universal, an appropriate set of inferential assumptions, whether explicit or implicit, must be in place: the assumptions underlying Pascal's work were those of the mathematician. Périer's adventure almost immediately became a kind of "Just So" story, a mythical tale of discovery like those catalogued in the sixteenth century by Polydore Vergil.[101] No tokens of good faith in the narrative needed to be made or implied, because the subsequent plausibility of the assertions that flowed from it did not rely on the story itself. Thus Périer, in his introduction to the *Traités* of 1663, gave an account of Pascal's actual performance of the experiences that the latter presented only in universalized theorem form in the 1647 pamphlet.[102] Périer apparently found such an account appropriate as part of a longer "histoire" of experiences on the void—that is to say, something other than philosophical or mathematical discourse. Similarly, Galileo claimed in 1624 to have tried (probably some twenty years earlier) the famous experiment of dropping a weight from the mast of a moving ship— and yet he clearly found it inappropriate to mention it in his discussion of the matter in the *Dialogo*.[103]

The cognitive practices represented by Pascal's tables were employed in the 1670s in another French study on pneumatic and atmospheric matters, their character underscored on this occasion by the direct contrast between them and a handling of the same subject within English "experimental philosophy." Robert Boyle's eponymous "law" had entered the world in 1662 as a detailed account of specific trials that included measurements sufficiently precise to bear witness to their slight deviation from a postulated ideal. They were the outcome of an attempt to demonstrate the power of the spring of the air. The account includes tables of actual measurements made on a specific occasion, unlike Pascal's tables of calculated figures developed from unspecified empirical sources. Boyle observes that "till further tryal hath more clearly informed me, I shall not venture to determine whether or no the intimated Theory will hold universally and precisely."[104] When, however, in 1676, Edme Mariotte published his ver-

---

101. See, for example, Emmanuel Maignan's retelling of it in his *Cursus philosophicus* (1653), vol. 4, pp. 1895–1986. On Vergil see Copenhaver, "The Historiography of Discovery in the Renaissance" (1978).
102. Pascal, *Œuvres* (1963), p. 235.
103. Drake, *Galileo at Work* (1978), pp. 84, 294.
104. Boyle, *A Defence of the Doctrine Touching the Spring and Weight of the Air* (1662), chap. 5, pp. 57–68, quote p. 62. For discussion see Webster, "The Discovery of Boyle's

sion of the same relationship and its justification, he proceeded in a quite different manner. Where Boyle had made experimental trials the core of his presentation, Mariotte set up his own account such that historical reportage of specific events was accorded minimal justificatory significance.

Mariotte starts out by postulating that the "condensation" of air occurs in direct proportion to the weight used to compress it. He gives an argument in support of that suggestion, drawn from considerations of the compression of the air by its own weight in the atmosphere—that is, he takes for granted the conclusions concerning the air's weight that Pascal had attempted to establish two decades or so earlier.[105] Only then does Mariotte proceed to discuss apparatus and its proper use. The discussion details the way in which a mercury barometer that also contains air, employed in a particular way, would manifest the postulated relationship between incumbent weight and compression.[106] Torricelli's device, in other words, is being used as a well-understood instrument to test a postulate: "To find out if this consequence was true, I tried it with Mr. Hubin, who is very expert at making barometers." Mariotte gives the measurements from this "experience," which accord with the postulate, gives another historical narration of a similar affair, and proceeds to set the seal of accomplishment on his account: "I furthermore had made [i.e., by Hubin] some other similar experiences, leaving more or less air in the same tube, or in others more or less large; and I always found that after the experience was made," the height of mercury rested where it ought according to the postulate. Notice that the specified trials have quickly given way to an unspecified multiplicity of trials. The overall result, he concludes, "makes sufficiently clear that one can take for a certain rule or law of nature that air condenses in proportion to the weight by which it is loaded."[107] A page of instructions follows, detailing ways to make "des

---

Law" (1965). Shapin, *A Social History of Truth*, chap. 7, esp. pp. 323–330, explains how, and why, Boyle did not actually enunciate "Boyle's law" qua "law of nature."

105. And cf. Pecquet, *Experimenta nova anatomica*, cited above.

106. Mariotte, "Discours de la nature de l'air," in *Œuvres* (1717), pp. 149–182 (originally published in 1676), on pp. 151–152. For details and diagrams of Mariotte's procedures, see also Costabel, "La loi de Boyle-Mariotte" (1986).

107. Mariotte, "Discours," p. 152: "Pour savoir si cette conséquence étoit véritable, j'en fis l'expérience avec le Sieur *Hubin*, qui est très-expert à faire les Baromètres & des Thermomètres de plusieurs sortes"; "Je fis faire encore quelques autres expériences semblables, laissant plus ou moins d'air dans le même tuyau, ou dans d'autres plus ou moins grands; et je trouvai toujours, qu'après l'expérience faite"; "ce qui fait connoître suffisamment, qu'on peut prendre pour une règle certaine ou loi de la nature, que l'air

expériences plus sensibles." It consists of tips and suggestions concerning apparatus and procedures.[108]

Mariotte finally rounds out the discussion with three problems based on his "law of nature." They represent, in effect, a version of the procedure adopted by Pascal in his unfinished treatise on the void. Mariotte's mathematical "problems" (this time explicitly labeled as such) show how to calculate such things as resultant heights of mercury columns from given initial conditions; the form is of a "problem" followed by a solution in the standard manner of geometrical problemata.[109] Once the general instructions, followed by worked problems, appear, historical reportage of specific events vanishes. Mariotte's treatise taken as a whole, in fact, deploys such reportage very sparingly: its historical accounts serve only to signify novelty, whereas the legitimation of that novelty consists in arguing for its likelihood in advance and rendering its production routine through the constitution of a "law of nature." Thus Mariotte describes procedures regularly manifesting that law, and even presents "problems" that treat the law as an unproblematic assumption. Mariotte tells his reader what *happens*, not what *happened*. We have seen exactly this procedure before, in Riccioli's "problems" concerning fall and pendulum motion.[110]

The meaning of a famous episode in the history of seventeenth-century science should now be evident. Robert Boyle, in his *Hydrostatical Paradoxes* of 1666, took Pascal to task for the form of presentation used in the latter's publication of 1663: "though the Experiments he mentions be delivered in such a manner, as is usual in mentioning matters of fact; yet I remember not that he expressly says that he actually try'd them, and therefore he might possibly have set them down as things that *must* happen, upon a just confidence that he was not mistaken in his Ratiocinations."[111] But Pascal, like Mariotte, and like any other practitioner of the mathematical sciences, behaved in this way because it was demanded by the form of the argument.[112] Scientific

---

se condense à proportion des poids dont il est chargé." Hubin's role is noteworthy: on "technicians" in this period, with especial reference to Robert Boyle's practice, see Shapin, *A Social History of Truth*, chap. 8.

108. Mariotte, *Discours*, p. 153.
109. Ibid., pp. 154–156.
110. See chapter 3, text to nn. 47, 60, above.
111. Boyle, *Hydrostatical Paradoxes* (1666), pp. 4–5, referring to Pascal's *Traité de l'équilibre des liqueurs*. See, for similar remarks, ibid., pp. 63–64; 141 (referring to Stevin); 171; 185; 243.
112. Pascal's form of presentation was used routinely in these matters, as in mixed mathematical treatises generally: compare, e.g., barometric material in Guiffart, *Discours du vuide*; Pecquet's *Experimenta nova anatomica*; Maignan's *Cursus philosophicus*. The occa-

experiences needed to be universal, true generally and for everyone at all times. A historical report of a specific event, of the kind that Boyle wrote endlessly, would have been scientifically meaningless; it would have been philosophical antiquarianism.

Newton's 1672 letter on light and colors represented a quite different situation. Jean Moss has explained the narrative structure with which Newton's piece begins by making reference to its rhetorical form: it is a letter, typical of the early *Philosophical Transactions*, and so conforms to the appropriate expectations for such a piece. It is this, she claims, rather than appeal to the Society's Baconian self-image, that accounts for its literary character.[113] But as Boyle's criticism of Pascal indicates, the very fact that the *Philosophical Transactions* itself encouraged use of such a form must be taken seriously: the close resemblance between the narratives of Boyle's books and those of the frequent epistolary contributions published by Henry Oldenburg is surely not accidental. The following chapter examines the peculiarity of the English case and considers Newton's correspondingly careful recasting of mathematical forms of argument. The experience of the mathematicians and the experience of Boylean "experimental philosophy" close the circle of the new scientific experience of the seventeenth century.

---

sional appearance of historical reportage in these texts should be compared with the analysis of Pascal's Puy-de-Dôme "experiment."

113. Moss, "Newton and the Jesuits" (1988), esp. p. 122.

Eight   BARROW, NEWTON, AND CONSTRUCTIVIST EXPERIMENT

I. Geometry and Mechanics

In the eighteenth century and beyond, Isaac Newton's name came to represent a form of empirical science that, in conjunction with the example and practice of celestial mechanics, elevated him to the status of a philosophical touchstone, comparable to Aristotle during previous centuries. But the meaning of Newton's practices regarding experiential forms of knowledge, and his expressed views on experience in natural philosophy (now, remarkably, of a mathematical kind), can be understood only with reference to the matters treated in previous chapters. The uses of mathematical argument in making accredited knowledge of nature show how the foundational assumptions of a mathematical science, and the kinds of experience that underwrote them, made it possible for Newton to announce a kind of declaration of independence for physico-mathematics. The adoption of his claims by the Royal Society then amounted to the reform of the radical and philosophically fruitless "experimental philosophy" of its Restoration period.[1]

Newton's *Principia* was criticized by Christiaan Huygens on the grounds that the doctrine of universal gravitation was not properly

---

1. Fruitless by seventeenth-century standards, of course. But when Shapin and Schaffer, *Leviathan and the Air-Pump* (1985), p. 344, say, albeit ironically, that "Hobbes was right" in pointing to the social permeability and inconclusiveness of Boylean experimentalism, the present argument sees Hobbes's judgment seconded, in effect, by the Royal Society itself, as well as by subsequent continental philosophers of nature, in adopting Newton's formulation of demonstrative experiment.

mechanical.² By that, Huygens meant that Newton failed to provide contact-action explanations for gravitational behavior couched in terms of the sorts of laws of collision developed by Newton and Huygens themselves, or else in acceptable fluid-dynamical terms. Huygens's use of the term "mechanical" in this context invoked the sort of explanations that Descartes had advocated or that he himself had provided in his own vortex theory of gravity. In that sense, Newton was certainly not providing mechanical explanations in the *Principia*. And yet in his preface to the *Principia*'s first edition, Newton had stressed the foundational role of mechanics in his mathematical natural philosophy.

The burden of Newton's famous preface is that geometry, the mathematical science with which he will explicate the "mathematical principles of natural philosophy," is rooted in mechanics, insofar as circles and right lines, the raw material of geometry, have to be drawn, or assumed as drawn, through motion: "Therefore geometry is founded in mechanical practice." This is evidently a different sense of the term "mechanical" from that of Huygens. The preface attempts to set the *Principia* as a whole into an intellectual context rooted in antiquity and sanctioned by modernity:

> Since the ancients (as we are told by *Pappus*) esteemed mechanics of greatest importance in the investigation of natural things, and the moderns, rejecting substantial forms and occult qualities, have endeavored to subject the phenomena of nature to mathematical laws, it has been thought in this treatise to cultivate mathematics as far as it relates to philosophy.[3]

Newton writes as if there were nothing of importance at issue; he will provide mathematical principles of natural philosophy because such an approach is, as it always has been, routinely understood as appropriate. The difficulties attending the art/nature distinction, so prominent earlier in the century, play no apparent role in the discussion. The rise of physico-mathematics bears witness to the rise of a sensibility such as Newton portrays; but his position cannot be simply reduced to an announcement of the triumph of demonstrative mixed mathematics over probabilistic physics in the study of nature. Mathematical knowledge has become a model for understanding in general, rather than merely a body of techniques capable of especially satisfying forms of

---

2. Koyré, "Huygens and Leibniz on Universal Attraction" (1965), esp. pp. 115–124; Martins, "Huygens's Reaction to Newton's Gravitational Theory" (1993).

3. Newton, *Principia* (1687/1972), vol. 1, p. 15; translation Cajori, *Principles* (1934), p. xvii.

explanation. Even for Descartes, whose mathematicism is notorious,[4] the use of quasi-mathematical structures of reasoning usually stands apart from the things reasoned about; mathematics is just another logic. In the fifth part of the *Discourse on Method*, for example, Descartes has been discussing the action of the heart, providing an account of the way in which blood is transferred from the veins to the arteries: the heart's innate heat rarefies the blood that enters from the veins, and the heart's arrangement of valves allows the expanded blood to escape only into the arteries, where it cools and condenses again. The account is cast in purely qualitative terms, and yet Descartes clarifies its explanatory virtue through reference to mathematics:

> Now those who are ignorant of the force of mathematical demonstrations and unaccustomed to distinguishing true reasons from probable may be tempted to reject this explanation without examining it. To prevent this, I would advise them that the movement I have just explained follows from the mere arrangement of the parts of the heart (which can be seen with the eye), from the heat in the heart (which can be felt with the fingers), and from the nature of the blood (which can be known through experience). This movement follows just as necessarily as the movement of a clock follows from the force, position, and shape of its counterweights and wheels.[5]

Descartes's explanation has "the force of mathematical demonstrations" even though it entirely ignores quantities. Mathematics is a convenient cultural association that lends an air of authority to an argument. Its specificity of reference has here vanished entirely.

For Newton, by contrast, mathematical knowledge (for which, typically, geometry stands as the exemplar) is about particular kinds of objects created by active construction—not contingently, but in their very essence. The name of that construction, from which mathematics derives, is *mechanics:*

> The ancients considered mechanics in a twofold respect; as rational, which proceeds accurately by demonstration, and practical. To practical mechanics all the manual arts belong, from which indeed mechanics took its name. But as artificers do not work with perfect accuracy, it comes to pass that mechanics is so distinguished from geometry that what is perfectly accurate is ascribed to geometry; what is less

---

4. Cf. Allard, *Le mathématisme de Descartes* (1963); Gaukroger, "Descartes' Project for a Mathematical Physics" (1980).

5. Descartes, *Œuvres* (1964–1976), vol. 6, p. 50; trans. slightly modified from Robert Stoothoff's in Descartes, *The Philosophical Writings,* vol. 1 (1985), p. 136.

so, is ascribed to mechanics. However, the errors do not belong to the art, but to the artificers.[6]

The geometrical is nothing but the perfectly mechanical. "He that works with less accuracy is an imperfect mechanic; and if any could work with perfect accuracy, he would be the most perfect mechanic of all, for the drawing of right lines and circles, upon which geometry is founded, belongs to mechanics." The "General Scholium," first appearing in 1713 in the *Principia*'s second edition, indicates the divine identity of such a perfect mechanic. Newton continues:

> Geometry does not teach us to draw these lines, but postulates them. For it postulates that the beginner first be taught to draw these accurately before he enters upon geometry, then it shows how by these operations problems may be solved. To draw right lines and circles are problems, but not geometrical problems. The solution of these problems is asked [postulated] from mechanics, and by geometry the use of them, when so solved, is shown. And it is the glory of geometry that from those few principles, brought from without, it is able to produce so many things. Therefore geometry is founded in mechanical practice, and is nothing but that part of universal mechanics which accurately proposes and demonstrates the art of measuring.[7]

The activity of measuring amounts to the establishment of spatial possibilities. Thomas Hobbes, famously, portrayed geometry as a human creation which, because made by man, could be known perfectly by man.[8] It is important to realize that Hobbes spoke of geometry and not of the space to which that science purports to refer. Newton's account is remarkably similar. However, Newton was not following Hobbes; instead, Hobbes's position reflects an understanding of geometry that had wide currency in the seventeenth century and that Newton himself surely knew well. Newton's remarks echo especially strongly those of Isaac Barrow, as we shall see below.[9]

---

6. Newton, *Principia*, vol. 1, p. 15; trans. modified from Cajori, *Principles*, p. xvii. On the "practical" versus "rational" mechanics dichotomy, see Gabbey, "The *Principia*" (1992).
7. Newton, *Principia*, vol. 1, p. 15; Cajori, *Principles*, p. xvii.
8. Hobbes, *English Works* (1839–1845), vol. 7, p. 184.
9. See especially Barrow, *Lectiones geometricae* (1670), reprinted in Barrow, *The Mathematical Works* (1860/1973), trans. as Barrow, *Geometrical Lectures* (1735), Lect. I; also Barrow, *The Mathematical Works*, pp. 25–26 (*Lectiones mathematicae*, Lect. II). The relation between Barrow's and Newton's views has been noted by Kargon, "Newton, Barrow and the Hypothetical Physics" (1965); idem, *Atomism in England* (1966), pp. 106–117; Guerlac, "Newton and the Method of Analysis" (1973); Garrison, "Newton and the Relation of Mathematics to Natural Philosophy" (1987); Malet, "Isaac Barrow on the Mathematiza-

Mechanics for Newton remained rooted in construction: it was, that is, represented by the "mechanical arts." The classical mathematical use of the word, as found, for example, in Pappus (as Newton knew), applied to the theoretical understanding of the five "simple machines," the lever, pulley, wedge, windlass, and screw.[10] But that science had to do with statics, the balancing of forces (always to be understood in terms of weights); it did not deal with the generation of motions. Newton, however, equated mechanics—"mechanical practice"—with the generation of those motions that realized the possibility of circles and right lines, the basic constituents of geometry. Amos Funkenstein has suggested that the classical tradition relating to the generation of different classes of curves may best be understood by seeing it as involving a distinction between generating motions that are made through "motion-in-time" and those that are not. By this, he means to distinguish between curves such as circular arcs and right lines, on the one hand, and so-called mechanical curves such as the spiral or the quadratrix on the other. The former can be conceived as the product of a given type of motion independent of time relations in the execution of those motions—mere displacement. The latter, by contrast, are traced out by joint motions that have a determinate temporal relation. Thus a straight line may be drawn quickly or slowly, and indeed might as well be drawn instantaneously, but a quadratrix, created by a combination of a rectilinear motion and a rotation about a point, can be made only if those two motions occur at the proper rates with respect to one another. The latter are therefore generated by "motion-in-time," since their correct relation can only be defined through temporal comparison: one cannot be allowed to go too fast or too slowly with respect to the other.[11] However, as Newton sees it, even the generation of straight lines and circles requires the motive action of an artificer and therefore counts as mechanical. His understanding of this key concept is thus

---

tion of Nature" (forthcoming); Gabbey, "The *Principia*"; A. Shapiro, *Fits, Passions, and Paroxysms* (1993), pp. 31–36. On Barrow's role in the development of the calculus, see the recent reassessment by Feingold, "Newton, Leibniz, and Barrow Too" (1993). See also Pycior, "Mathematics and Philosophy" (1987). On Newton's early acquaintance with Hobbes's writings, see McGuire and Tamny, *Certain Philosophical Questions* (1983), esp. pp. 219–221.

10. Garrison, "Newton and the Relation of Mathematics to Natural Philosophy," p. 624, misunderstands Newton's reference in this preface to "the five powers that relate to the manual arts," taking these to be the five senses. They are, of course, the five simple machines.

11. Funkenstein, *Theology and the Scientific Imagination* (1986), pp. 301–303; see also pp. 315–316. Descartes to some extent transcended the older categorizations through his proportional compass: see text at nn. 31–33 below.

closer to the notion of the mechanical arts as exemplified by the classical science of machines than it is to the modern, seventeenth-century view championed by Huygens. In his unpublished *Geometria*, dating from the 1690s, Newton goes so far as to say that "geometry was devised, not for the purposes of bare speculation, but for workaday use, and the reason for its first institution must be preserved."[12] Newton appears in this sense to resemble the mathematical practitioners of whom J. A. Bennett has written in connection with the emergence of a distinctively English form of experimental mechanical philosophy—Robert Hooke would stand as a canonical example, but Newton seems to represent a similar cluster of cultural elements.[13]

In a discussion focused on Newton's preface, James W. Garrison has pointed out, in connection with other writings of Newton (especially the *Geometria*),[14] that Newton's talk of how geometry requires circles and right lines to be drawn properly relates to the category of geometrical *postulates*.[15] However, Garrison's argument has several difficulties. First of all, he imagines that Newton's position implies a crucial distinction between the mechanical foundations of geometry—mechanically rooted postulates—that are accomplished by human technicians and those achieved by nature or God. In consequence of this view, Garrison further maintains that Newton effectively acknowledged a category of "natural geometry" that flows from postulates created by natural motions rather than human artifice, and that knowledge of such a form of geometry would be imperfect because the foundations of natural geometry, not being man-made, would be exempt from the privileged kind of "maker's knowledge" available for ordinary geometry.[16]

However, natural geometry appears to be a pure invention of Garrison's and makes no sense in the context of Newton's presentations of the matter. In the preface Newton explicitly connected the activities of

12. Newton, *The Mathematical Papers*, vol. 7 (1976), p. 291.
13. Bennett, "The Mechanics' Philosophy and the Mechanical Philosophy" (1986); idem, "Robert Hooke as Mechanic and Natural Philosopher" (1980); and further references in chapter 5, n. 66, above. Hooke's *Micrographia* may plausibly be seen as written along constructivist geometrical lines: Dennis, "Graphic Understanding" (1989), on p. 330. Gabbey, "The *Principia*," pp. 313–314, gives a quote from Robert Boyle that works along similar lines, although Gabbey does not invoke the context of the mathematical practitioner stressed by Bennett.
14. Book I of the first version of the *Geometria* begins with an expanded discussion of the issues that he had treated in the part of the *Principia*'s preface that we have been considering: Newton, *The Mathematical Papers*, pp. 286–298 (Latin).
15. Garrison, "Newton and the Relation of Mathematics to Natural Philosophy."
16. Ibid., pp. 618–619.

imperfect human artificers with that of some "most perfect mechanic of all"; there is no implication that the unavoidable imperfections of human technicians compromise their geometrical knowledge. Geometrical postulates assume those things that must be taken as done or made in geometry; the level of perfection attaching to their actual realization is irrelevant to their epistemological status. So too, therefore, whether God or Newton is considered to be able to draw a perfect circle around any given point makes no difference to the resulting geometrical reasoning and its legitimate steps. If something is conceived of as if it were possible for a human agent, it is also ipso facto possible for God or for nature.[17] The resulting geometry is literally identical, as are the foundations of its knowledge. Motions and the forces that create them are, according to Newton, a part of "universal mechanics" and are to be distinguished from "rational mechanics" associated with the simple machines and human artifice; but no such distinction applies, or would make sense, to geometry.

Garrison, being interested primarily in Newton's views and their development, treats Newton's own understanding of the matter as if it were sui generis. However, an examination of the meaning and use of the classical category "postulate" in the seventeenth century reveals that the apparently "constructivist" conception of it found in Newton's writings had achieved widespread acceptance and had even breached the boundaries of mathematics itself.

II. Postulates and Constructivism

Commentators on Euclid have long puzzled over the categories employed at the beginning of the *Elements* for accommodating the basic principles, or assumptions, from which Euclid builds his demonstrations. Comprising "definitions," "postulates," and "axioms," they distribute statements according to criteria that are not always evident. Only the category "definitions" appears to have a degree of coherence, insofar as all its contained items clearly conform to the label, whether

17. In the *Geometria* Newton writes thus of postulates: "Any plane figures executed by God, nature or any technician [Artifex] you will are measured by geometry in the hypothesis that they are exactly constructed.... Geometry does not posit modes of description [i.e., drawing]: we are free to describe them [plane figures] by moving rulers around, using optical rays, taut threads, compasses, the angle given in a circumference, points separately ascertained, the unfettered motion of a careful hand, or finally any mechanical means whatever. Geometry makes the unique demand they be described exactly." Newton, *The Mathematical Papers*, vol. 7, p. 289.

satisfactorily or not. The other two present severe difficulties, which appear most starkly in a consideration of the label "postulates."[18]

Euclid lists five postulates. The first three are about the possibility of certain constructions: the constructibility of a straight line between any two given points; the indefinite producibility of a straight line in either direction; and the constructibility of a circle of any given radius around any point. The fourth and fifth, however, follow a different pattern. The fourth states that all right angles are equal, while the fifth is the so-called "parallel postulate." It says that if two straight lines in the same plane, when cut by a third, make interior angles on the same side equal to less than two right angles, then those two lines, if produced, will eventually meet on that side—that is, in reference to Euclid's earlier definition, will not be parallel. In a sense, therefore, Euclid does little more in his postulates than ask the reader to accept certain propositions before getting on with the job; there seems to be no greater conceptual coherence than that.[19]

The various editions and translations of the *Elements* that appeared during the first century or so of printing retained Euclid's formulation of the category.[20] The first significant variation occurred in the late sixteenth century, in a commentary version of the *Elements* by Christopher Clavius that served as the standard text in Jesuit colleges through most of the seventeenth century. Descartes would undoubtedly have used the commentary, as he did other mathematical texts by Clavius, during his schooldays at the college of La Flèche.[21] In the "Prolegomena" pre-

---

18. See Heath's comments in Euclid, *The Thirteen Books of Euclid's Elements* (1956), vol. 1, pp. 119–120, 195. Lloyd, *Magic, Reason and Experience* (1979), chap. 2, discusses the whole issue of the foundations of Greek deductive systems.

19. Euclid, *The Thirteen Books of Euclid's Elements*, pp. 195–220.

20. E.g., the printed version of the medieval Campanus, [Euclid] *Opus elementorum Euclidis Megarensis* (1482), 2d page; Campanus's postulates are not identical to Euclid's, combining Euclid's first two and adding one, perhaps implied by the first (and presented by Pappus as an axiom: see Euclid, *The Thirteen Books of Euclid's Elements*, pp. 195–196, 232), to the effect that two straight lines cannot enclose a surface. Commandino gives Euclid's five in his Latin and Italian versions, of which *Elementorum Euclidis libri tredecim* (1620), which I have used, is based upon his Latin translation (see Euclid, *The Thirteen Books of Euclid's Elements*, p. 102); the edition of the Italian version that I have used is Commandino, *Degli Elementi d'Euclide libri quindici* (1619).

21. Clavius, *Opera mathematica* (1611–1612), vol. 1, "Commentaria in Euclidis Elementa Geometrica." For a general synthetic account of Jesuit education in Descartes's time see Shea, *The Magic of Numbers and Motion* (1991), pp. 4–6. Hervey, "Hobbes and Descartes" (1952), on p. 78, quotes Pell to Cavendish, 2/12 March 1646, on a conversation with Descartes: "He says he had no other instructor for Algebra than ye reading of Clavy Algebra above 30 yeares agoe"—which was precisely when Descartes was at La Flèche.

fixed to the commentary,[22] Clavius explains the various kinds of "principles" used by mathematicians as the foundation for their demonstrations. In particular, he draws on alternative definitions of the difference between axioms and postulates given in Proclus's *Commentary on the First Book of Euclid's Elements*.

"*Petitiones*, or *Postulata*," says Clavius, "are very clear and perspicuous in the science that is under consideration, so that they need no confirmation, but merely demand the assent of the hearer, and neither is there any hesitation or difficulty in explaining [them]." He then touches on "*Axiomata*, or common notions of the mind, which are so manifest and evident, not only in the science in question but also in all others, that he can dissent from them with no cause who will rightly understand the very words."[23] The only real difference between axioms and postulates on this view, then, is that axioms should be common to all the sciences whereas each science has its own proper postulates.

Clavius's definitions in the "Prolegomena" are taken from Proclus's *Commentary*, in which they are given as one of a number of possible alternative approaches. However, an examination of the items that Clavius actually lists under the heading of "postulates" in his Euclidean geometry itself reveals an entirely different picture. First of all, Clavius does not list the five postulates of Euclid: both the right-angle postulate and the parallel postulate are missing, having been transplanted into the "axioms."[24] Instead, Clavius's postulates consist of Euclid's first three together with a fourth stating that a magnitude can always be taken that is greater or less than any given magnitude.[25] Clavius's procedure breaks the rules of the definitions he had given in the "Prolegomena" in at least two ways: first, the statements concerning right angles and parallel lines ought not to have counted as axioms

---

Cf. Descartes, *Œuvres*, vol. 10, p. 156 and n., on Descartes's use of Clavius's algebraic terminology in 1619.

22. In all editions from 1574 onwards, including *Opera mathematica*.

23. Clavius, "In disciplinas mathematicas prolegomena," in *Opera mathematica*, vol. 1, p. 9: "Secundum genus complectitur Petitiones, sive Postulata, quae quidem adeo clara sunt, & perspicua in illa scientia, quae in manibus habetur ut nulla indigeant confirmatione, sed auditoris duntaxat assensum exposcant, ne ulla sit in demonstrando haesitatio, aut difficultas. Ad tertium genus referuntur Axiomata, seu communes animi notiones, quae non solum in scientia proposita, sed etiam in omnibus alijs ita manifesta sunt & evidentia, ut ab eis nulla ratione dissentire queat is, qui ipsa vocabula recte perceperit." Cf. the very similar (probably derivative) account by Otto Cattenius in his 1610–1611 mathematical lectures, printed in Krayer, *Mathematik im Studienplan der Jesuiten* (1991), pp. 181–360, on p. 184.

24. Clavius, "Commentaria," pp. 22–23; axioms on p. 25.

25. This latter as an axiom is attributed to Pappus: see Euclid, *The Thirteen Books of Euclid's Elements*, p. 232.

since they do not transcend their particular subject matter, being specific to geometry; second, the additional fourth postulate ought not to have counted as a postulate since it was *not* specific to a particular subject matter, applying equally well to arithmetic.

In fact, after having listed his definitions, postulates, and axioms, Clavius acknowledges his use of a different definition of these categories, one also recorded by Proclus and championed by the Greek geometer Geminus.[26] Here, axioms and postulates are differentiated by defining axioms as unproven propositions the truth of which is easily grasped by the understanding, such as "the whole is greater than its proper part," and postulates as unproven propositions that assert the possibility of *doing* something—that is, performing constructions. Proclus himself had pointed out that if this characterization is accepted, Euclid's fourth and fifth postulates will not count as true postulates because they have nothing to do with constructibility.[27] Clavius's classification tacitly accepts this alternative Proclean scheme by removing Euclid's final two postulates and putting them under "axioms," a heading for which they now qualify. Clavius's new postulate, concerning the possibility of addition to or subtraction from any given magnitude, is, like the first three, about constructibility, or the ability to *do* something.

Clavius's version of Euclid, therefore, in effect presents postulates as assertions of existence based on an operational or constructivist criterion. Constructibility, in geometry, asserts the potential existence of a figure by maintaining the possibility of actualizing that existence through construction. Thus a universal concept could be rooted in an empiricist ontology by making a universal claim of constructibility: the universal "circle" is real because circles can *always* be drawn.

The Aristotelian word that came to be used for this kind of postulate was "hypothesis."[28] In its nonmathematical usage, the term was usually translated into the Latin of the scholastics by *suppositio* rather than *postulatum;* Aristotle's use of the term in the *Posterior Analytics* invoked the sense of a statement of existence corresponding to the constructivist understanding in geometry.[29] "Postulates," therefore, were equiva-

26. Clavius, "Commentaria," p. 26.
27. Proclus, *A Commentary on the First Book of Euclid's Elements* (1970), pp. 140–143.
28. The picture is slightly confused by Aristotle's additional distinction between "hypotheses" ("suppositions") and "postulates" strictly speaking, the latter being assertions that are simply assumed without being demonstrated even though a demonstration might properly be demanded. See Euclid, *The Thirteen Books of Euclid's Elements*, pp. 117–122.
29. Aristotle characterized "hypotheses" as statements of existence in *Posterior Analytics* I.10, which corresponds closely to the assertions of constructibility that Clavius presents as his postulates: see discussion in Lloyd, *Magic, Reason and Experience*, chap. 2.

lent to "mathematical suppositions," and in their constructivist sense could be compared to the "suppositions" of empirical sciences. As we have seen in chapter 2, the latter had ultimately to rest, for an Aristotelian, in sensory experience. An empirical supposition typically was established by appeal to uncontested experience, whence it could be used as raw material in constructing scientific, causal demonstrations. In the case of *mathematical* suppositions, that is, "postulates," therefore, the "experience" to which Clavius makes appeal is the practical experience of geometrical construction. This is experiential knowledge that something can always be done; in other words, that the outcome is always possible. That possibility thus constitutes a universally true statement. Clavius's constructivism, evidenced by his handling of geometrical postulates at the beginning of his Euclid edition, means that geometry remains true to experience because mathematics is itself, like all branches of knowledge, rooted in the senses. The only proper accounts of geometrical figures and concepts are operational and constructivist, whereby the objects of geometry are made according to the possibilities codified in operationally rooted geometrical postulates.[30]

Descartes's well-known use of mechanical constructibility in his classification of mathematical curves is thus unsurprising in the work of a onetime student of Clavius's textbooks, despite the non-Aristotelian aspects of Descartes's metaphysics and epistemology.[31] It goes along with his use of a kind of "proportional compass" to establish a new concept of general magnitude that would exclude dimensionality from the operation of multiplication applied to geometrical magnitudes; founding geometrical objects in the nature of their constructibility was a central part of his understanding of geometry.[32] Accordingly, in his treatise *La géométrie* Descartes conceptualizes the category of "postulates," or *demandes*, as statements of constructional techniques: he criticizes the ancients for refusing to admit into geometry curves more complex than the conic sections, since, he says, the same exactitude can be attained with instruments more complex than ruler and com-

---

30. For more on Clavius and geometrical constructivism, see Dear, "Mersenne's Suggestion" (1995).

31. Molland, "Shifting the Foundations" (1976); Bos, "On the Representation of Curves in Descartes' *Géométrie*" (1981); Lenoir, "Descartes and the Geometrization of Thought" (1979). See also Shea, *The Magic of Numbers and Motion*, chap. 3, for a good synthetic treatment. Gaukroger, "The Nature of Abstract Reasoning" (1992), has recently stressed a slightly different perspective. Funkenstein, *Theology and the Scientific Imagination* (1986), pp. 315–316, sets the point within a wider discussion of "knowing by doing" in the seventeenth century, for which Hobbes may stand as the exemplar.

32. See Schuster, "Descartes' *Mathesis universalis*" (1980), esp. pp. 47–51; also Shea, *The Magic of Numbers and Motion*, chap. 3.

pass, such as when two straight lines are moved with respect to one another so that their intersection traces a curve.[33]

The mathematical, constructivist definition of "postulate" found in Proclus appears not just in mathematical texts, but also in seventeenth-century philosophical lexicons, in preference to the rather different definition found in Aristotle.[34] The latter understanding, which Proclus also discusses but does not favor, and that Clavius mentions but proceeds to ignore, has postulates as things that strictly are amenable to proof, although the hearer is asked simply to grant them; they are distinguished from the "hypotheses" or "suppositions" that came to be identified with postulates on the constructivist view.[35] Aristotle's discussion concerned the principles of formal deductive demonstration in general, and the fact that a specifically mathematical understanding that departed from Aristotle's own text could find philosophical favor in the seventeenth century reflects the increasing prestige of mathematics as a cognitive model, as we have seen elsewhere. The mathematical usage, extended in this way beyond the subject matter of mathematics itself, could even reach into natural philosophy.

In a widely known logic textbook that first appeared in 1638, the German pedagogue Joachim Jung defines "postulate" as a supposition that something can be done or brought about. He gives three standard examples from geometry (such as the indefinite producibility of a straight line) and two from arithmetic; he then proceeds to add to the list as follows: "Thus the *natural philosopher* postulates, 'fire can be struck from flint,' 'water can be evaporated by heat.'"[36] The kinds of empirical generalizations that Jung identifies with postulates were in exactly the form appropriate for the foundations of any Aristotelian science of nature: an empirical supposition for a scholastic Aristotelian was established by just such uncontested common experience, which rendered the supposition suitable for use in the construction of scientific, causal demonstrations. In Clavius's reading of Euclid, geometrical postulates were rooted in the practical experience of geometrical construction; Jung roots physical postulates in practical, active experience

---

33. Descartes, *Œuvres*, vol. 6; Descartes, *The Geometry* (1952), pp. 41–42 (pp. 315–316 of first edition; my translation). Molland, "Implicit versus Explicit Geometrical Methodologies" (1991), on p. 196, makes an explicit comparison between Descartes's instrumental constructivist view of geometrical curves and Newton's preface to the *Principia*.

34. See, e.g., Goclenius, *Lexicon philosophicum* (1613/1964), pp. 836–837; Aristotle, *Posterior Analytics* I.10.

35. See n. 28, above.

36. Jung, *Logica Hamburgensis* (1657/1681), p. 228. On Jung, see Meinel, "*In physicis futurum saeculum respicio*" (1984).

of the physical world. The extension of mathematical forms of argument beyond their proper subject matter thus rested on an appeal to a particular conception of both mathematical and nonmathematical experience.[37]

Isaac Barrow underwrote a constructivist conception of knowledge in his own discussion of postulates in the *Lectiones mathematicae*, originally delivered in the mid-1660s. Hypotheses, or postulates, he says (using the two words indifferently) "are Propositions assuming or affirming some evidently possible Mode, Action, or Motion of a Thing."[38] After giving some geometrical and optical examples, he continues: "Of such Positions the Sense it self proves, and Experience clearly attests that they are not rashly assumed, but ought to be admitted, *i.e.* that they contain nothing repugnant to Possibility. From whence it appears by the Way, that as there is a very near Affinity between *Axioms* and *Theorems*, which was observed by the Ancients, so there is the same between *Hypotheses* and *Problems*."[39]

III. Isaac Barrow and the Scope of Physico-Mathematics

Alan Gabbey has recently considered the meaning of the term "mechanics" in the seventeenth century in order to discover what Newton and his contemporaries meant by it. Gabbey stresses the distinction then made between practical and "rational" mechanics to show how Newton saw himself as a practitioner of the latter, but in a novel form sanctioned by the new assertions made by such as Isaac Barrow and John Wallis on the relationship between mechanics and natural philosophy.[40] Our previous considerations of physico-mathematics, in chapter 6, and geometrical constructivism, in the last section, now allow a fuller understanding of the conceptual elements on which Newton drew.

Barrow, Newton's immediate predecessor as Lucasian professor of

---

37. M. R. Reif, "Natural Philosophy in Some Early Seventeenth Century Scholastic Textbooks" (1962), p. 309, notes an increasing tendency among the early seventeenth-century textbook writers whose works she examines to "pattern the format of their manuals on that of a geometrical treatise."

38. Barrow, *The Mathematical Works*, p. 125, trans. in idem, *The Usefulness of Mathematical Learning*, trans. John Kirkby (1734), p. 128.

39. Barrow, *The Mathematical Works*, p. 126, trans. in idem, *The Usefulness of Mathematical Learning*, p. 129. Barrow here cites Proclus following Geminus. See also Barrow, *The Mathematical Works*, p. 69, trans. in idem, *The Usefulness of Mathematical Learning*, p. 57: "Who will deny that these Things [some geometrical constructional postulates] may be done or conceived to be done?"

40. Gabbey, "The *Principia*"; see also idem, "Between *ars* and *philosophia naturalis*" (1993).

mathematics at Cambridge University, knew a lot about contemporary methodological and epistemological issues concerning the status and nature of mathematical knowledge. He was particularly well informed about the discussions of the century's preeminent theorists on the subject, the Jesuit mathematicians. The *Lectiones mathematicae* of the mid-1660s, his major discussion of epistemological issues, mention especially Pereira and Blancanus, and his position owed much to each. Barrow's views demonstrate the emergence of physico-mathematics as a methodologically superior "successor discipline" to Aristotelian natural philosophy.

Barrow's methodological position regarding mathematics, including the nature of postulates, was part of a more general conception of knowledge in which the mathematical sciences tended to assume priority in the understanding of all branches of natural philosophy.[41] He regarded what he called either "physico-mathematics" or "mixed mathematics" as applicable to all areas of natural philosophy, insofar as all parts of physics implicated considerations of quantity.[42] His view was not that physics reduces to mathematics, but, rather like Clavius, he maintained that mathematics is of use in all areas of physics: "there is no branch of natural Science that may not arrogate the Title ["mathematical"] to itself; since there is really none, from which the Consideration of Quantity is wholly excluded, and consequently to which some Light or Assistance may not be fetched from Geometry."[43] Barrow clearly perceived a role for the mathematical sciences in which properly physical questions may eventually become sidelined in favor of those which can be answered using mathematics: "as to what belongs

---

41. Both Garrison and Gabbey draw a casual, but erroneous, contrast between Newton and Barrow on this question. Whereas Newton, according to Garrison, regarded geometry and mechanics as clearly distinct, although inseparable, "for Barrow the two could not be discriminated" (Garrison, "Newton," p. 615). His evidence is a remark from Barrow's *Lectiones mathematicae* (Barrow, *The Mathematical Works*, p. 44) translated as "mathematics is commonly held to be, so to speak, coextensive with physics." But Barrow does not here say, as Garrison claims, that mathematics and physics are the *same*; instead, he says that they are coextensive, that is, are each applicable to the same array of objects. Gabbey, "The *Principia*," p. 312, makes a similar argument to that of Garrison on the basis of similar material; once again, however, the textual evidence only supports the claim that Barrow saw all areas of physics as amenable to mathematical treatment, not that, as Gabbey says, "the foundations of all practical and theoretical mechanical disciplines ... become identical with the principles of natural philosophy" (ibid.; cf. Gabbey, "Between *ars* and *philosophia naturalis*," pp. 138–139). See the somewhat more nuanced discussion of this matter in A. Shapiro, *Fits, Passions, and Paroxysms*, pp. 32–33.

42. E.g., Barrow, *The Mathematical Works*, pp. 31, 89.

43. Ibid., p. 40 (see also p. 41); trans. in idem, *The Usefulness of Mathematical Learning*, p. 21. Cf. the discussion by Guerlac, "Newton and the Method of Analysis," esp. p. 388.

to the Sciences termed *Mixed Mathematics,* I suppose they ought all to be taken as Parts of *Natural Science,* being the same in Number with the Branches of *Physics.*"[44]

It is symptomatic of his position that Barrow elucidates his epistemology of mathematics through frequent reference to Aristotle's *Posterior Analytics* and the wider issues of axiom systems in knowledge-making with which it is concerned. One might argue that Barrow's is in spirit an authentically Aristotelian approach, although it involves some quite radical criticisms of the generally accepted understanding of the Aristotelian text. These criticisms rest on considerations of practice in the mathematical sciences that seem to make them more different from physical sciences than Aristotle had allowed. But since Barrow continued (much as had Pascal) to cleave to the view of the *Posterior Analytics* that the logical structures of both mathematical and physical sciences were really the same, the net result was to pull the methodological form of the latter over towards that of the former, in their new guise as physico-mathematics. Barrow addresses the central problem of the establishment of foundational principles in a science by a flat rejection of Aristotle's position: "the Truth of Principles does not solely depend on *Induction,* or a perpetual Observation of Particulars, as *Aristotle* seems to have thought."[45] The importance of Barrow's point for the legitimation of the singular experiment is considerable. He says that the Aristotelian "perpetual Observation of Particulars" that provided the warrant for empirical principles is unnecessary "since only one Experiment will suffice (provided it be sufficiently clear and indubitable) to establish a true Hypothesis [i.e., supposition], to form a true Definition; and consequently to constitute true Principles. I own the Perfection of Sense is in some Measure required to establish the Truth of Hypotheses, but the Universality or Frequency of Observation is not so."[46]

Henry Guerlac has noted this last passage in connection with Newton's assertion, made variously to Hooke and to Lucas during his optical controversies of the 1670s, that the making of many experiments is irrelevant when one (here, the *experimentum crucis*) will suffice to conclude the point.[47] Newton's declaration was no doubt born of frustra-

---

44. Barrow, *The Mathematical Works,* p. 40; trans. in idem, *The Usefulness of Mathematical Learning,* p. 20.
45. See chapter 1, section IV, above, for other remarks by Barrow on induction.
46. Barrow, *The Mathematical Works,* pp. 116–117; trans. in idem, *The Usefulness of Mathematical Learning,* p. 116.
47. Guerlac, "Newton and the Method of Analysis," p. 387, mentions Hooke; see also A. Shapiro, *Fits, Passions, and Paroxysms,* pp. 34–35. For incisive discussion of Newton on this point, Feyerabend, "Classical Empiricism" (1970), esp. p. 162 n. 10.

tion as well as methodological conviction, but it echoes Barrow's argument quite well. Nonetheless, there is a crucial difference. Newton's point was that a particular experimental design yields a result that makes a certain conclusion logically unavoidable. Hence any other experimental designs are irrelevant to establishing or refuting that conclusion, once so established. By contrast, Barrow had argued, from a related but importantly different perspective, that a single experimental *event* could suffice to establish a "true principle" in a science. It is, of course, plausible that Newton believed the same thing, but Barrow not only made the claim explicitly, but also explained its rationale. "Sometimes," he wrote, "from the Constancy of Nature, we may prudently infer an universal Proposition even by one Experiment alone." That constancy was underwritten by the same metaphysical/theological conviction that served to warrant faith in frequently experienced regularities:

> ... when we still find our Expectations most accurately answered, after a thousand Researches; and especially when we have the constant Agreement of Nature to confirm our Assent, and the immutable Wisdom of the first Cause forming all Things according to simple Ideas, and directing them to certain Ends: Which Consideration alone is almost sufficient to make us look upon any Proposition confirmed with frequent Experiments, as universally true, and not suspect that Nature is inconstant, and the great Author of the Universe unlike himself.[48]

Barrow here incorporates into his scheme precisely the issues that the Jesuit mathematicians at the beginning of the century had juggled with, but now they are organized into a much clearer picture. The regular nature perceived by the natural philosopher had not been so bound by immutable law as to fail to produce occasional anomalies;[49] thus the representativeness of any single experience was always in question, and, as we saw in chapters 3 and 5, a multiplicity of instances was routinely adduced. For Barrow, however, the mathematical sciences invited the perception of a different universe: if light rays can always be represented by straight lines, then geometrical inferences about their behavior are certain. There is no guarantee, of course, that the experience underwriting this universal characterization (namely,

48. Barrow, *The Mathematical Works*, p. 83; trans. in idem, *The Usefulness of Mathematical Learning*, p. 74. Barrow here cites Aristotle, *Posterior Analytics* I.31, on how a universal can sometimes be grasped from a single perception: if (says Aristotle) one saw that glass was perforated and light came through it, one would grasp the universal reason for the transparency of glass from that one perception.

49. See chapter 5 and chapter 1, above, on "laws of nature" and on *ex suppositione* argument and monsters.

that one cannot see around corners) must always hold. Indeed, of course, it does not, and that is where the concept of refraction comes from. But the codification of departures from the basic rule, ultimately giving rise to Descartes's rule of sines for refraction, only served to confirm the mathematical conviction that behaviors in nature, separated from the essentialist kinds of causal explanation in which the natural philosopher dealt, could be calculated.

Newton's geometry was based on postulates provided by mechanics; they were not arbitrary postulates, but were rooted in the empirical constituents of space and time. Barrow too, at the start of his *Lectiones geometricae* (1670), proclaimed local motion to be the primary, underlying way of generating magnitude, "without which nothing can be produced."[50] He means here to make a strong correlation between the foundations of mathematics and physical nature itself: he goes on to cite the authority of Aristotle, who says in *Physics* III, 1, that to be ignorant of motion is necessarily to be ignorant of nature.[51] But since the true nature and causes of motion are such fraught matters, Barrow says that "'Tis enough that [mathematicians] take for granted what is allow'd by common Sense, and proved by obvious Experience, *viz.* That any Magnitude ... is moveable."[52] Mathematicians assume motion, he continues, and then investigate and demonstrate what follows from it. This is really Newton's argument: it discards the concerns of natural philosophers as inconclusive and uninteresting, replacing them with a "mathematical" approach that is characterized by its reliance on postulates—that is, assumptions of possibility rooted in experience.

Not everyone, including Barrow's English contemporaries, saw matters in this way. Robert Boyle criticized the practices of mathematical writers in his *Hydrostatical Paradoxes* of 1666. In doing so, he adverted to the central feature of the physics/mathematics distinction: "Those Mathematicians, that, (like *Marinus Ghetaldus, Stevinus,* and *Galileo*) have added anything considerable to the Hydrostaticks ... have been wont to handle them, rather as Geometricians, then as Philosophers, and without referring them to the explication of the Phaenomena of Nature."[53] For Boyle, the disadvantages of the mathematical disciplin-

---

50. Barrow, *The Mathematical Works*, p. 159: "sine quo nil procreari potest."
51. Ibid.
52. Ibid., p. 160: "Sufficere potest his, quae communis sensus agnoscit, et obvia comprobant experimenta pro concessis arripere; hoc imprimis generale, Quamvis magnitudinem ... mobilem esse," trans. in Barrow, *Geometrical Lectures*, p. 3. See also the comments on Barrow's discussions of this subject in Garrison, "Newton and the Relation of Mathematics to Natural Philosophy," pp. 614–615.
53. Boyle, *Hydrostatical Paradoxes* (1666), Preface (u.p.), 6th p. On other aspects of Boyle's antipathy, see Shapin, *A Social History of Truth,* chap. 7.

ary model far outweighed its advantages in demonstrative certainty over his own kind of experimental natural philosophy. The latter fulfilled the role of "physics" to the extent that it sought "causes" or "reasons," albeit only probable ones.[54] However, the contradiction between the suspicion of mathematics manifested by Boyle and the central place accorded it by Barrow is by no means complete. Both approaches converge in what John Wilkins had, only a few years earlier, announced as the mission of the new Royal Society: "Physico-Mathematicall-Experimentall Learning."[55] The most famous exponent of that enterprise, and one who reshaped it, was Barrow's protégé Isaac Newton.

IV. Experimental Events in the Early Royal Society

The mathematical sciences occupied a fitful place in the doings of the Royal Society during its earliest decades. Early numbers of the *Philosophical Transactions* contain items on mathematical subjects such as optical lenses and astronomical observations of cometary positions, and in 1667 the mathematical-experimental work of John Wallis and Christopher Wren on the rules of collision appeared in the journal's pages.[56] Robert Hooke routinely produced for the entertainment of the Fellows mechanical devices of the sort generally associated with the more practical, instrumental mathematical sciences.[57] In the main, however, the attention of the Society, and the burden of its public self-presentation, concerned qualitative observational and experimental reporting of the kind celebrated by Sprat in his eulogistic *History of the Royal Society* (1667), and exemplified in the many writings of Robert Boyle. When Newton was elected a Fellow in early 1672 it was in recognition of his achievement in producing a reflecting telescope (and presenting it to the Society), rather than as the due of a Cambridge professor of mathematics.[58]

As is well known, especially through the work of Steven Shapin and Simon Schaffer, experience appears in the writings of the early Royal Society Fellows (in contrast to much of the material considered in this book) in the form of the singular experienced event: a Fellow of

---

54. Cf. B. Shapiro, *Probability and Certainty in Seventeenth-Century England* (1983), chap. 2; Van Leeuwen, *The Problem of Certainty in English Thought* (1963).

55. B. Shapiro, *John Wilkins* (1969), p. 192. A. Shapiro, *Fits, Passions, and Paroxysms*, p. 31, contrasts Boyle's views on the solidity of "physico-mathematical" demonstrations to those of Christopher Wren, a Fellow who was himself a mathematician.

56. *Philosophical Transactions* 3 (1669), pp. 864, 867.

57. Cf. Bennett, "The Mechanics' Philosophy and the Mechanical Philosophy"; also essays in Hunter and Schaffer, *Robert Hooke* (1989).

58. Westfall, *Never at Rest* (1980), pp. 234–237.

the Royal Society made a contribution to its cooperative project by reporting an experience.[59] Unlike a typical experience found in scholastic-Aristotelian practice—that is, a universal statement of how some aspect of the world behaves—it was a report of how, on one specific occasion, the world had *behaved*. An example from a Boylean air-pump trial conveys the flavor:

> We took a slender and very curiously blown cylinder of glass, of nearly three foot in length, and whose bore had in diameter a quarter of an inch, wanting a hair's breadth: this pipe, being hermetically sealed at one end was, at the other, filled with quicksilver, care being taken in the filling, that as few bubbles as was possible should be left in the mercury. Then the tube being stopt with the finger and inverted, was opened, according to the manner of the experiment, into a somewhat long and slender cylindrical box (instead of which we are now wont to use a glass of the same form) half filled with quicksilver: and so, the liquid metal being suffered to subside, and a piece of paper being pasted on level with its upper surface, the box and the tube and all were by strings carefully let down into the receiver.[60]

Boyle's report gives the impression of a discrete historical event and of the observer's central role within it, both by his detailed recounting of what happened, and by his frequent use of the first person, active voice. This situated form of experience is overwhelmingly the rule in the writings of the Fellows, regardless of subject matter.[61]

The experiences of others routinely appear in the same form as those reported at first hand, but, again, the veracity of the report clearly depends on the original experience of a specified person on a particular occasion. Even reports from people who had no pretensions to being philosophers were grist to the mill, as Sprat famously claimed: "We find many Noble Rarities to be every day given in, not onely by the hands of Learned and profess'd Philosophers; but from the Shops of *Mechanicks;* from the Voyages of *Merchants;* from the Ploughs of *Husbandmen;* from the Sports, the Fishponds, the Parks, the Gardens of *Gentlemen.*"[62] Thus: "Here follows a Relation ... which is about the

---

59. Shapin and Schaffer, *Leviathan and the Air-Pump*, chap. 2; Shapin, "Pump and Circumstance" (1984). See also Dear, *"Totius in verba"* (1985).

60. Quoted in M. B. Hall, *Robert Boyle on Natural Philosophy* (1966), pp. 329–330. Shapin, "Pump and Circumstance," discusses the function of Robert Boyle's use of detailed accounts as a form of "virtual witnessing" designed to establish "matters of fact." The importance of the latter as a form of social boundary marker is discussed in Schaffer, "Making Certain" (1984).

61. For other examples, see Dear, *"Totius in verba."*

62. Sprat, *History of the Royal Society* (1667), p. 72. Note the relevance of Sprat's listing to the issue of "social topography," considered in chapter 5, above; it is here integrated with a demarcation of "fact" from "hypothesis."

new *Whale-fishing* in the *West-Indies* about the *Bermudas,* as it was delivered by an understanding & hardy Sea-man, who affirmed to have been at the killing work himself,"⁶³ reads an item in the *Philosophical Transactions.* The crucial element in these reports is the rhetorical establishment of the actuality of a discrete event. Such assertion of a specific trial begs comparison with the nearest parallels we have seen in scholastic texts: Riccioli's occasional accounts of specific trials stress his personal credentials and expertise, which are converted into knowledge-claims on the basis of indefinitely many other single events. The Fellows of the Royal Society, by contrast, stress the particularities of the singular instance itself, as if the establishment of belief in the historical event was itself sufficient.

An instructional, or recipe-like, format of the kind that we have seen repeatedly in earlier chapters often has its place too in writings by Fellows of the Royal Society, but it is almost invariably accompanied by a historical description of what had ensued on the prototypical occasion on which the instructions had been followed. The procedure could always be repeated; the historical event could not. Thus "An extract of a letter lately written by an observing friend of the *Publisher,* concerning the vertue of Antimony," printed in the *Philosophical Transactions* in 1668, sets up the presentation with a historical assertion: "I tried that a Boare, to whom I had given an ounce of crude Antimony at a time, putting him into the Sty, would be fat a fortnight before another, having no Antimony, upon the like feeding." Then follows the recipe that had tested the alleged effect: "The manner of using it, is this. Take one drachme of crude Antimony powder'd for one Horse, and when you give him his Oats in a morning. . . ."⁶⁴

Examples could be chosen ad nauseam to show how the presentation of experiences (whether "experiments" or "observations") by those connected with the early Royal Society put the narrator at the center of a historical event. This was an approach to the making of knowledge that entailed the rooting of claims about the natural world in discrete events rather than universal experiences. Robert Boyle criticized those who "neither themselves ever took the pains to make trial of [experiments], nor received from any credible persons that professed themselves to have tried them";⁶⁵ knowledge about the world had to rely on discrete experiences happening to particular persons. The credentials that established the actuality of the event were provided by surrounding the description by a wealth of circumstantial

---

63. *Philosophical Transactions* 1 (1665/1666), p. 11.
64. Ibid. 3 (1668), p. 774.
65. Quoted in M. B. Hall, *Robert Boyle on Natural Philosophy,* p. 131.

detail. This detail generally included information regarding time, place, and participants, together with extraneous remarks about the experience, all serving to add verisimilitude.[66] The literary form constituting a publicly available singular experience represented a credential formerly supplied by different literary forms—in natural philosophy, the commentary on an authoritative text; in the mathematical sciences, the axiomatic-deductive treatise.[67]

John Wallis's theory of the tides, read to the Society in 1666 and printed in the *Philosophical Transactions*, provides a crucial instance of the representation of a mathematical scientific argument as an empirical investigation rooted in reportable singular experience. Wallis, a mathematician, had invented an explanation of the tides, based on a model of the motion of the earth, that closely resembled Galileo's famous account presented in the *Dialogo*.[68] The chief difference was its use of a monthly rotation around the center of gravity of the earth-moon system, in addition to the earth's daily and annual motions, to account for the apparent lunar correlation of the tides that Galileo had neglected. Wallis's mathematical argument was not the sort of thing that the Royal Society usually treated as grist to its collective mill, since it was not predicated upon warranted singular experiences (beyond its intention to explain the routinely experienced behavior of the tides). Indeed, its status as an explanatory theory or hypothesis placed it in dangerous territory; Fellows of the Society habitually disclaimed speculative hypothesizing.[69] However, a consequence of Wallis's theory was that the highest tides should occur not at the equinoxes, as was popularly held to be the case, but during February and November. This

66. For much detail on the social placement of trust in such reportage, see Shapin, *A Social History of Truth*.
67. On the creation of a publicly available experience through such means, see Shapin, "Pump and Circumstance."
68. See T. Birch, *The History of the Royal Society* (1756–1757), vol. I, pp. 88–89; *Philosophical Transactions* 1 (1666), pp. 263–289, esp. pp. 271–275. For a brief secondary treatment, see Aiton, "Galileo's Theory of the Tides" (1954), on pp. 50–54. The methodological form of Galileo's argument has been discussed by Wisan, "Galileo's Scientific Method" (1978), on pp. 35–36, in terms of its conformity to a mathematical structure of the kind considered in earlier chapters.
69. Wood, "Methodology and Apologetics" (1980); Shapin and Schaffer, *Leviathan and the Air-Pump*; Henry, "The Scientific Revolution in England" (1992), gives an overview of possible religious cultural correlates of the Royal Society's distrust of hypotheses. Robert Hooke's dedicatory preface to his *Micrographia* (1665) is one of the most famous expressions of this position: Hooke, "To the Royal Society," preface to *Micrographia*; Birch, *The History of the Royal Society* I, pp. 490–491, gives the Council's directions to Hooke to make this disclaimer. See also Harwood, "Rhetoric and Graphics in *Micrographia*" (1989); Dennis, "Graphic Understanding."

circumstance provided him with an opportunity to justify his presentation by reporting firsthand experiences. Wallis first gave the testimony of the people of Romney Marsh that tides tend to be higher in February and November and made it clear that he had heard this view before he had ever considered questioning the usual opinion. This at least served to make the point that his theory was not a post hoc contrivance. But Wallis then proceeded to bolster his claim in a fashion appropriate for his audience:

> And since that time, I have my self very frequently observed (both at *London* and elsewhere, as I have had occasion) that in those months of *February* and *November,* (especially *November*) the Tides have run much higher, than those at other times: Though I confess, I have not been so diligent to set down those Observations, as I should have done. Yet this I do particularly very well remember, that in *November* 1660 (the same year that his Majesty returned) having occasion to go by Coach from the *Strand* to *Westminster,* I found the Water so high in the middle of King-street, that it came up, not onely to the Boots, but into the body of the Coach; and the *Pallace-yard* (all save a little place near the *West-End*) overflow'd; as likewise the Market-place; and many other places; and their Cellars generally filled up with Water.[70]

Wallis, a founding Fellow of the Royal Society, thus used singular experience in presenting material conceived within the context of the mathematical sciences. How this procedure could establish universal knowledge-claims is less clear, however, than in the cases of the mathematicians discussed in earlier chapters. There, the use of historical reports served not to underwrite a claim about nature, but to warrant the expertise, the fitness to speak, of the author. But the protocol employed in the early Royal Society shifted the meaning of such reports firmly towards the "matter of fact" itself. The use of singular experience in this way tended to atomize the world that it purported to reveal; experimental philosophy thus called into question the knowability of universals and enshrined a form of nominalism. Boyle and Joseph Glanvill sometimes fell back on the position that, if nothing else, at least they could claim to be assembling a stockpile of "facts"—thereby deferring indefinitely the necessity of showing how such "facts" could ever be turned into universal philosophical truths.[71]

---

70. *Philosophical Transactions* 1 (1665/1666), pp. 275–276.
71. See especially the discussion on "hypothesis" and its problems in the early Royal Society in B. Shapiro, *Probability and Certainty in Seventeenth-Century England,* chap. 2, esp. pp. 49, 53; cf. Dear, *"Totius in verba."* The difficulties attendant on turning experimental reports into philosophical truths was the main burden of Hobbes's criticisms of Boyle's work on the air-pump: Shapin and Schaffer, *Leviathan and the Air-Pump.*

However, when Newton himself employed reportorial tactics in 1672, he did so within a physico-mathematical context that allowed the development of arguments to warrant just such a philosophical move.

V. Newton's Mathematical-Experimental Philosophy of Light

Newton's earliest published work, on light and colors, was based on material that he had worked up for his first series of lectures (1670–1672) as the new Lucasian professor at Cambridge, succeeding Barrow.[72] As the holder of a chair in mathematics, Newton developed lectures that adhered closely to the usual form for geometrical optics. The texts proceed discursively from point to point, but Newton nonetheless presents formal demonstrations of his "propositions" as he goes along. Exactly like the mathematical texts in optics or astronomy or mechanics that were produced throughout the seventeenth century, Newton's optical lectures, while presenting a large number of set-piece experimental contrivances (a characteristic feature of optical writings), do so in the universalized instructional form with which we are already very familiar. On only a handful of occasions does the use of the first person intrude in respect of experimental trials.[73]

Thus Newton presents his account of differential refrangibility with a demonstration of the assertion that light rays "differ from one another with respect to the quantity of refraction: Of those rays that all have the same angle of incidence, some will have an angle of refraction somewhat larger, and others a smaller one." This universal proposition is further specified in relation to a geometrical diagram, commencing: "For the sake of a fuller illustration, let EFG be any refracting surface, for example, glass, and draw any line OF meeting it at F and making with it an acute angle OFE."[74] Once having provided a precise, formal

---

72. Newton, *The Optical Papers* (1984). See also A. R. Hall, *All Was Light* (1993), pp. 45–59.

73. On my count, there are four first-person experimental narratives in what appears to be the earlier of the two surviving drafts of the lectures (called "Lectiones opticae" in Shapiro's edition, pp. 64, 66, 70–72, 186 of the Latin text; this edition provides facing-page English translation). There are in addition five more such passages in the extra material of the longer draft (the "Optica," apparently a version closer to an intended but aborted publication: see Newton, *The Optical Papers*, pp. 16–20), although one such new passage, a formally stated "proposition" on the immutability by refraction of the colors of spectrally produced rays, runs for about five pages (ibid., pp. 452–462, 512, 538–540, 542–544, 552, 554).

74. Ibid., p. 283 (continuing to p. 285 of the facing-page translation); this is the version "Optica."

exposition of the proposition, Newton now offers a proof in similar vein, together with another diagram:

> Having thus briefly explained my [our] view on this subject, I will at once present the reasoning and experiments that support these things, lest you think that I have set forth fables instead of the truth. Since a certain very commonly encountered experiment with a prism first presented me the opportunity to think out the rest, I will explain that first. In the wall or window of a room let F be some hole through which solar rays OF are transmitted, while other holes elsewhere have been carefully sealed off so that no light enters from any other place.... Then place at that hole a triangular glass prism AαBβCk that refracts the rays OF transmitted through it toward PYTZ.[75]

The reader is assured that "you will see these rays ... formed into a very oblong figure." After providing a fuller description of what the reader will see, and concluding that the result establishes that some rays are refracted more than others, Newton confides that "in whatever position I placed the prism, I nonetheless could never make it happen that the image's length was not more than four times its breadth."[76] This concluding historical remark stands somewhat at odds with the presentation immediately preceding it, although it alludes back to the introductory observation of Newton's initial encounter with the described procedure. It is followed, however, with a formal examination of the same point (that a band appears when the received rules of refraction would predict a circular disk) that adopts the usual instructional form.

At about the same time as the text of the lectures was being prepared, Newton wrote his other, more famous account of these matters in a letter to Henry Oldenburg, to be presented to the Royal Society. The letter starts with an account paralleling, but with a different formal structure, that just examined from the lectures. Its famous opening takes on more than a mere dramatic quality when one recognizes its crafted character—its careful difference from the usual construction/instructions found in the mathematical texts of optics.

> To perform my late promise to you, I shall without further ceremony acquaint you, that in the beginning of the Year 1666 (at which time I applied my self to the grinding of Optick glasses of other figures than *Spherical*,) I procured me a Triangular glass-Prisme, to try therewith the celebrated *Phaenomena* of *Colours*. And in order thereto hav-

---

75. Ibid., p. 285.
76. Ibid.

ing darkened my chamber, and made a small hole in my window-shuts, to let in a convenient quantity of the Suns light, I placed my Prisme at his entrance, that it might be thereby refracted to the opposite wall.[77]

For several pages Newton continues in this explicitly historical vein, towards the end of which he tells the reader that "Amidst these thoughts I was forced from *Cambridge* by the Intervening Plague, and it was more then two years, before I proceeded further."[78] His determinedly anecdotal, event-focused style restructures the material of the lectures into a story of the discovery of universal properties of light rather than an account of the means whereby those properties could be made manifest (although the latter function is also served thereby). It is, perhaps, of secondary interest that the story is not literally true: Newton's original research on light and colors was conducted over a period of several years, and his presentation to the Royal Society is an idealized version representing a whole series of investigations.[79]

The letter changes form towards the end, its historical narrative replaced by a statement of thirteen numbered propositions of the usual mathematical kind. They are each universal (such as number two: "To the same degree of Refrangibility ever belongs the same colour"); none is provided with a formal demonstration such as is given in the optical lectures. Having established the difformity of refrangibility in the rays composing white light, Newton introduces these propositions with the promise that they will reveal the origin of colors. "I shall lay down the *Doctrine* first, and then, for its examination, give you an instance or two of the *Experiments*, as a specimen of the rest."[80] Newton's lack of

77. Newton, "New Theory about *Light* and *Colours*" (1672), on pp. 3075–3076. This paper has recently been examined by writers interested in scientific rhetoric: Bazerman, *Shaping Written Knowledge* (1988), chap. 4; Gross, *The Rhetoric of Science* (1990), chap. 8; D. Locke, *Science as Writing* (1992). None, however, has been concerned with its fuller historical meaning.

78. Newton, "New Theory about *Light* and *Colours*," p. 3080.

79. See Newton, *The Optical Papers*, "Introduction" by Alan Shapiro, pp. 2–15; Westfall, "The Development of Newton's Theory of Color" (1962), esp. pp. 351–352; Westfall, *Never at Rest*, pp. 156–174, 211–222, esp. pp. 156–164; Kuhn, "Newton's Optical Papers" (1958), esp. pp. 33–34 n. 11. Blay, "Remarques sur l'influence de la pensée baconienne à la Royal Society" (1985), concludes that, since Newton's apparently "Baconian" letter was a fabrication, that he was therefore not a true Baconian. It is unclear what being a Baconian really amounted to (e.g., Webster, *The Great Instauration* [1975]); Newton's procedure exploits different philosophical genres within a basically physico-mathematical enterprise.

80. Newton, "New Theory about *Light* and *Colours*," p. 3081. In the original letter to Oldenburg, but omitted in the published version, Newton had commenced this sentence with the words: "To continue the historicall narration of these experiments would make

concern to provide formal demonstrations runs directly counter to his procedure in the lectures, of course; his attempt to concentrate the essence of his views into so short a span must be accounted part of the reason. He presents his approach, however, as if it simply conformed to the logic of the Royal Society's experimental philosophy.[81] "Reviewing what I have written, I see the discourse it self will lead to divers Experiments sufficient for its examination: And therefore I shall not trouble you further, than to describe one of those, which I have already insinuated." There follows his experiment for recombining the spectrum with a lens, so as to produce the appearance of whiteness at the focus.[82] The presentation is now instructional, not historical, and refers to a lettered diagram. Investigative experimental philosophy has been supplanted by confident, predetermined mathematical science.[83]

This mathematical science must be understood, however, in the physico-mathematical terms of Barrow and others. Newton attempts to step beyond the apparently strict boundaries of classical mixed mathematics when he presumes to speak to unequivocally natural-philosophical questions on the basis of his mathematical conclusions concerning color. "These things being so, it can no longer be disputed, whether there be colours in the dark, nor whether they be the qualities of the objects we see, no nor perhaps whether Light be a Body."[84] Newton dares to answer, as a matter of at least moral certainty ("it can no longer be disputed"), questions concerning the nature of colors, a question that arises from fundamentally Aristotelian philosophical categories. His mathematical science of optics, in other words, adopts the presumptions that had driven the increasing use of the label "physico-mathematics" throughout the seventeenth century—it takes over topics from natural philosophy itself. Others had recently discussed the question of the nature of light. Grimaldi's *Physico-mathesis de lumine*

---

a discourse too tedious & confused, & therefore. . . ." Newton, *The Correspondence*, vol. 1 (1959), p. 97.

81. Newton's account book records that he had purchased the *Philosophical Transactions* in 1667; he was already very familiar with the style of philosophy approved by the Royal Society: Newton, *The Optical Papers*, "Introduction," p. 14.

82. Newton, "New Theory about *Light* and *Colours*," p. 3085.

83. Moss, "Newton and the Jesuits" (1988), characterizes Newton's overall argument as exemplifying a scholastic-Aristotelian scientific methodology, quoting on pp. 120–121 William A. Wallace, *The Scientific Methodology of Theodoric of Freiburg* (Fribourg, Switzerland: The University Press, 1959), pp. 268–272. Aside from the judgments of the absolute demonstrative worth of Newton's argument, there seems much worth attending to in this view, when Newton's letter is properly seen in reference to the mathematical sciences.

84. Newton, "New Theory about *Light* and *Colours*," p. 3085.

of 1665 had focused, in an explicitly physico-mathematical vein, on whether light is substantial or only accidental (that is, a property of something else). He argued in the first part of the work for the former conclusion, balancing it in the second with a purported dissolution of the former's arguments and a reaffirmation of the peripatetic doctrine of light as an accident.[85] Honoré Fabri, in 1669, had published *Dialogi physici quorum primus est de lumine,* an explicitly physical work that nonetheless incorporated arguments and topics from the mixed mathematical sciences (the work's second and third dialogues concern motion and the force of percussion, good Galilean mathematical topics).[86] In it he considers Grimaldi's account of diffraction as evidence for light as a body (it is not approved), and uses an elaborate geometrical diagram of the experimental setup (tried a number of times, with never a variation in the result) in mathematical rather than traditional physical style.[87] Newton knew Fabri's discussion, although not directly Grimaldi's, in 1675, and possibly earlier.[88] He appears in his own laconic discussion of it to regard it as a matter of little dispute, given his findings: "For, since Colours are the *qualities* of Light, having its Rays for their intire and immediate subject, how can we think those Rays *qualities* also, unless one quality may be the subject of and sustain another; which in effect is to call it *Substance.* We should not know Bodies for substances, were it not for their sensible qualities, and the Principal of those being now found due to something else, we have as good reason to believe that to be a Substance also."[89] The argument is both physicomathematical and, on its physical side, deeply scholastic.

Newton quickly regretted this adventure, retreating in some haste back to the security of mathematical optics when challenged by Robert Hooke. In the original letter Newton had included a passage that Oldenburg had judged fit to be expunged from the printed version. "A naturalist would scearce expect to see ye science of those [i.e., colors] become mathematicall, & yet I dare affirm that there is as much cer-

---

85. Grimaldi, *Physico-mathesis de lumine* (1665). Ziggelaar, *Le physicien Ignace Gaston Pardies* (1971), pp. 176–177, suggests that Grimaldi's heart was really with the proposition of substantiality, the argument of the work's second part being a pro forma adherence to official orthodoxy.
86. Fabri, *Dialogi physici* (1669).
87. Ibid, dialogue 1, pp. 4–10, esp. pp. 6–7; the diagram is fig. I of Tab. I^a.
88. See A. R. Hall, "Beyond the Fringe" (1990), on pp. 17–19. Hall suggests (ibid., n. 1) that Newton's remarks on the nature of light in his 1672 letter may have been brought to mind by a review of Grimaldi's book given in the preceding issue of the *Philosophical Transactions;* Newton's verdict is the opposite of that reported for Grimaldi.
89. Newton, "New Theory about *Light* and *Colours,"* p. 3085.

tainty in it as in any other part of Opticks."⁹⁰ Hooke saw the original and considered the apparent claim to absolute certainty unwarranted; in response to Hooke's criticism, Newton tried to clarify his position in another letter to Oldenburg. "I said indeed that the *Science of Colours was Mathematicall & as certain as any other part of Optiques;* but who knows not that Optiques & many other Mathematicall Sciences depend as well on Physicall Principles as on Mathematicall Demonstrations: And the absolute certainty of a Science cannot exceed the certainty of its Principles."⁹¹ Empirically validated propositions relied in part on the physical natures of things, and uncertainty over such matters meant that the propositions lacked absolute mathematical certainty. Newton's appeal to mixed mathematics underscored the extent of his claims: if the mixed mathematical sciences were valid, then so was his science of colors.

Physical causes, "hypotheses," then, were to be kept apart from the mathematical sciences, and claims to any degree of certainty were to be warranted through the secure exemplars of mathematics.⁹² This was the line adopted in the optical lectures. Newton's remarks there are worth examination:

> ... the generation of colors includes so much geometry, and the understanding of colors is supported by so much evidence [*evidentiâ:* "evidentness"], that for their sake I can thus attempt to extend the bounds of mathematics somewhat, just as astronomy, geography, navigation, optics, and mechanics are truly considered mathematical sci-

---

90. Newton, *The Correspondence,* p. 96. Henry Oldenburg judiciously omitted this passage when Newton's letter was published in the *Philosophical Transactions.* See, for the role such claims played in the subsequent controversy over Newton's ideas, Bechler, "Newton's 1672 Optical Controversies" (1974); Bechler judges that the objections of critics such as Hooke and Huygens derived from their perception of Newton as transgressing the proper bounds of knowledge-claims in natural philosophy, insofar as he claimed certainty for his assertions rather than the probability they deemed appropriate. Bechler's persuasive account would appear to require, in addition, consideration of the disciplinary relationship between natural philosophy and mathematics.
91. Newton, *The Correspondence,* p. 187; cf. A. Shapiro, *Fits, Passions, and Paroxysms,* pp. 37–38.
92. Newton used the term "theory" in a manner quite distinct from the term "hypothesis." A theory was a generalized descriptive account of some aspect of the physical world discovered through phenomena; a hypothesis was a conjectural explanation for such an account, as one sees also in the *Principia,* which presents various theories while eschewing hypotheses: see in this connection Feyerabend, "Classical Empiricism," p. 159 n. 7. Newton's "theory of comets," for example (as mentioned in his preface to the *Principia's* second edition), no doubt owes its name to conventional astronomical (mathematical) usage, although he employs the term widely for nonastronomical matters.

ences even if they deal with physical things: the heavens, earth, seas, light, and local motion. Thus although colors may belong to physics, the science of them must nevertheless be considered mathematical, insofar as they are treated by mathematical reasoning.[93]

Newton urges geometers "to investigate nature more rigorously, and those devoted to natural science to learn geometry first." If this is done, "with the help of philosophical geometers and geometrical philosophers, instead of the conjectures and probabilities that are being blazoned about everywhere, we shall finally achieve a natural science supported by the greatest evidence [i.e., "evidentness"]."[94] Newton's philosophical geometers, like Galileo's philosophical astronomers, herald the new mathematical study of nature that others had come to call "physico-mathematics."

Regarding the physical question of the nature of colors, Newton had concluded in the 1672 letter that "to determine more absolutely, what Light is, after what manner refracted, and by what modes or actions it produceth in our minds the Phantasms of Colours, is not so easie." And so, he says, "I shall not mingle conjectures with certainties."[95] In another letter to Oldenburg, published in the *Philosophical Transactions* as a reply to the Jesuit mathematics professor Ignace Gaston Pardies,[96] Newton asserted that the "best and safest way of philosophizing" was the establishment of the properties of things by experiments, only then moving towards a consideration of hypotheses for explaining those properties.[97] In effect, the establishment of properties such as Newton claimed to have done belonged to mathematics, whereas hypotheses for their causal explanation belonged to natural philosophy.

But the cognitive style of Newton's letter to the Royal Society, however mathematical, represents more than classical mathematical science: it is an example of what Wilkins had called "Physico-Mathematicall-Experimentall Learning," and the experimental component was of great importance. Newton's use of first-person historical narration in the letter was not an isolated occurrence bearing little rela-

93. Newton, *The Optical Papers*, p. 439 (from "Optica").
94. Ibid.
95. Newton, "New Theory about *Light* and *Colours*," p. 3085.
96. *Philosophical Transactions* 7 (1672), pp. 5014–5018. On Pardies and optics see Ziggelaar, *Le physicien Ignace Gaston Pardies*, chap. 11, esp. pp. 182–184, for the exchange with Newton.
97. Newton, *The Correspondence*, p. 164. See also, on the issues of "hypotheses" and causal explanations, McMullin, "Conceptions of Science in the Scientific Revolution" (1990), on pp. 67–74; McMullin does not, however, use a "mathematics/natural philosophy" distinction in his explication of these questions.

tion to his other work. When, in 1704 and as president of the Royal Society, he finally came to publish a full treatise on light and colors, the differences between it and the earlier, still unpublished, lectures emphasized precisely those features that distinguish the famous letter. Like his Royal Society predecessors and colleagues, Newton places emphasis throughout the *Opticks* on the foundational status of events, weaving them into a structure that claims to be as firmly established as any science could ever be. Hypothetical causes are to be avoided in favor of demonstrable behaviors, but the model for that distinction clearly remains the older division between physics and mathematics. "My Design in this Book," Newton begins, "is not to explain the Properties of Light by Hypotheses, but to propose and prove them by Reason and Experiments." Which means to follow the procedures of the mathematician: "In order to which I shall premise the following Definitions and Axioms."[98] Colors, he concludes later in the work, "are derived, not from any physical Change caused in Light by Refraction or Reflexion, but only from the various Mixtures or Separations of Rays, by virtue of their different Refrangibility or Reflexibility." They are thus conformable to the geometry of light rays. "And in this respect the Science of Colours becomes a Speculation as truly Mathematical as any other part of Opticks."[99]

In the *Opticks*, Newton presents himself as having experimentalized a mathematical science in such a way as to allow its development along the procedural lines promulgated in the Royal Society's early statements: "matters of fact" rooted in firsthand experience. The *Opticks* is written chiefly as a structured, organized demonstrative compendium of Newton's experiments, accounts usually detailing, literally, "what I did." Interspersed instructional passages clarify details or variants, but the framework is provided by historical stories of Newton performing specific experiments. Unlike the story that begins the letter of 1672, these are not given specific locations in time but sit in a temporal limbo: things that happened once upon a time. The geometrical demonstrations that form so much of the optical lectures have been discarded (references to some of these, available in the editions of the lectures that finally appeared in 1728 and 1729, after Newton's death, are made in the posthumous 1730 edition of the *Opticks*).[100] But the treatise still remains a mathematical scientific text in optics. It begins with two foundational sections that contain "definitions" and

98. Newton, *Opticks* (1730/1952), p. 1. See A. R. Hall, *All Was Light*, for an overview.
99. Newton, *Opticks*, p. 244.
100. See Newton, *The Optical Papers*, "Introduction," pp. 21–23.

"axioms," the latter presented as "the sum of what hath hitherto been treated of in Opticks" and has "been generally agreed on," and therefore assumed "under the notion of Principles."[101] It proceeds through the subsequent presentation of "propositions," "theorems," and the occasional "problem," as well as experiments (often numbered) that provide the "proof." Newton's commitment to these forms represents his celebrated "mathematical way," codified in his occasional discussions of the method of analysis and synthesis in the study of nature.[102]

Newton's most famous pronouncement on method appears in the third edition of the *Opticks* (1717), in an addition to what appears there as Query 31:

> As in Mathematicks, so in Natural Philosophy, the Investigation of difficult Things by the Method of Analysis, ought ever to precede the Method of Composition. This Analysis consists in making Experiments and Observations, and in drawing general Conclusions from them by Induction, and admitting of no Objections against the Conclusions, but such as are taken from Experiments, or other certain Truths. For Hypotheses are not to be regarded in experimental Philosophy. And although the arguing from Experiments and Observations by Induction be no Demonstration of general Conclusions; yet it is the best way of arguing which the Nature of Things admits of, and may be looked upon as so much the stronger, by how much the Induction is more general. And if no Exception occur from Phaenomena, the Conclusion may be pronounced generally.[103]

Newton's talk of the method of analysis and synthesis (or "resolution and composition") had a long pedigree, derived from the ancient writings of Pappus on mathematical procedure. The latter seem to have fed into Galileo's similar terminological and methodological usage.[104] In relation to Newton's talk of "experiments," however, it also uses what looks like an idiosyncratic notion of "induction." Newton's advo-

---

101. Newton, *Opticks*, pp. 19–20.
102. Guerlac, "'Newton's Mathematical Way'" (1984), details Newton's use of the expression against I. Bernard Cohen's characterization of it as a spurious "Newtonism" in his *The Newtonian Revolution* (1980), p. 95. For a general discussion of method in the seventeenth century, see Dear, "Method in the Study of Nature" (forthcoming).
103. Newton, *Opticks*, p. 404. Henry Guerlac indicates which parts of this passage and its continuation, quoted below, first appeared in the 1717 edition in "Newton and the Method of Analysis," p. 379.
104. See for a general discussion Guerlac, "Newton and the Method of Analysis" (see also Newton's manuscript remarks quoted in ibid., p. 385); Birch, "The Problem of Method in Newton's Natural Philosophy" (1991). For the background as it pertains especially to Galileo, N. Jardine, "Galileo's Road to Truth" (1976). See also Hintikka and Remes, *The Method of Analysis* (1974), and Dear, "Method in the Study of Nature."

cated procedure in 1717, to which he seems to have adhered at least in practice in the optical arguments of the 1670s, and which is strongly echoed in the fourth of the "Rules of Philosophizing" found in the *Principia's* third edition of 1726, takes the following form. An experimental or observational situation is conceptually analyzed to reveal the necessary features of its behavior and properties, just as a particular kind of geometrical figure might be analyzed to determine its logically required corollary properties.[105] Induction then serves to generalize those features to all other situations deemed similar to the experimental or observational exemplar. Induction can never yield conclusions later found to be false in light of new experiments; its conclusions are subject only to possible qualification: "But if at any time afterwards any Exception shall occur from Experiments, it may then begin to be pronounced with such Exceptions as occur."[106]

The mathematical prototype of induction in Newton's usage reminds us immediately of Barrow's views on that question. Echoing one of Aristotle's diverse illustrations, Barrow, like others, had allowed that knowledge of a universal in geometry could be acquired through experience of a single example.[107] Similarly, Newton allows the formation,

---

105. Hintikka and Remes, *The Method of Analysis*, p. 110, valuably summarizes "the Newtonian method" as follows:
  (i) an analysis of a certain [experimental or observational] situation into its ingredients and factors →
  (ii) an examination of the interdependencies between these factors →
  (iii) a generalization of the relationships so discovered to all similar situations ["induction"] →
  (iv) deductive applications of these general laws to explain and to predict other situations ["synthesis"].
106. Newton, *Opticks*, p. 404; see the insightful discussion in Feyerabend, "Classical Empiricism," esp. p. 166. The fourth "Rule of Philosophizing," appearing in 1626, states: "*In experimental philosophy we are to look upon propositions inferred by general induction from phenomena as accurately or very nearly true, notwithstanding any contrary hypotheses that may be imagined, till such time as other phenomena occur, by which they may either be made more accurate, or liable to exceptions. This rule we must follow, that the argument of induction may not be evaded by hypotheses*" (Cajori, *Principles*, p. 400). This is the same notion of induction as in Query 31, denying in effect the possibility of any properly constituted inductive generalization ever being falsified. The matter of sufficient similarity left a lot of room, of course; the second of Newton's "Rules of Philosophizing" (or "Hypotheses" in the first edition) appears to recognize the problem: "Therefore to the same natural effects we must, as far as possible, assign the same causes" (ibid., p. 398). On the development of the "Rules" in successive editions see Koyré, "Newton's 'Regulae Philosophandi'" (1965), and annotations in Newton, *Principia*, vol. 2, pp. 550–555. On Newton's use of the term *phaenomena* in such passages, and its contemporary meaning, see Baroncini, *Forme di esperienza e rivoluzione scientifica* (1992), chap. 4.
107. See above, section III, and chapter 1, section IV.

also "inductive," of a universal from a single experimental setup, properly considered.[108] In effect, Newton experimentalizes for the physico-mathematical realm the particular notion of making a universal from a singular that had been discussed by Barrow. However, Barrow had also, as we saw in section III, asserted that frequency of observation was not necessary to establish certain kinds of supposition. This followed from Barrow's mathematical-scientific perspective: the necessary relations of the mathematician were not compromised by qualitative essences incapable of certain determination. Of course, as Newton observed to Oldenburg, a mixed mathematical science (interpreted here physico-mathematically) relied on physical principles as well as mathematical demonstrations and to that degree wanted absolute certainty. But that was unavoidable; Newton had found, he thought, the way of making the most secure knowledge that could be had.

Hence Newton's version of "experimental philosophy" postulated the actual production of particular phenomena so as to allow the formation of a universal science from singulars; the trick lay in that final stage and was accomplished, to the extent that it could be accomplished, by framing the issues in terms of physico-mathematics. The difficulties in attributing to contrived events a philosophical meaning that enabled them to have relevance to the establishment of universal knowledge-claims had left the Royal Society's enterprise at something of an impasse; Newton's work retrospectively validated the experimental program that Boyle had advocated and that the Royal Society had largely exemplified. It placed event experiments into the frame of the expanding mathematical sciences by providing them with a new meaning constructed from the methodological language that those sciences had already developed for themselves. The event experiments of a Robert Boyle aimed at disclosing localized natural behavior; those occasionally found in mathematical practice had aimed at the vindication of the physico-mathematician's competence (as a repository of legitimate experience) to speak for nature.[109] The difference between the two collapsed when Newton imported the former into the latter, while the Royal Society's experimental philosophers imported the latter into the former. The Newtonian rhetoric of inductive generalization, purportedly based on the mathematical method of analysis and synthesis, gave event experiments a philosophical respectability that they had formerly lacked.

108. See also text to n. 47, above.
109. Cf. Shapin, *A Social History of Truth*, chap. 7, esp. pp. 347–349, on Boyle's views of the variation in physical properties of the "same" chemical substances obtained from different localities; chapter 3, above, on the establishment of expertise.

The very vagueness of Newton's "Rules of Philosophizing" (which are really tips or maxims, like those of scholastic philosophy), reveal the point that David Hume was later to make about the character of the natural science produced by this kind of experimental philosophy.[110] Taking the latter for granted as the proper way to look at how human knowledge is made, Hume saw what Newton expressed in terms of prudential maxims as amounting to a radical empiricism that disavowed the possibility of access to causes and connections behind observable objects and events. The Humean "problem of induction" arises only for those who require guarantees for inductive moves. Otherwise, Hume is simply concerned with how these things are actually done; their legitimacy is never in doubt. And that legitimacy, as we have seen, derived from the practices of the essence-disavowing mathematical sciences.

110. See chapter 1, above. Capaldi, *David Hume* (1975), plays up Hume's debt to Newton, although he works with a somewhat idealized version of Newtonian philosophy in doing so. On "maxims," see chapter 7, section I, above.

CONCLUSION: A MATHEMATICAL NATURAL PHILOSOPHY?

In *Leviathan and the Air-Pump*, Shapin and Schaffer have provided a local reading of the meaning of experimental philosophy, one that focuses on the program laid forth by Robert Boyle. While their account suggests that Boyle's ideal corresponded in some way to the realities of the early Royal Society, they succeed above all in unpacking certain contingent meanings of experimentalism as they were seen by Boyle himself and, from a different perspective, by Thomas Hobbes. What they do not do is examine the sociocultural conventions that allowed the attribution of certain meanings to experiment and disallowed others.[1] This book, by contrast, has considered socially embedded genres of argument in philosophy so as to understand what inferential moves were taken for granted or contested within particular knowledge-producing communities. Those groups, typically trained within the universities and colleges, prosecuted the literary endeavors that constituted dominant seventeenth-century natural philosophy and mathematical science.

By examining widely shared modes of argument, and understanding them in relation to the academic disciplinary constraints that shaped and enabled them, it becomes possible to see how the phenomenon that Steven Shapin has called "virtual witnessing" relates to the presumed reading practices of a seventeenth-century philosophical public.[2] It not only shows us what that public would have expected,

1. Shapin, *A Social History of Truth* (1994), addresses some of the relevant issues in regard to truth-telling and trust.
2. Shapin, "Pump and Circumstance" (1984); Shapin and Schaffer, *Leviathan and the Air-Pump* (1985), chap. 2.

however; it also reveals the sources of philosophical legitimacy that attached to the form. Boyle, in encouraging virtual witnessing by his readers, effectively appealed to increasingly respectable philosophical precedent. That was how he could portray his work as "philosophy."

The story outlined in the foregoing pages indicates an intimate relationship between what Thomas Kuhn described as largely independent "mathematical" and "experimental" traditions in the seventeenth century.[3] The mathematical sciences spilled over into the domain of natural philosophy, carrying over practices that, with altered connotations, quickly became the hallmarks of experimental inquiry.

The new scientific experience of the seventeenth century was characterized by the singular, historical event experiment, which acted as a surrogate for universal experience. The latter had routinely been regarded as the proper grounding for philosophically legitimate knowledge-statements about nature; the advent of event experiments was a practical response within the mixed mathematical sciences to a confrontation between such Aristotelian methodological demands and the practical exigencies of making knowledge that would be acceptable to all relevant judges. Trusting the author of a scientific text, especially on a mathematical subject such as astronomy or optics, had always been an implicit necessity. In the seventeenth century, however, the weakening social structures of academic knowledge increasingly promoted the explicit courting of trust; event experiments served that function.

The reported historical event thus began by serving as a token of the mathematical philosopher's expertise.[4] Firsthand reports underwrote the authority of the text so that assertions of universal truths might be rendered acceptable. At the same time, event-centered experience became contrived experience, again following a pattern familiar in the mathematical sciences. Experimental science was about *making* experience, even as its wider philosophical legitimacy involved the extension of mathematical procedures into the realm formerly reserved for a demarcated natural philosophy.

The experimental philosophy of the early Royal Society, unusually, accorded event experiments independent, rather than token, philosophical significance. This can be understood partly through the local reasons that have been detailed by Shapin and Schaffer and partly through more general considerations relating to the social place of craft

---

3. Kuhn, "Mathematical versus Experimental Traditions" (1976).

4. Recall that mathematical scientists often stressed that mathematics was one of Aristotle's three branches of philosophy: chapter 2, section I, above.

knowledge, as entertained most recently by William Eamon.[5] But the larger significance of the Royal Society's enterprise seems to derive retrospectively from what it became during a slightly later period. After the appearance of Newton's *Principia*, the Society began to associate its experimental practice with Newton's physico-mathematical justifications. Only then (for all its earlier European fame)[6] did it begin to appear as an original mover of a new kind of natural philosophy. In the early eighteenth century the *Dictionnaire de Trévoux* could speak of a new physics, one in which "the ancients made mediocre progress, whether because they neglected the aid of experiments [*expériences*] or because of the humor of the Greeks, incapable of application." This applied, experimental physics was one that "observations made by the Royal Society [now commanded by Newton] have carried to a high point of perfection."[7]

When the Fellows adopted Newton as their champion, the mathematical sciences achieved their final triumph over scholastic natural philosophy. But in the process, those sciences had themselves mutated into something new, because explicitly experimental. When John Wilkins spoke of "Physico-Mathematicall-Experimentall Learning" as the intended business of the new Royal Society, he invoked a form of knowledge that Newton's work would be the first fully to exemplify. Wilkins could coin the expression without needing explanation because so many of his peers now grasped the connotations of the term "physico-mathematics" as it had evolved over the preceding decades. With Jesuit mathematicians in the vanguard (together with often combative, competing intellectual soulmates such as Galileo and Pascal), mathematical philosophers had pushed claims for their way of doing things that forced a new label to characterize their ambitions. Discarding apologetic deference to natural philosophy, the mathematicians were now also philosophers of nature.

The kind of scientific experience in which these mathematical philosophers dealt was less overtly reliant on patterns of trust regarding em-

---

5. Shapin and Schaffer, *Leviathan and the Air-Pump*; Eamon, *Science and the Secrets of Nature* (1994).

6. See M. B. Hall, *Promoting Experimental Learning* (1991), chap. 9.

7. *Dictionnaire de Trévoux*, (1704; 2d edition 1771), s.v. "Philosophie," quoted in Salomon-Bayet, *L'institution de la science et l'expérience du vivant* (1978), p. 42: the physics that the "observations faites par la Société d'Angleterre ont portée à un haut point de perfection, les Anciens ont fait de médiocres progrès, soit qu'ils aient négligé le secours des expériences, soit par l'humeur des Grecs incapables d'application." On the activities of the Royal Society under Newton, see Heilbron, *Physics at the Royal Society* (1983); for interaction with Continental philosophers, Guerlac, *Newton on the Continent* (1981), esp. chaps. 3 and 5; also Schaffer, "Glass Works" (1989).

pirical knowledge-claims than the English experimentalists of the Royal Society.[8] Rather than relying on patterns of gentlemanly conduct for its integrity, such experience relied on associated academic status or on established disciplinary practices (or both) for its integrity. It is a story of trust that is much more mediated than in the English case offered by Shapin (and perhaps closer to his portrayal of more recent science). The dispute over sunspots between Scheiner and Galileo, considered in chapter 4, was concerned more with the validity of mathematical arguments and their relation to physics than with the truth of appearances. Scheiner's endeavor can be seen as an attempt to integrate the phenomena into an extension of accepted astronomical practice—a continuation of the tradition—different from the extension proposed by Galileo. But neither argued about the foundations of astronomical knowledge ("phenomena" and "observations") or the formal conduct of observational work. When Riccioli and Arriaga wrangled over the behavior of falling bodies, the pointedness of their dispute was, again, permitted by just those assumptions that they shared. A common intellectual culture, one that understood the uses of disciplinary boundaries, set the ground rules for debate.

Finishing this book with a discussion of Isaac Newton is more than just a standard move in a study of the Scientific Revolution. The significance of Newton's handling of experience makes sense only when seen in light of the story that leads up to it. Mathematical sciences and the experiential conventions that attached to them show Newton making a kind of mathematical natural philosophy that would, not long before, have been unthinkable. His was not, therefore, the triumph of a lone figure, but the achievement of one who stood on the shoulders of many physico-mathematicians. And it was not an uncontested triumph. Newton acknowledged that, in dealing with the physical world, the techniques of the mixed mathematical sciences could not yield principles that would be absolutely certain: Newton's "mathematical way" acknowledged its own limitations. A number of English writers in the early eighteenth century were to display the paradox of Newton's position.

George Gordon's hostile *Remarks upon the Newtonian Philosophy* (1719) advised that "[i]t is most certain that Mathematics give very great Assistence to Natural Philosophy; but I cannot see the Advantage of handling Natural Philosophy itself in the same Method as Mathematics."[9] Henry Pemberton, by contrast a self-avowed Newtonian, was

8. Shapin, *A Social History of Truth*.
9. Quoted in Risse, *Die Logik der Neuzeit*, 2. Band (1970), p. 140 n. 578.

more explicit as well as more sympathetic in his *A View of Sir Isaac Newton's Philosophy* (1728): "The proofs in natural Philosophy cannot be so absolutely conclusive, as in mathematics. For the subjects of that science are purely the ideas of our minds. . . . But in natural knowledge the subject of our contemplation is without us, and not so compleatly to be known: therefore our method of arguing must fall a little short of absolute perfection."[10] Isaac Watts's *Logick* (1726) made this general pronouncement:

> The *antient scholastic* Writers have taken a great deal of Pains, and engaged in useless Disputes about these two Methods [analytic and synthetic], and after all have not been able to give such an Account of them as to keep them entirely distinct from each other, neither in the Theory or in the Practice. Some of the *Moderns* have avoided this Confusion in some Measure by confining themselves to describe almost nothing else but the *synthetic* and *analytic* Methods of *Geometricians* and *Algebraists*, whereby they have too much narrowed the Nature and Rules of Method, as tho' every thing were to be treated in *mathematical* Forms.[11]

Louis Bertrand Castel, the eighteenth-century Jesuit scholastic who, as we saw in chapter 1, resisted the new experimental philosophy because of its reliance on singular ("monstrous") experiences, was more forthright than Watts.[12] Newton, he asserted, "believes everything done in physics when he has represented nature, phenomena, observations, by figures and by calculations. For that is all he does. Until him, that was just called astronomy, mechanics, or physico-mathematical geometry at the very most. He calls it simply physics."[13] This scandalous physics, of first importance in the appearance of modern science, attempted to inventory the universe.

10. Ibid., n. 579.
11. Watts, *Logick: or, The Right Use of Reason* (1726/1984), p. 346.
12. See chapter 1, section II, above.
13. My translation from Castel, *Le vrai système de physique générale de M. Isaac Newton* (Paris, 1743), p. 369, as quoted in Schier, *Louis Bertrand Castel* (1941), p. 103: "Newton croit tout fait en physique lorsqu'il a représenté la nature, les phénomènes, les observations, par des figures et par des calculs. Car il ne fait que cela. Jusqu'à lui cela ne s'appellait qu'astronomie, mécanique, ou géométrie physico-mathématique tout au plus. Il l'appelle physique tout court." Unpopular as Castel's view became in many quarters during the century of the Enlightenment, its echoes, transforming into romanticism, reverberate even in the pages of the *Encyclopédie*, as well as in Diderot's *Rêve d'Alembert*: Gillispie, "The *Encyclopédie* and the Jacobin Philosophy of Science" (1959); essays in Cunningham and Jardine, *Romanticism and the Sciences* (1990), Poggi and Bossi, *Romanticism in Science* (1994).

BIBLIOGRAPHY

Aguilonius, Franciscus. *Opticorum libri sex.* Antwerp, 1613.
Aiton, E. J. "Galileo's Theory of the Tides." *Annals of Science* 10 (1954): 44–57.
Akagi, Shozo. "Pascal et le problème du vide." *Osaka Daigaku Kyoyobu. Kenkyu Shoroku: Gaikokugo Gaikoku Bungaku* 3, 4, 5 (1967, '68, '69): 185–202, 170–184, 109–149.
Allard, Jean-Louis. *Le mathématisme de Descartes.* Ottawa: Éditions de l'Université d'Ottawa, 1963.
Aristotle. *The Complete Works of Aristotle: The Revised Oxford Translation,* ed. Jonathan Barnes. Princeton: Princeton University Press, 1984.
[Arnauld, Antoine, and Pierre Nicole] *La logique ou l'art de penser.* Paris, 1662. Facsimile reprint in Bruno Baron von Freytag Löringhoff and Herbert Brekle (eds.), *L'art de penser: La logique de Port-Royal,* 3 vols. Stuttgart–Bad Cannstatt: Friedrich Frommann Verlag, 1965–1967. Vol. 1.
Arriaga, Roderigo de. *Cursus philosophicus,* 1st edition. Antwerp, 1632.
———. *Cursus philosophicus,* 4th edition. Paris, 1647.
Auger, Léon. *Un savant méconnu: Gilles Personne de Roberval (1602–1675), son activité intellectuelle dans les domaines mathématique, physique, mécanique et philosophique.* Paris: Blanchard, 1962.
Aversa, Rafael. *Philosophia metaphysicam physicamque complectens.* Rome, 1625.
Bachelard, Gaston. *La formation de l'esprit scientifique.* Paris: J. Vrin, 1938.
Bacon, Francis. *The Works of Francis Bacon,* 8 vols. Ed. and trans. J. Spedding, R. L. Ellis, and D. D. Heath. Boston: Brown and Taggart, 1860–1864.
Baldini, Ugo. *"Additamenta Galilaeana:* I. Galileo, la nuova astronomia e la critica all'Aristotelismo nel dialogo epistolare tra Giuseppe Biancani e i revisori romani della Compagnia di Gesu." *Annali dell'Istituto e Museo di Storia della Scienza di Firenze* 9 (1984), fasc. 2:13–43.
———. "Christoph Clavius and the Scientific Scene in Rome." In G. V. Coyne, S. J., M. A. Hoskin, and O. Pedersen (eds.), *Gregorian Reform of the Calendar:*

Proceedings of the Vatican Conference to Commemorate its 400th Anniversary 1582–1982, pp. 137–169. Vatican City: Specola Vaticana, 1983.

———. *Legem impone subactis: Studi su filosofia e scienza dei gesuiti in Italia 1540–1632*. Rome: Bulzoni, 1992.

———. "La nova del 1604 e i matematici e filosofi del Collegio Romano: Nota su un testo inedito." *Annali dell'Istituto e Museo di Storia della Scienza di Firenze* 6 (1981), fasc. 2:63–98.

Baldini, Ugo, and George V. Coyne, eds. *The Louvain Lectures (Lectiones Lovanienses) of Bellarmine and the Autograph Copy of his 1616 Declaration to Galileo*. Vatican City: Vatican Observatory, 1984.

Baliani, J. B. *De motu naturali gravium solidorum et liquidorum*. Genoa, 1646.

Barker, Peter, and Roger Ariew, eds. *Revolution and Continuity: Essays in the History and Philosophy of Early Modern Science*. Washington, DC: Catholic University of America Press, 1991.

Barnes, Barry. "On the Conventional Character of Knowledge and Cognition." In Karin D. Knorr-Cetina and Michael Mulkay (eds.), *Science Observed: Perspectives on the Social Study of Science*, pp. 19–51. London: Sage, 1983.

Baroncini, Gabriele. *Forme di esperienza e rivoluzione scientifica* (Bibliotheca di Nuncius, studi e testi IX). Florence: Leo S. Olschki, 1992.

Barrow, Isaac. *Geometrical Lectures . . . translated from the Latin Edition, revised, corrected and amended by the late Sir Isaac Newton. by Edmund Stone, F.R.S.* London, 1735.

———. *Lectiones geometricae*. London, 1670.

———. *The Mathematical Works of Isaac Barrow, D.D.* Ed. William Whewell. Cambridge, 1860; facs. reprint in 1 vol., Hildesheim: Georg Olms, 1973.

———. *The Usefulness of Mathematical Learning Explained and Demonstrated*. Trans. John Kirkby. London, 1734.

Bazerman, Charles. *Shaping Written Knowledge: The Genre and Activity of the Experimental Article in Science*. Madison: University of Wisconsin Press, 1988.

Bechler, Zev. "Newton's 1672 Optical Controversies: A Study in the Grammar of Scientific Dissent." In Yehuda Elkana (ed.), *The Interaction Between Science and Philosophy*, pp. 115–142. Atlantic Highlands, NJ: Humanities Press, 1974.

Beeckman, Isaac. *Journal tenu par Isaac Beeckman de 1604 à 1634*, 4 vols. Ed. C. de Waard. The Hague: Martinus Nijhoff, 1939–1953.

Bennett, J. A. "The Challenge of Practical Mathematics." In Stephen Pumfrey, Paolo L. Rossi, and Maurice Slawinski (eds.), *Science, Culture and Popular Belief in Renaissance Europe*, pp. 176–190. Manchester: Manchester University Press, 1991.

———. "The Mechanics' Philosophy and the Mechanical Philosophy." *History of Science* 24 (1986): 1–28.

———. "Robert Hooke as Mechanic and Natural Philosopher." *Notes and Records of the Royal Society* 35 (1980): 33–48.

Berkel, Klaas van. "Beeckman, Descartes et 'la philosophie physico-mathématique'." *Archives de philosophie* 46 (1983): 620–626.

———. *Isaac Beeckman (1588–1637) en de mechanisering van het wereldbeeld.* Amsterdam: Rodopi, 1983.
Bertoloni Meli, Domenico. "Guidobaldo dal Monte and the Archimedean Revival." *Nuncius* 7 (1992): fasc. 1:3–34.
Biagioli, Mario. *Galileo, Courtier: The Practice of Science in the Culture of Absolutism.* Chicago: University of Chicago Press, 1993.
———. "The Social Status of Italian Mathematicians 1450–1600." *History of Science* 27 (1989): 41–95.
Birch, Andrea Croce. "The Problem of Method in Newton's Natural Philosophy." In Daniel O. Dahlstrom (ed.), *Nature and Scientific Method* (Studies in Philosophy and the History of Philosophy, vol. 22), pp. 253–270. Washington, DC: Catholic University of America Press, 1991.
Birch, Thomas. *The History of the Royal Society of London.* 4 vols. London, 1756–1757.
Black, Max. *Models and Metaphors: Studies in Language and Philosophy.* Ithaca: Cornell University Press, 1962.
Blackwell, Richard J. *Galileo, Bellarmine, and the Bible.* Notre Dame: University of Notre Dame Press, 1991.
Blake, Ralph M. "Isaac Newton and the Hypothetico-Deductive Method." In Edward H. Madden (ed.), *Theories of Scientific Method: The Renaissance Through the Nineteenth Century,* chap. 6. Seattle: University of Washington Press, 1960.
———. "Theory of Hypothesis among Renaissance Astronomers." In Edward H. Madden (ed.), *Theories of Scientific Method: The Renaissance Through the Nineteenth Century,* chap. 2. Seattle: University of Washington Press, 1960.
Blancanus, Josephus. *Aristotelis loca mathematica ex universis ipsius operibus collecta, & explicata.* Bologna, 1615.
———. *De mathematicarum natura dissertatio. Una cum clarorum mathematicorum chronologia.* Bologna, 1615.
———. *Sphaera mundi.* Bologna, 1620.
Blay, Michel. "Remarques sur l'influence de la pensée baconienne à la Royal Society: pratique et discours scientifique dans l'étude des phénomènes de la couleur." *Les études philosophiques* 1985 [no vol. #]:359–373.
Bloor, David. "The Dialectics of Metaphor." *Inquiry* 14 (1971): 430–444.
———. *Wittgenstein: A Social Theory of Knowledge.* London: Macmillan, 1983.
Boehm, A. "L'aristotélisme d'Honoré Fabri (1607–1688)." *Revue des sciences religieuses* 39 (1965): 305–360.
Bos, H. J. M. "On the Representation of Curves in Descartes' *Géométrie.*" *Archive for History of Exact Sciences* 24 (1981): 295–338.
Boyle, Robert. *A Defence of the Doctrine Touching the Spring and Weight of the Air . . . against the Objections of Franciscus Linus.* London, 1662.
———. *Hydrostatical Paradoxes, Made Out by New Experiments, (For the Most Part Physical and Easie).* Oxford, 1666.
Brahe, Tycho. *Tychonis Brahe Dani opera omnia.* J. L. E. Dreyer et al. (eds.). Reprint, Amsterdam: Swets & Zeitlinger, 1972.

Brockliss, L. W. B. *French Higher Education in the Seventeenth and Eighteenth Centuries: A Cultural History.* Oxford: Clarendon Press, 1987.
Butts, Robert E., and Joseph C. Pitt, eds. *New Perspectives on Galileo.* Dordrecht: D. Reidel, 1978.
Buzon, Frédéric de. "Descartes, Beeckman et l'acoustique." *Archives de philosophie* 44 (1981), *Bulletin cartésien* 10:1–8.
———. "Science de la nature et théorie musicale chez Isaac Beeckman (1588–1637)." *Revue d'histoire des sciences* 38 (1985): 97–120.
Cabeo, Niccolò. *In quatuor libros Meteorologicorum Aristotelis commentaria.* Rome, 1646.
———. *Philosophia magnetica.* Cologne, 1629.
Cajori, Florian. *Principles.* See Newton, Isaac, *Mathematical Principles of Natural Philosophy.*
Cannon, Susan Faye. "Humboldtian Science." Chap. 3 in Cannon, *Science in Culture: The Early Victorian Period.* New York: Science History Publications, 1978.
Cantor, Geoffrey. "Light and Enlightenment: An Exploration of Mid-Eighteenth Century Modes of Discourse." In David C. Lindberg and Geoffrey Cantor (eds.), *The Discourse of Light from the Middle Ages to the Enlightenment,* pp. 69–106. Los Angeles: Clark Memorial Library, 1985.
Capaldi, Nicholas. *David Hume: The Newtonian Philosopher.* Boston: G. K. Hall (Twayne Publishers), 1975.
Carugo, Adriano. "Giuseppe Moleto: Mathematics and the Aristotelian Theory of Science at Padua in the Second Half of the 16th Century." In Luigi Olivieri (ed.), *Aristotelismo veneto e scienza moderna,* vol. 1, pp. 509–517. Padua: Editrice Antenore, 1983.
———. "Les Jésuites et la philosophie naturelle de Galilée: Benedictus Pereira et le *De motu gravium* de Galilée." *History and Technology* 4 (1987): 321–333.
Carugo, Adriano, and Alistair C. Crombie. "The Jesuits and Galileo's Ideas of Science and of Nature." *Annali dell'Istituto e Museo di Storia della Scienza di Firenze* 8 (1983): fasc. 2:3–68.
Céard, Jean. *La nature et les prodiges: L'insolite au XVIe siècle, en France.* Geneva: Librairie Droz, 1977.
Chartier, Roger, Dominique Julia and Marie-Madeleine Compère. *L'éducation en France du XVIe au XVIIIe siècle.* Paris: Société d'Édition d'Enseignment Supérieur, 1976.
Chauvin, Stephanus. *Lexicon Philosophicum.* Leeuwarden, 1713; facsimile reprint Düsseldorf: Stern-Verlag Janssen, 1967.
Christianson, J. R. "Tycho Brahe's German Treatise on the Comet of 1577: A Study in Scientific Patronage." *Isis* 70 (1979): 110–140.
Clagett, Marshall. *The Science of Mechanics in the Middle Ages.* Madison: University of Wisconsin Press, 1961.
Clarke, Desmond. *Descartes' Philosophy of Science.* Manchester: Manchester University Press, 1982.
Clavius, Christopher. *Opera mathematica,* 5 vols. in 4. Mainz, 1611–1612.

Cochrane, Eric. "Science and Humanism in the Italian Renaissance." *American Historical Review* 81 (1976): 1039–1057.

Codina Mir, Gabriel. *Aux sources de la pédagogie des Jésuites: Le "Modus Parisiensis."* Rome: Institutum Historicum S.I., 1968.

Cohen, I. Bernard. *Isaac Newton's Papers and Letters on Natural Philosophy.* Cambridge, MA: Harvard University Press, 1958.

———. *The Newtonian Revolution: With Illustrations of the Transformation of Scientific Ideas.* Cambridge: Cambridge University Press, 1980.

Collingwood, R. G. *The Idea of Nature.* Oxford: Clarendon Press, 1945.

Collins, H. M. *Artificial Experts: Social Knowledge and Intelligent Machines.* Cambridge, MA: MIT Press, 1990.

———. *Changing Order: Replication and Induction in Scientific Practice,* 2d edition. Chicago: University of Chicago Press, 1991.

———. "Public Experiments and Displays of Virtuosity: The Core-Set Revisited." *Social Studies of Science* 18 (1988): 725–748.

———. "The Structure of Knowledge." *Social Research* 60 (1993): 95–116.

Commandino, Federigo. *Degli Elementi d'Euclide libri quindici.* Pesaro, 1619.

Copenhaver, Brian P. "The Historiography of Discovery in the Renaissance: The Sources and Composition of Polydore Vergil's *De inventoribus rerum.*" *Journal of the Warburg and Courtauld Institutes* 41 (1978): 192–214.

Copernicus, Nicholas. *On the Revolutions.* Trans. and commentary by Edward Rosen. Baltimore: Johns Hopkins University Press, 1992.

Cosentino, Giuseppe. "L'insegnamento delle matematiche nei collegi Gesuitici nell'Italia settentrionale: Nota introduttiva." *Physis* 13 (1971): 205–217.

———. "Le matematiche nella 'Ratio studiorum' della Compagnia di Gesu." *Miscellanea Storica Ligure* 2.2 (1970): 171–213.

Costabel, Pierre. "La loi de Boyle-Mariotte." In Costabel (ed.), *Mariotte, savant et philosophe (d. 1684): Analyse d'une renommée,* pp. 65–73. Paris: J. Vrin, 1986.

Costantini, Claudio. *Baliani e i Gesuiti: Annotazione in margine alla corrispondenza del Baliani con Gio Luigi Confalonieri e Orazio Grassi* [= Istituto Italiano per la storia della tecnica, Sezione IV, vol. 3]. Milan: Giunti, 1969.

Coyne, George V. "Bellarmino e la nuova astronomia nell'età della Controriforma." In *Bellarmino e la Controriforma: Atti del simposio internazionale di studi,* pp. 571–577. Sora: Centro di Studi Sorani "V. Patriarca," 1990.

Crombie, Alistair C. "Mathematics and Platonism in the Sixteenth-Century Italian Universities and in Jesuit Educational Policy." In Y. Maeyama and W. G. Salzer (eds.), *Prismata: Naturwissenschaftsgeschichtliche Studien (Festschrift für Willy Hartner),* pp. 63–94. Wiesbaden: Franz Steiner, 1977.

———. *Robert Grosseteste and the Origins of Experimental Science 1100–1700.* Oxford: Clarendon Press, 1953.

———. "Sources of Galileo's Early Natural Philosophy." In M. L. Righini Bonelli and W. R. Shea (eds.), *Reason, Experiment and Mysticism in the Scientific Revolution,* pp. 157–175. New York: Science History Publications, 1975.

Cunningham, Andrew. "Getting the Game Right: Some Plain Words on the

Identity and Invention of Science." *Studies in History and Philosophy of Science* 19 (1988): 365–389.
Cunningham, Andrew, and Nicholas Jardine (eds.). *Romanticism and the Sciences*. Cambridge: Cambridge University Press, 1990.
Cunningham, Andrew, and Perry Williams. "De-centring the 'Big Picture': *The Origins of Modern Science* and the Modern Origins of Science." *British Journal for the History of Science* 26 (1993): 407–432.
Dainville, François de. *L'Éducation des Jésuites XVIe–XVIIIe siècles*. Paris: Éditions de Minuit, 1978.
———. *La géographie des humanistes*. Paris: Beauchesne, 1940.
———. *La naissance de l'humanisme moderne*. Paris: Beauchesne, 1940.
Daston, Lorraine J. "Baconian Facts, Academic Civility, and the Prehistory of Objectivity." *Annals of Scholarship* 8 (1991): 337–363.
———. *Classical Probability in the Enlightenment*. Princeton: Princeton University Press, 1988.
———. "The Factual Sensibility." Essay-review of Oliver Impey and Arthur MacGregor (eds.), *The Origins of Museums: The Cabinet of Curiosities in Sixteenth and Seventeenth-Century Europe* (Oxford: Clarendon Press, 1985), and R. F. Ovenell, *The Ashmolean Museum, 1683–1894* (Oxford: Clarendon Press, 1986). *Isis* 79 (1988): 452–467.
———. "Marvelous Facts and Miraculous Evidence in Early Modern Europe." *Critical Inquiry* 18 (1991): 93–124.
Daston, Lorraine, J., and Katherine Park. "Unnatural Conceptions: The Study of Monsters in Sixteenth- and Seventeenth-Century France and England." *Past and Present*, no. 92 (1981): 20–54.
De Angelis, Enrico. *Il metodo geometrico nella filosofia del Seicento*. Pisa: Istituto di Filosofia, 1964.
De Gandt, François. "Cavalieri's Indivisibles and Euclid's Canons." In Barker and Ariew, *Revolution and Continuity*, pp. 157–182.
———. "Les *Mécaniques* attribuées à Aristote et le renouveau de la science des machines au XVIe siècle." *Les études philosophiques* 1986 (no vol. #): 391–405.
De Pace, Anna. *Le matematiche e il mondo: Ricerche su un dibattito in Italia nella seconda metà del Cinquecento*. Milan: FrancoAngeli, 1993.
De Waard, Cornélis. *L'expérience barométrique*. Thouars, 1936.
Dear, Peter. "From Truth to Disinterestedness in the Seventeenth Century." *Social Studies of Science* 22 (1992): 619–631.
———. *Mersenne and the Learning of the Schools*. Ithaca: Cornell University Press, 1988.
———. "Mersenne's Suggestion: Cartesian Meditation and the Mathematical Model of Knowledge in the Seventeenth Century." In Roger Ariew and Marjorie Grene (eds.), *Descartes and His Contemporaries: Objections and Replies*. Chicago: University of Chicago Press, 1995.
———. "Method in the Study of Nature." In Michael Ayers and Daniel Garber (eds.), *The Cambridge History of Seventeenth-Century Philosophy*. Cambridge: Cambridge University Press, forthcoming.

———. "Miracles, Experiments, and the Ordinary Course of Nature." *Isis* 81 (1990): 663–683.

———. "*Totius in verba*: Rhetoric and Authority in the Early Royal Society." *Isis* 76 (1985): 145–161.

Della Porta, J. B. *Natural Magick.* London, 1658; facsimile reprint, New York: Basic Books, 1957.

Dennis, Michael Aaron. "Graphic Understanding: Instruments and Interpretation in Robert Hooke's *Micrographia.*" *Science in Context* 3 (1989): 309–364.

Descartes, René. *Discours de la méthode,* texte et commentaire par Étienne Gilson. Paris: J. Vrin, 1930.

———. *The Geometry of René Descartes.* Ed. and trans. David Eugene Smith and Marcia L. Latham. La Salle, IL: Open Court, 1952.

———. *Œuvres de Descartes.* Eds. Charles Adam and Paul Tannery. Paris: J. Vrin, 1964–1976.

———. *The Philosophical Writings of Descartes.* Trans. and ed. John Cottingham, Robert Stoothoff, and Dugald Murdoch. Cambridge: Cambridge University Press, 1985–1991.

———. *Regles utiles et claires pour la direction de l'esprit en la recherche de la vérité.* Trans. and annotated by Jean-Luc Marion. La Haye: Martinus Nijhoff, 1977.

Dijksterhuis, E. J. *Archimedes.* Trans. C. Dikshoorn, with a new bibliographical essay by Wilbur R. Knorr. Princeton: Princeton University Press, 1987.

———. *The Mechanization of the World Picture: Pythagoras to Newton.* Trans. C. Dikshoorn. Princeton: Princeton University Press, 1986.

Drake, Stillman. *Cause, Experiment and Science.* Chicago: University of Chicago Press, 1981.

———. "Free Fall from Albert of Saxony to Honoré Fabri." *Studies in History and Philosophy of Science* 5 (1975): 347–366.

———. "Galileo and Satellite Prediction." *Journal for the History of Astronomy* 10 (1979): 75–95.

———. *Galileo at Work: His Scientific Biography.* Chicago: University of Chicago Press, 1978.

———. "Galileo's First Telescopic Observations." *Journal for the History of Astronomy* 7 (1976): 153–168.

———. *Galileo's Notes on Motion.* Florence: Istituto e Museo di Storia della Scienza, 1979.

———. "Impetus Theory and Quanta of Speed." *Physis* 16 (1974): 47–65.

———. "Ptolemy, Galileo, and Scientific Method." *Studies in History and Philosophy of Science* 9 (1978): 99–115.

Drake, Stillman, and C. D. O'Malley, eds. and trans. *The Controversy on the Comets of 1618.* Philadelphia: University of Pennsylvania Press, 1960.

Ducreux, Francis. *Le collège des Jésuites de Tulle.* [Tulle?]: Éditions des Champs, 1981.

Duhem, Pierre. *To Save the Phenomena: An Essay on the Idea of Physical Theory from Plato to Galileo.* Trans. Edmund Doland and Chaninah Maschler. Chicago: University of Chicago Press, 1969.

Eamon, William. *Science and the Secrets of Nature: Books of Secrets in Medieval and Early Modern Culture.* Princeton: Princeton University Press, 1994.

Eastwood, Bruce S. "Descartes on Refraction: Scientific versus Rhetorical Method." *Isis* 75 (1984): 481–502.

———. "Medieval Empiricism: The Case of Grosseteste's Optics." *Speculum* 43 (1968): 306–321.

———. "On the Continuity of Western Science from the Middle Ages: A. C. Crombie's *Augustine to Galileo.*" *Isis* 83 (1992): 84–99.

Edelstein, Ludwig. "Vesalius, the Humanist." *Bulletin of the History of Medicine* 14 (1943): 547–561.

Edgerton, Samuel Y., Jr. *The Heritage of Giotto's Geometry: Art and Science on the Eve of the Scientific Revolution.* Ithaca: Cornell University Press, 1991.

Edwards, William F. "Randall on the Development of Scientific Method in the School of Padua—A Continuing Reappraisal." In John P. Anton (ed.), *Naturalism and Historical Understanding: Essays on the Philosophy of John Herman Randall, Jr.*, pp. 53–68. Albany: State University of New York Press, 1967.

Elias, Norbert. *The Civilizing Process: The History of Manners and State Formation and Civilization.* Translated by Edmund Jephcott. Oxford, U.K., and Cambridge, MA: Blackwell, 1994.

Ernst, Germana. "Astrology, Religion and Politics in Counter-Reformation Rome." In Pumfrey et al. (eds.), *Science, Culture and Popular Belief*, 249–273.

Eschinardus, Francescus. *Microcosmi physicomathematici, seu compendii . . . tomus primus.* Perugia, 1658.

Euclid. *Elementorum Euclidis libri tredecim.* London, 1620.

———. *Opus elementorum Euclidis Megarensis in geometriam artem.* Ed. Campanus of Novara. Venice, 1482.

Euclid. *The Thirteen Books of Euclid's Elements*, 2 vols. Ed. and trans. Thomas L. Heath. New York: Dover, 1956.

Fabri, Honoré. *Controversiae logicae.* Lyon, [1646?]. (Bound with separate pagination between Fabri, *Philosophiae tomus,* and idem, *Tractatus physicus de motu locali.*)

———. *Dialogi physici.* Lyon, 1669.

———. *Philosophiae tomus primus: qui complectitur scientiarum methodum sex libris explicatam: Logicam analyticam, duodecim libris demonstratam & aliquot controversias logicas, breviter disputatas. Auctore Petro Mosnerio Doctore Medico. Cuncta excerpta ex praelectionibus R. P. Hon. Fabry. Soc. Iesu.* Lyon, 1646.

———. *Tractatus physicus de motu locali, in quo effectus omnes, qui ad impetum, motum naturalem, violentum, & mixtum pertinent, explicantur, & ex principiis Physicis demonstrantur.* Lyon, 1646.

Fanton d'Andon, Jean-Pierre. *L'horreur du vide: Expérience et raison dans la physique pascalienne.* Paris: Éditions du CNRS, 1978.

Feingold, Mordechai. *The Mathematician's Apprenticeship: Science, Universities and Society in England, 1560–1640.* Cambridge: Cambridge University Press, 1984.

———. "Newton, Leibniz, and Barrow Too." *Isis* 84 (1993): 310–338.

Feldhay, Rivka. "Knowledge and Salvation in Jesuit Culture." *Science in Context* 1 (1987): 195–213.
Feldhay, Rivka, and Michael Heyd. "The Discourse of Pious Science." *Science in Context* 3 (1989): 109–142.
Feyerabend, Paul K. "Classical Empiricism." In Robert E. Butts and John W. Davis (eds.), *The Methodological Heritage of Newton*, pp. 150–170. Toronto: University of Toronto Press, 1970.
———. "Explanation, Reduction, and Empiricism." *Minnesota Studies in the Philosophy of Science* 3 (1962): 28–97.
Field, J. V., and Frank A. J. L. James, eds. *Renaissance and Revolution: Humanists, Scholars, Craftsmen and Natural Philosophers in Early Modern Europe*. Cambridge: Cambridge University Press, 1993.
Findlen, Paula. "Jokes of Nature and Jokes of Knowledge: The Playfulness of Scientific Discourse in Early Modern Europe." *Renaissance Quarterly* 43 (1990): 292–331.
Fischer, Karl Adolf Franz. "Die Jesuiten-mathematiker des nordostdeutschen Kulturgebietes." *Archives internationales d'histoire des sciences*, no. 112 (1984): 124–162.
———. "Jesuiten Mathematiker in der deutschen Assistenz bis 1773." *Archivum Historicum Societatis Iesu* 47 (1978): 159–224.
———. "Jesuiten Mathematiker in der französischen und italienischen Assistenz bis 1762, bzw. 1773." *Archivum Historicum Societatis Iesu* 52 (1983): 52–92.
Foucault, Michel. "The Discourse on Language." In Foucault, *The Archaeology of Knowledge and the Discourse on Language*, trans. A. M. Sheridan Smith, pp. 215–237. New York: Pantheon Books, 1972.
Frängsmyr, Tore, J. L. Heilbron, and Robin E. Rider, eds. *The Quantifying Spirit in the 18th Century*. Berkeley: University of California Press, 1990.
Funkenstein, Amos. *Theology and the Scientific Imagination from the Middle Ages to the Seventeenth Century*. Princeton: Princeton University Press, 1986.
Gabbey, Alan. "Between *ars* and *philosophia naturalis*: Reflections on the Historiography of Early Modern Mechanics." In Field and James, *Renaissance and Revolution*, pp. 133–145.
———. "Cudworth, More, and the Mechanical Analogy." In Richard Kroll, Richard Ashcraft and Perez Zagorin (eds.), *Philosophy, Science, and Religion in England 1640–1700*, pp. 109–127. Cambridge: Cambridge University Press, 1992.
———. "Henry More and the Limits of Mechanism." In Sarah Hutton (ed.), *Henry More (1614–1687): Tercentenary Studies*, pp. 19–35. Dordrecht: Kluwer, 1990.
———. "Mariotte et Roberval, son collaborateur involontaire." In Pierre Costabel (ed.), *Mariotte, savant et philosophe (d. 1684): Analyse d'une renommée*, pp. 205–244. Paris: J. Vrin, 1986.
———. "The Mechanical Philosophy and Its Problems: Mechanical Explanations, Impenetrability, and Perpetual Motion." In Joseph C. Pitt (ed.), *Change and Progress in Modern Science*, pp. 9–84. Dordrecht: Reidel, 1985.

———. "The *Principia*: A Treatise on 'Mechanics'?" In P. M. Harman and Alan E. Shapiro (eds.), *The Investigation of Difficult Things: Essays on Newton and the History of the Exact Sciences in Honour of D. T. Whiteside*, pp. 305–322. Cambridge: Cambridge University Press, 1992.

Gagné, Jean. "Du *quadrivium* aux *scientiae mediae*." In *Arts libéraux et philosophie au moyen âge* [= *Actes du quatrième congrès international de philosophie mediévale*], pp. 975–986. Montreal: Institut d'Études Médiévales, 1969.

Galen. *Three Treatises on the Nature of Science*. Trans. R. Walzer and M. Frede. Indianapolis: Hackett, 1985.

Galileo Galilei. *Dialogue Concerning the Two Chief World Systems—Ptolemaic & Copernican*, 2d edition. Trans. Stillman Drake. Berkeley: University of California Press, 1967.

———. *Discourses and Demonstrations Concerning Two New Sciences*. Trans. Stillman Drake. Madison: University of Wisconsin Press, 1974.

———. *Le opere di Galileo Galilei: Edizione Nazionale*, 20 vols. Ed. Antonio Favaro. Florence: G. Barbèra, 1890–1909.

———. *Sidereus nuncius, or The Sidereal Messenger*. Ed. and trans. Albert Van Helden. Chicago: University of Chicago Press, 1989.

Galison, Peter, and Alexi Assmus. "Artificial Clouds, Real Particles." In Gooding, Pinch, and Schaffer, *The Uses of Experiment*, pp. 225–274.

Galluzzi, Paolo. "Il 'Platonismo' del tardo Cinquecento e la filosofia di Galileo." In Paola Zambelli (ed.), *Ricerche sulla cultura dell'Italia moderna*, pp. 37–79. Bari: Editore Laterza, 1973.

Garber, Daniel. "Descartes and Experiment in the *Discourse* and the *Essays*." In Stephen Voss (ed.), *Essays on the Philosophy and Science of René Descartes*, pp. 288–310. Oxford: Oxford University Press, 1993.

———. "Descartes et la méthode en 1637." In Nicolas Grimaldi and Jean-Luc Marion (eds.), *Le Discours et sa méthode: Colloque pour le 350e anniversaire du "Discours de la Méthode,"* pp. 65–87. Paris: Presses Universitaires de France, 1987.

———. *Descartes' Metaphysical Physics*. Chicago: University of Chicago Press, 1992.

Garrison, James W. "Newton and the Relation of Mathematics to Natural Philosophy." *Journal of the History of Ideas* 48 (1987): 609–627.

Gascoigne, John. "A Reappraisal of the Role of the Universities in the Scientific Revolution." In Lindberg and Westman, *Reappraisals of the Scientific Revolution*, pp. 207–260.

Gaukroger, Stephen. "Bachelard and the Problem of Epistemological Analysis." *Studies in History and Philosophy of Science* 7 (1976): 189–244.

———. "Descartes' Project for a Mathematical Physics." In Gaukroger (ed.), *Descartes: Philosophy, Mathematics and Physics*, pp. 97–140. Brighton, Sussex: Harvester, 1980.

———. *Explanatory Structures: A Study of Concepts of Explanation in Early Physics and Philosophy*. Atlantic Highlands, NJ: Humanities Press, 1978.

———. "The Nature of Abstract Reasoning: Philosophical Aspects of Descartes' Work in Algebra." In John Cottingham (ed.), *The Cambridge Companion to Descartes*, pp. 91–114. Cambridge: Cambridge University Press, 1992.

Giacobbe, Giulio Cesare. *Alle radici della rivoluzione scientifica rinascimentale: Le opere di Pietro Catena sui rapporti tra matematica e logica.* Pisa: Domus Galilaeana, 1981.

———. "Epigone nel Seicento della 'Quaestio de certitudine mathematicarum': Giuseppe Biancani." *Physis* 18 (1976): 5–40.

Gilbert, Neal W. *Renaissance Concepts of Method.* New York: Columbia University Press, 1961.

Gilbert, William. *De magnete.* London, 1600.

———. *On the Loadstone.* Trans. P. Fleury Mottelay. Chicago: Encyclopaedia Britannica Great Books, 1952.

Gillispie, Charles C. "The *Encyclopédie* and the Jacobin Philosophy of Science: A Study in Ideas and Consequences." In Marshall Clagett (ed.), *Critical Problems in the History of Science*, pp. 255–289. Madison: University of Wisconsin Press, 1959.

Gilson, Étienne. *Études sur le rôle de la pensée médiévale dans la formation du système cartésien* (1921). Paris: J. Vrin, 1951.

———. *Index scolastico-cartésien.* Paris, 1913.

Goclenius, Rodolphus. *Lexicon philosophicum.* Frankfurt, 1613; facsimile reprint Hildesheim: Georg Olms, 1964.

Golinski, Jan. "The Theory of Practice and the Practice of Theory: Sociological Approaches in the History of Science." *Isis* 81 (1990): 492–505.

Gooding, David, Trevor Pinch, and Simon Schaffer, eds. *The Uses of Experiment: Studies in the Natural Sciences.* Cambridge: Cambridge University Press, 1989.

Grafton, Anthony. *Defenders of the Text.* Cambridge, MA: Harvard University Press, 1991.

———. "Humanism, Magic and Science." In Anthony Goodman and Angus MacKay (eds.), *The Impact of Humanism on Western Europe*, pp. 99–117. London and New York: Longman, 1990.

———. *New Worlds, Ancient Texts: The Power of Tradition and the Shock of Discovery.* Cambridge, MA: Harvard University Press, 1992.

Grant, Edward. "Aristotelianism and the Longevity of the Medieval World View." *History of Science* 16 (1978): 93–106.

———. "In Defense of the Earth's Centrality and Immobility: Scholastic Reaction to Copernicanism in the Seventeenth Century." *Transactions of the American Philosophical Society*, n.s., 74 (1984), pt. 4.

——— (ed.). *A Source Book in Medieval Science.* Cambridge, MA: Harvard University Press, 1974.

Grassi, Orazio [Lothario Sarsi, pseud.]. *Libra astronomica ac philosophica* (1619). In Galileo, *Opere*, vol. 6, pp. 109–179.

———. *Ratio ponderum librae et simbellae* (1626). In Galileo, *Opere*, vol. 6, pp. 373–500.

Gray, Hanna H. "Renaissance Humanism: The Pursuit of Eloquence." *Journal of the History of Ideas* 24 (1963): 497–514. Reprinted in P. O. Kristeller and P. Wiener (eds.), *Renaissance Essays*. New York: Harper Torchbooks, 1968.

Grene, Marjorie. *A Portrait of Aristotle*. Chicago: University of Chicago Press, 1963.

Grimaldi, Francesco Maria. *Physico-mathesis de lumine, coloribus, et iride*. Bologna, 1665.

Gross, Alan G. *The Rhetoric of Science*. Cambridge, MA: Harvard University Press, 1990.

Guerlac, Henry. "Newton and the Method of Analysis." In P. P. Wiener (ed.), *Dictionary of the History of Ideas*, vol. 3, pp. 378–391. New York: Charles Scribner's Sons, 1973.

———. *Newton on the Continent*. Ithaca, NY: Cornell University Press, 1981.

———. "'Newton's Mathematical Way': Another Look." *British Journal for the History of Science* 17 (1984): 61–63.

Guiffart, Pierre. *Discours du vuide*. Rouen, 1647.

Hacking, Ian. *The Emergence of Probability: A Philosophical Study of Early Ideas about Probability, Induction and Statistical Inference*. Cambridge: Cambridge University Press, 1975.

Hall, A. Rupert. *All Was Light: An Introduction to Newton's "Opticks."* Oxford: Clarendon Press, 1993.

———. *Ballistics in the Seventeenth Century: A Study in the Relations of Science and War with Reference Principally to England*. Cambridge: Cambridge University Press, 1952.

———. "Beyond the Fringe: Diffraction as seen by Grimaldi, Fabri, Hooke and Newton." *Notes and Records of the Royal Society* 44 (1990): 13–23.

Hall, A. Rupert, and Marie Boas Hall. "Philosophy and Natural Philosophy: Boyle and Spinoza." In *Mélanges Alexandre Koyré* [= *Histoire de la pensée*, nos. 12, 13], vol. 2, pp. 241–256. Paris: École Pratique des Hautes Études, 1964.

Hall, Marie Boas. *Nature and Nature's Laws: Documents of the Scientific Revolution*. New York: Walker and Co., 1970.

———. *Promoting Experimental Learning: Experiment and the Royal Society, 1660–1727*. Cambridge: Cambridge University Press, 1991.

———. *Robert Boyle on Natural Philosophy*. Bloomington: Indiana University Press, 1966.

Hanafi, Zakiya Asha Jenan. "Matter, Machines, and Metaphor: Monstrosity in the *Seicento*." Ph.D. diss., Stanford University, 1991.

Hankins, Thomas L. "The Ocular Harpsichord of Louis-Bertrand Castel; or, The Instrument that Wasn't." *Osiris* n.s. 9 (1994): 141–156.

Hannaway, Owen. "Laboratory Design and the Aims of Science: Andreas Libavius versus Tycho Brahe." *Isis* 77 (1986): 585–610.

Hanson, Norwood Russell. *Patterns of Discovery*. Cambridge: Cambridge University Press, 1958.

Harrington, Thomas More. *Pascal philosophe: Une étude unitaire de la pensée de Pascal*. Paris: Société d'Édition d'Enseignement Supérieur, 1982.

Harris, Steven J. "Transposing the Merton Thesis: Apostolic Spirituality and the Establishment of the Jesuit Scientific Tradition." *Science in Context* 3 (1989): 29–65.

Harwood, John T. "Rhetoric and Graphics in *Micrographia*." In Hunter and Schaffer (eds.), *Robert Hooke*, pp. 119–147.

Heath, T. L. (ed.). *The Works of Archimedes*. New York: Dover, 1953.

Heilbron, John L. *Electricity in the 17th and 18th Centuries: A Study in Early Modern Physics*. Berkeley: University of California Press, 1979.

———. *Physics at the Royal Society during Newton's Presidency*. Los Angeles: William Andrews Clark Memorial Library, 1983.

Hemmendinger, David. "Galileo and the Phenomena: On Making the Evidence Visible." In R. S. Cohen and M. W. Wartofsky (eds.), *Physical Sciences and History of Physics*, pp. 115–143. Dordrecht: D. Reidel, 1984.

Henry, John. "Occult Qualities and the Experimental Philosophy: Active Principles in Pre-Newtonian Matter Theory." *History of Science* 24 (1986): 335–381.

———. "The Scientific Revolution in England." In Porter and Teich (eds.), *The Scientific Revolution in National Context*, pp. 178–209.

Hervey, Helen. "Hobbes and Descartes in the Light of Some Unpublished Letters of the Correspondence between Sir Charles Cavendish and Dr. John Pell." *Osiris* 10 (1952): 67–90.

Hesse, Mary. *The Structure of Scientific Inference*. Berkeley: University of California Press, 1974.

Hesse, Mary, and Michael A. Arbib. *The Construction of Reality*. Cambridge: Cambridge University Press, 1986.

Hetherington, Norriss S. *Science and Objectivity*. Ames: Iowa State University Press, 1988.

Hintikka, Jaako, and Unto Remes. *The Method of Analysis: Its Geometrical Origin and its General Significance* [= Boston Studies in the Philosophy of Science, vol. 25]. Dordrecht: D. Reidel, 1974.

Hobbes, Thomas. *English Works*, 11 vols. Ed. Sir William Molesworth. London, 1839–1845.

Hobsbawm, Eric, and Terence Ranger. *The Invention of Tradition*. Cambridge: Cambridge University Press, 1983.

Homann, Frederick A. "Christophorus Clavius and the Renaissance of Euclidean Geometry." *Archivum Historicum Societatis Iesu* 52 (1983): 233–246.

Hooke, Robert. *Micrographia or Some Physiological Descriptions of Minute Bodies Made by Magnifying Glasses with Observations and Inquiries Thereupon*. London, 1665; facsimile reprint New York: Dover, 1962.

Hooykaas, Rejner. "The Discrimination between 'Natural' and 'Artificial' Substances and the Development of Corpuscular Theory." *Archives internationales d'histoire des sciences* 1 (1948): 640–651.

———. "Isaac Beeckman." In Charles C. Gillispie (ed.), *Dictionary of Scientific Biography*, vol. 1, pp. 566–568. New York: Scribner's, 1970.

Humbert, Pierre. *Cet effrayant génie . . . L'œuvre scientifique de Blaise Pascal*. Paris: Éditions Albin Michel, 1947.

Hume, David. *Enquiries Concerning Human Understanding and Concerning the Principles of Morals.* Ed. L. A. Selby-Bigge, 3d edition with text revised and notes by P. H. Nidditch. Oxford: Clarendon Press, 1975.

Hunter, Michael, and Simon Schaffer (eds.). *Robert Hooke: New Studies.* Woodbridge, Suffolk: The Boydell Press, 1989.

Hutchison, Keith. "What Happened to Occult Qualities in the Scientific Revolution?" *Isis* 73 (1982): 233–253.

Iliffe, Robert. "'In the Warehouse': Privacy, Property and Priority in the Early Royal Society." *History of Science* 30 (1992): 29–68.

Jacob, James R. *Henry Stubbe, Radical Protestantism and the Early Enlightenment.* Cambridge: Cambridge University Press, 1983.

Jardine, Lisa. *Francis Bacon: Discovery and the Art of Discourse.* Cambridge: Cambridge University Press, 1974.

Jardine, Nicholas. *The Birth of History and Philosophy of Science: Kepler's "A Defence of Tycho against Ursus" with Essays on Its Provenance and Significance.* Cambridge: Cambridge University Press, 1984.

———. "Epistemology of the Sciences." In Charles B. Schmitt et al. (eds.), *The Cambridge History of Renaissance Philosophy,* pp. 685–711. Cambridge: Cambridge University Press, 1988.

———. "The Forging of Modern Realism: Clavius and Kepler against the Sceptics." *Studies in History and Philosophy of Science* 10 (1979): 141–173.

———. "Galileo's Road to Truth and the Demonstrative Regress." *Studies in History and Philosophy of Science* 7 (1976): 277–318.

Joy, Lynn Sumida. *Gassendi the Atomist: Advocate of History in an Age of Science.* Cambridge: Cambridge University Press, 1987.

———. "Humanism and the Problem of Traditions in Seventeenth-Century Natural Philosophy." In Patricia Cook (ed.), *Philosophical Imagination and Cultural Memory: Appropriating Historical Traditions,* pp. 139–148. Durham, NC: Duke University Press, 1993.

Jung, Joachim. *Logica Hamburgensis.* Original version Hamburg, 1638; modern edition, from the 1681 version, by Rudolf W. Meyer. Hamburg: J. J. Augustin, 1957.

Kargon, Robert. *Atomism in England from Hariot to Newton.* Oxford: Clarendon Press, 1966.

———. "Newton, Barrow and the Hypothetical Physics." *Centaurus* 11 (1965): 46–56.

Kemp, Martin. *The Science of Art: Optical Themes in Western Art from Brunelleschi to Seurat.* New Haven and London: Yale University Press, 1990.

Kepler, Johannes. *Dissertatio cum nuncio sidereo.* Reprinted in Galileo, *Opere,* vol. 3, pt. 1, pp. 96–126.

Kircher, Athanasius. *Ars magnesia, hoc est, disquisitio bipartita-empeirica seu experimentalis, physico-mathematica de natura, viribus, et prodigiosis effectibus magnetis.* Würzburg, 1631.

———. *Musurgia universalis.* Rome, 1650; facsimile reprint Hildesheim and New York: Georg Olms, 1970.

Knobloch, Eberhard. "Christoph Clavius—Ein Astronom zwischen Antike und

Kopernikus." In Klaus Döring and Georg Wöhrle (eds.), *Vorträge des ersten Symposions des Bamberger Arbeitskreises "Antike Naturwissenschaft und ihre Rezeption" (AKAN)*, pp. 113–140. Wiesbaden: Otto Harrassowitz, 1990.

———. "Christoph Clavius: Ein Namen- und Schriftenverzeichnis zu seinen *Opera mathematica*." *Bolletino di storia delle scienze matematiche* 10 (1990): 135–189.

———. "Sur la vie et l'œuvre de Christophore Clavius." *Revue d'histoire des sciences* 41 (1988): 331–356.

Koutná-Karg, Dana. "Experientia fuit, mathematicum paucos discipulos habere . . . : Zu den Naturwissenschaften und der Mathematik an der Universität Dillingen zwischen 1563 und 1632." In Martin Kintzinger, Wolfgang Stürner, and Johannes Zahlten (eds.), *Das andere Wahrnehmen: Beiträge zur europäischen Geschichte. August Nitschke zum 65. Geburtstag gewidmet*, pp. 451–466. Cologne and Weimar: Böhlau Verlag, 1991.

Koyré, Alexandre. "A Documentary History of the Problem of Fall from Kepler to Newton: De motu gravium naturaliter cadentium in hypothesi terrae motae." *Transactions of the American Philosophical Society*, n.s., 45 (1955), pt. 4.

———. "An Experiment in Measurement." In Koyré, *Metaphysics and Measurement*, pp. 89–117.

———. "Galileo and Plato." In Koyré, *Metaphysics and Measurement*, pp. 16–43.

———. *Galileo Studies*, trans. John Mepham of *Études galiléennes*, 1939. Brighton, Sussex: Harvester Press, 1978.

———. "Galileo's Treatise *De motu gravium*: The Use and Abuse of Imaginary Experiment." In Koyré, *Metaphysics and Measurement*, pp. 44–88.

———. "Huygens and Leibniz on Universal Attraction." In Koyré, *Newtonian Studies*, pp. 115–138.

———. *Metaphysics and Measurement: Essays in Scientific Revolution*. Cambridge, MA: Harvard University Press, 1968.

———. *Newtonian Studies*. Chicago: University of Chicago Press, 1965.

———. "Newton's 'Regulae philosophandi'." In Koyré, *Newtonian Studies*, pp. 261–272.

———. "Pascal savant." In Koyré, *Metaphysics and Measurement*, pp. 131–156.

Krayer, Albert. *Mathematik im Studienplan der Jesuiten: Die Vorlesung von Otto Cattenius an der Universität Mainz (1610–1611)*. Stuttgart: Franz Steiner, 1991.

Kristeller, Paul Oskar. "Humanism and Scholasticism in the Italian Renaissance." In Kristeller, *Renaissance Thought*, pp. 92–119.

———. "The Humanist Movement." In Kristeller, *Renaissance Thought*, pp. 3–23.

———. *Renaissance Thought: The Classic, Scholastic, and Humanist Strains*. New York: Harper Torchbooks, 1961.

Kuhn, Thomas S. "Alexandre Koyré and the History of Science: On an Intellectual Revolution." *Encounter* (Jan. 1970): 67–69.

———. "The Essential Tension: Tradition and Innovation in Scientific Research." In Kuhn, *The Essential Tension: Selected Studies in Scientific Tradition and Change*, pp. 225–239. Chicago: University of Chicago Press, 1977.

———. "Mathematical versus Experimental Traditions in the Development of

Physical Science." *Journal of Interdisciplinary History* 7 (1976): 1–21. Reprinted in Kuhn, *The Essential Tension*, pp. 31–65.

———. "Newton's Optical Papers." In I. Bernard Cohen (ed.), *Isaac Newton's Papers and Letters on Natural Philosophy*, pp. 27–45. Cambridge, MA: Harvard University Press, 1958.

———. *The Structure of Scientific Revolutions*, 2d edition. Chicago: University of Chicago Press, 1970.

Laird, W. R. "Archimedes among the Humanists." *Isis* 82 (1991): 629–638.

———. "The *Scientiae mediae* in Medieval Commentaries on Aristotle's *Posterior Analytics*." Ph.D. diss., University of Toronto, 1983.

———. "The Scope of Renaissance Mechanics." *Osiris*, n.s., 2 (1986): 43–68.

Lakatos, Imre, and Elie Zahar. "Why did Copernicus's Research Programme Supersede Ptolemy's?" In Lakatos, *The Methodology of Scientific Research Programmes (Philosophical Papers vol. 1)*, ed. John Worrall and Gregory Currie, pp. 168–192. Cambridge: Cambridge University Press, 1978.

Lakoff, George, and Mark Johnson, *Metaphors We Live By*. Chicago: University of Chicago Press, 1980.

Latour, Bruno. "Postmodern? No, Simply Amodern: Steps Towards an Anthropology of Science." Essay-review of Shapin and Schaffer, *Leviathan and the Air-Pump*, Michel Serres, *Statues* (Paris: Bourin, 1987), and Sharon Traweek, *Beam Times and Life Times: The World of High Energy Physicists* (Cambridge, MA: Harvard University Press, 1988). *Studies in History and Philosophy of Science* 21 (1990): 145–171.

———. *Science in Action: How to Follow Scientists and Engineers Through Society*. Cambridge, MA: Harvard University Press, 1987.

———. *We Have Never Been Modern*. Trans. Catherine Porter. Cambridge, MA: Harvard University Press, 1993.

Latour, Bruno, and Steve Woolgar. *Laboratory Life: The [Social] Construction of Scientific Facts*. Beverley Hills and London: Sage, 1979; 2d edition, Princeton: Princeton University Press, 1986.

Lattis, James M. *Between Copernicus and Galileo: Christoph Clavius and the Collapse of Ptolemaic Cosmology*. Chicago: University of Chicago Press, 1994.

———. "Christoph Clavius and the *Sphere* of Sacrobosco: The Roots of Jesuit Astronomy on the Eve of the Copernican Revolution." Ph.D. diss., University of Wisconsin at Madison, 1989.

Lejeune, Albert. *L'Optique de Claude Ptolémée, dans la version latine d'après l'arabe de l'emir Eugène de Sicile*. Leiden: E. J. Brill, 1989.

———. "Recherches sur la catoptrique grecque d'après les sources antiques et médiévales." Académie Royale de Belgique *Mémoires*, Classe des lettres, 2d series, vol. 52 (1957–1958), fasc. 2.

Lennox, James G. "Aristotle, Galileo, and 'Mixed Sciences'." In William A. Wallace (ed.), *Reinterpreting Galileo*, pp. 29–51. Washington, DC: Catholic University of America Press, 1986.

Lenoble, Robert. *Esquisse d'une histoire de l'idée de nature*. Paris: Albin Michel, 1969.

———. *Mersenne ou la naissance du mécanisme*. Paris: J. Vrin, 1943, 1971.

Lenoir, Timothy. "Descartes and the Geometrization of Thought: The Method-

ological Foundation of Descartes' *Géométrie.*" *Historia Mathematica* 6 (1979): 355–379.
Lindberg, David C. *Theories of Vision from Al-Kindi to Kepler.* Chicago: University of Chicago Press, 1976.
Lindberg, David C., and Robert S. Westman, eds. *Reappraisals of the Scientific Revolution.* Cambridge: Cambridge University Press, 1990.
Linemannus, Albertus. *Disputatio ordinaria continens controversias physicomathematicas quam publicae ventilatione subjicit M. Albertus Linemannus mathem: in Acad: Regiom: Professor Publicus.* Königsberg, 1636.
Lloyd, G. E. R. *Magic, Reason and Experience: Studies in the Origin and Development of Greek Science.* Cambridge: Cambridge University Press, 1979.
Locke, David. *Science as Writing.* New Haven and London: Yale University Press, 1992.
Locke, John. *Essay Concerning Human Understanding.* Ed. Peter H. Nidditch. Oxford: Clarendon Press, 1975.
Loeck, Gisela. *Der cartesische Materialismus: Maschine, Gesetz und Simulation.* Frankfurt am Main, Bern, and New York: Peter Lang, 1986.
Lohr, Charles H. "Renaissance Latin Aristotle Commentaries: Authors A–B." *Studies in the Renaissance* 21 (1974): 228–289.
Long, Pamela O. "Humanism and Science." In Albert Rabil Jr. (ed.), *Humanism and the Disciplines,* vol. 3 of *Renaissance Humanism: Foundations, Forms, and Legacy,* pp. 486–512. Philadelphia: University of Pennsylvania Press, 1988.
Lukens, David Clough. "An Aristotelian Response to Galileo: Honoré Fabri, S.J. (1608–1688) on the Causal Analysis of Motion." Ph.D. diss., University of Toronto, 1979.
Lux, David S. *Patronage and Royal Science in Seventeenth-Century France: The Académie de physique in Caen.* Ithaca: Cornell University Press, 1989.
Lynch, Michael. *Scientific Practice and Ordinary Action: Ethnomethodology and Social Studies of Science.* Cambridge: Cambridge University Press, 1993.
Machamer, Peter. "Galileo and the Causes." In Robert E. Butts and Joseph C. Pitt (eds.), *New Perspectives on Galileo,* pp. 161–180. Dordrecht: D. Reidel, 1978.
MacLachlan, James. "A Test of an 'Imaginary' Experiment of Galileo's." *Isis* 64 (1973): 374–379.
Magnus, Valerianus. *Demonstratio ocularis. Loci sine locato: Corporis successivè moti in vacuo: Luminis nulli corpori inhaerentis. A Valeriano Magno Fratre Capuccino, exhibita.* Part of Petit, *Observation.*
Maignan, Emmanuel. *Cursus philosophicus.* Toulouse, 1653.
Makdisi, George. "The Scholastic Method in Medieval Education: An Inquiry into its Origins in Law and Theology." *Speculum* 49 (1974): 640–661.
Malet, Antoni. "Isaac Barrow on the Mathematization of Nature: Theological Voluntarism and the Rise of Geometrical Optics." Forthcoming.
Mali, Joseph. *The Rehabilitation of Myth: Vico's "New Science."* Cambridge: Cambridge University Press, 1992.
———. "Science, Tradition, and the Science of Tradition." *Science in Context* 3 (1989): 143–173.
Mancosu, Paolo. "Aristotelean Logic and Euclidean Mathematics: Seventeenth-

Century Developments of the *Quaestio de certitudine mathematicarum.*" *Studies in History and Philosophy of Science* 23 (1992): 241–265.

Marenbon, John. *Later Medieval Philosophy (1150–1350): An Introduction.* London: Routledge and Kegan Paul, 1987.

Mariotte, Edme. "Discours de la nature de l'air" (1676). In *Œuvres de Mr. Mariotte,* pp. 149–182. Leyden, 1717.

Martin, Julian. *Francis Bacon, the State, and the Reform of Natural Philosophy.* Cambridge: Cambridge University Press, 1992.

Martins, Roberto de A. "Huygens's Reaction to Newton's Gravitational Theory." In J. V. Field and Frank A. J. L. James (eds.), *Renaissance and Revolution: Humanists, Scholars, Craftsmen and Natural Philosophers in Early Modern Europe,* pp. 203–213. Cambridge: Cambridge University Press, 1993.

Mastrius, Bartholomaeus, and Bonaventura Bellutus. *Disputationes in libros de celo et metheoris.* Venice, 1640.

Mayr, Otto. *Authority, Liberty and Automatic Machinery in Early Modern Europe.* Baltimore and London: Johns Hopkins University Press, 1986.

McEvoy, James. *The Philosophy of Robert Grosseteste.* Oxford: Clarendon Press, 1983.

McGuire, J. E., and Martin Tamny. *Certain Philosophical Questions: Newton's Trinity Notebook.* Cambridge: Cambridge University Press, 1983.

McKirahan, Richard D. "Aristotle's Subordinate Sciences." *British Journal for the History of Science* 11 (1978): 197–220.

———. *Principles and Proofs: Aristotle's Theory of Demonstrative Science.* Princeton: Princeton University Press, 1992.

McKirahan, Richard D., Jr. "Aristotelian Epagoge in *Prior Analytics* 2.21 and *Posterior Analytics* 1.1." *Journal of the History of Philosophy* 21 (1983): 1–13.

McMullin, Ernan. "The Conception of Science in Galileo's Work." In Butts and Pitt, *New Perspectives on Galileo,* pp. 209–257.

———. "Conceptions of Science in the Scientific Revolution." In Lindberg and Westman, *Reappraisals of the Scientific Revolution,* pp. 27–92.

Meinel, Christoph. "Early Seventeenth-Century Atomism: Theory, Epistemology, and the Insufficiency of Experiment." *Isis* 79 (1988): 68–103.

Mersenne, Marin. *Cogitata physico-mathematica.* Paris, 1644.

———. *Correspondance du P. Marin Mersenne, religieux minime,* 17 vols. Ed. C. de Waard et al. Paris: Beauchesne (vol. 1); Presses Universitaires de France (vols. 2–4); Centre National de la Recherche Scientifique (vols. 5–17), 1932–1988.

———. *Harmonie universelle.* Paris, 1636–1637; facsimile reprint Paris: CNRS, 1963.

———. *Novarum observationum . . . tomus III.* Paris, 1647.

———. *Les preludes de l'harmonie universelle.* Paris, 1634.

———. *Traité des mouvemens.* Paris, 1634; facsimile reprint in *Corpus: Revue de philosophie,* no. 2 (1986): 25–58.

Micraelius, Johannes. *Lexicon philosophicum,* 2d edition. Stetini, 1662; facsimile reprint Düsseldorf: Stern-Verlag Janssen, 1966.

Middleton, W. E. Knowles. *The History of the Barometer.* Baltimore: Johns Hopkins Press, 1964.

———. "Science in Rome, 1675–1700, and the Accademia Fisicomatematica of Giovanni Giustino Ciampini." *British Journal for the History of Science* 8 (1975): 138–154.

Millen, Ron. "The Manifestation of Occult Qualities in the Scientific Revolution." In Margaret J. Osler and Paul L. Farber (eds.), *Religion, Science and Worldview: Essays in Honor of Richard S. Westfall*, pp. 185–216. Cambridge: Cambridge University Press, 1985.

Milton, J. R. "Induction before Hume." *British Journal for the Philosophy of Science* 38 (1987): 49–74.

———. "Laws of Nature." In Michael Ayers and Daniel Garber (eds.), *The Cambridge History of Seventeenth-Century Philosophy*. Cambridge: Cambridge University Press, forthcoming.

———. "The Origin and Development of the Concept of the 'Laws of Nature'." *Archives européennes de sociologie* 22 (1981): 173–195.

Molland, A. G. "Implicit versus Explicit Geometrical Methodologies: The Case of Construction." In Roshdi Rashed (ed.), *Mathématiques et philosophie de l'antiquité à l'âge classique: Hommage à Jules Vuillemin*, pp. 181–196. Paris: CNRS, 1991.

———. "Shifting the Foundations: Descartes's Transformation of Ancient Geometry." *Historia Mathematica* 3 (1976): 21–49.

Montaigne, Michel de. *Les essais de Montaigne*. Ed. Pierre Villey. Paris: Nizet, 1972.

Monumenta Paedagogica Societatis Jesu quae Primam Rationem Studiorum anno 1586 praecessere. Madrid, 1901.

Morse, JoAnn S. "The Reception of Diophantus' *Arithmetic* in the Renaissance." Ph.D. diss., Princeton University, 1981.

Moscovici, Serge. *L'expérience du mouvement: Jean-Baptiste Baliani disciple et critique de Galilée*. Paris: Hermann, 1967.

Moss, Jean Dietz. "Newton and the Jesuits in the *Philosophical Transactions*." In G. V. Coyne, M. Heller and J. Zycinski (eds.), *Newton and the New Direction in Science: Proceedings of the Cracow Conference 25 to 28 May 1987*, pp. 117–134. Vatican City: Vatican Observatory, 1988.

———. *Novelties in the Heavens: Rhetoric and Science in the Copernican Controversy*. Chicago: University of Chicago Press, 1993.

Motley, Mark. *Becoming a French Aristocrat: The Education of the Court Nobility 1580–1715*. Princeton: Princeton University Press, 1990.

Murdoch, John E. "The Analytic Character of Late Medieval Learning: Natural Philosophy without Nature." In Lawrence D. Roberts (ed.), *Approaches to Nature in the Middle Ages*, pp. 171–213. Binghamton, NY: Center for Medieval and Early Renaissance Studies, 1982.

Mydorge, Claude. *Examen du livre des recreations mathematiques: et de ses problemes en geometrie, mechanique, optique, & catoptrique*. Paris, 1630.

———. *La seconde partie des recreations mathematiques. Composée de plusieurs problemes plaisans & facetieux*. Paris, 1630.

Nagel, Ernest. *The Structure of Science*. New York: Harcourt, Brace, 1961.

Naux, Charles. "Le père Christophore Clavius (1573–1612), sa vie et son œuvre." *Revue des questions scientifiques* 154 (1983): 55–67, 181–193, 325–347.

Naylor, R. H. "Galileo's Experimental Discourse." In David Gooding, Trevor Pinch, and Simon Schaffer (eds.), *The Uses of Experiment: Studies in the Natural Sciences*, pp. 117–134. Cambridge: Cambridge University Press, 1989.

Newton, Isaac. *The Correspondence of Isaac Newton*, vol. 1. Ed. H. W. Turnbull. Cambridge: Cambridge University Press, 1959.

———. *The Mathematical Papers of Isaac Newton*, vol. 7. Ed. and trans. D. T. Whiteside. Cambridge: Cambridge University Press, 1976.

———. *Mathematical Principles of Natural Philosophy*. Trans. Andrew Motte, trans. revised by Florian Cajori. Berkeley: University of California Press, 1934.

———. "New Theory about *Light* and *Colours*." *Philosophical Transactions* 6 (1672): 3075–3087.

———. *The Optical Papers of Isaac Newton*, vol. 1: *The Optical Lectures, 1670–72*. Ed. Alan E. Shapiro. Cambridge: Cambridge University Press, 1984.

———. [*Principia.*] *Isaac Newton's Philosophiae naturalis principia mathematica: The Third Edition with Variant Readings*, 2 vols. Ed. Alexandre Koyré and I. Bernard Cohen. Cambridge, MA: Harvard University Press, 1972.

Noël, Étienne. *Stephani Natalis . . . gravitas comparata*. Paris, 1648.

Nutton, Vivian. "Greek Science in the Sixteenth-Century Renaissance." In Field and James, *Renaissance and Revolution*, pp. 15–28.

———. "Humanistic Surgery." In Andrew Wear, Roger K. French, and Ian M. Lonie (eds.), *The Medical Renaissance of the Sixteenth Century*, pp. 75–99. Cambridge: Cambridge University Press, 1985.

Oakley, Francis. *Omnipotence, Covenant, and Order: An Excursion in the History of Ideas from Abelard to Leibniz*. Ithaca: Cornell University Press, 1984.

O'Malley, C. D. *Andreas Vesalius of Brussels (1514–1564)*. Berkeley: University of California Press, 1964.

Omar, Saleh Beshara. *Ibn al-Haytham's "Optics": A Study of the Origins of Experimental Science*. Minneapolis and Chicago: Bibliotheca Islamica, 1977.

Ophir, Adi, and Steven Shapin. "The Place of Knowledge: A Methodological Survey." *Science in Context* 4 (1991): 3–21.

Ortony, Andrew, ed. *Metaphor and Thought*. Cambridge: Cambridge University Press, 1979.

Owen, G. E. L. "Aristotle: Method, Physics and Cosmology." In Charles C. Gillispie (ed.), *Dictionary of Scientific Biography*, vol. 1, pp. 250–258. New York: Scribner's, 1970–1980.

———. "Tithenai ta phainomena." In Owen, *Logic, Science and Dialectic: Collected Papers in Greek Philosophy*, pp. 239–251. Ithaca: Cornell University Press, 1986.

Pagden, Anthony. *European Encounters with the New World: From Renaissance to Romanticism*. New Haven: Yale University Press, 1993.

Parker, Gary D. "Galileo and Optical Illusion." *American Journal of Physics* 54 (1986): 248–252.

Pascal, Blaise. *Œuvres complètes II: Texte établi, présenté et annoté par Jean Mesnard*. Paris: Desclée de Brouwer, 1970.

———. *Œuvres complètes, présentation et notes de Louis Lafuma*. Paris: Éditions du Seuil, 1963.

———. *Œuvres de Blaise Pascal*, 14 vols. Ed. Léon Brunschvicg and Pierre Boutroux. Paris, 1904–1923.
Pecquet, Jean. *Experimenta nova anatomica*. Paris, 1651.
Pedersen, Olaf. "Astronomy." In David C. Lindberg (ed.), *Science in the Middle Ages*, pp. 303–337. Chicago: University of Chicago Press, 1978.
Pérez-Ramos, Antonio. *Francis Bacon's Idea of Science and the Maker's Knowledge Tradition*. Oxford: Clarendon Press, 1988.
Petit, Pierre. *Observation touchant le vuide, faite pour la premiere fois en France: Contenuë en une lettre écrite à Monsieur Chanut Resident pour sa Majesté en Suede. Par Monsieur Petit Intendant des fortifications, le 10. Novembre 1646*. Paris, 1647.
Pickering, Andrew, ed. *Science as Practice and Culture*. Chicago: University of Chicago Press, 1992.
[Pierius, Jacobus]. *An detur vacuum in rerum natura*. Rouen, [1646?].
Pierius, Jacobus. *Iacobi Pierii Doctoris Medici et Philophiae [sic] Professoris, ad experientiam nuperam circa vacuum. R.P. Valeriani Magni demonstrationem ocularem. Et mathematicorum quorumdam nova cogitata. Responsio. . . .* Paris, 1648.
Pitt, Joseph. "The Heavens and Earth: Bellarmine and Galileo." In Peter Barker and Roger Ariew (eds.), *Revolution and Continuity: Essays in the History and Philosophy of Early Modern Science*, pp. 131–142. Washington DC: Catholic University of America Press, 1991.
Pocock, J. G. A. "Time, Institutions and Action: An Essay on Traditions and Their Understanding." In Pocock, *Politics, Language and Time: Essays on Political Thought and History*, pp. 232–272. New York: Atheneum, 1971.
Poggi, Stefano, and Maurizio Bossi, eds. *Romanticism in Science: Science in Europe, 1790–1840*. Boston Studies in the Philosophy of Science, vol. 152. Dordrecht: Kluwer, 1994.
Popkin, Richard H. *The History of Scepticism from Erasmus to Spinoza*. Berkeley: University of California Press, 1979.
———. "Scepticism, Theology and the Scientific Revolution in the Seventeenth Century." In Imre Lakatos and Alan Musgrave (eds.), *Problems in the Philosophy of Science*, pp. 1–28. Amsterdam: North Holland Publishing, 1968.
———. "Theories of Knowledge." In Charles B. Schmitt, Quentin Skinner, Eckhard Kessler, and Jill Kraye (eds.), *The Cambridge History of Renaissance Philosophy*, pp. 668–684. Cambridge: Cambridge University Press, 1988.
Popper, Karl. *The Logic of Scientific Discovery*. London: Hutchinson, 1959.
Poppi, Antonino. *La dottrina della scienza in Giacomo Zabarella*. Padua: Antenore, 1972.
Porter, Roy, and Mikuláš Teich, eds. *The Scientific Revolution in National Context*. Cambridge: Cambridge University Press, 1992.
Proclus. *A Commentary on the First Book of Euclid's Elements*. Trans. Glenn R. Morrow. Princeton: Princeton University Press, 1970.
Ptolemy. *Ptolemy's Almagest*. Trans. G. J. Toomer. London: Duckworth, 1984.
Pumfrey, Stephen. "The History of Science and the Renaissance Science of History." In Pumfrey et al. (eds.), *Science, Culture and Popular Belief*, pp. 48–70.
———. "Neo-Aristotelianism and the Magnetic Philosophy." In John Henry and Sarah Hutton (eds.), *New Perspectives on Renaissance Thought: Essays in*

the *History of Science, Education and Philosophy in Memory of Charles B. Schmitt*, pp. 177–189. London: Duckworth, 1990.

Pumfrey, Stephen, Paolo L. Rossi, and Maurice Slawinski (eds.). *Science, Culture and Popular Belief in Renaissance Europe*. Manchester: Manchester University Press, 1991.

Pycior, Helena M. "Mathematics and Philosophy: Wallis, Hobbes, Barrow, and Berkeley." *Journal of the History of Ideas* 48 (1987): 265–286.

Pyenson, Lewis. *Civilizing Mission: Exact Sciences and French Overseas Expansion, 1830–1940*. Baltimore: Johns Hopkins University Press, 1993.

———. *Cultural Imperialism and Exact Sciences: German Expansion Overseas, 1900–1930*. New York: P. Lang, 1985.

———. *Empire of Reason: Exact Sciences in Indonesia, 1840–1940*. Leiden: E. J. Brill, 1989.

Raeder, Hans, Elis Stromgren, and Bengt Stromgren, eds. *Tycho Brahe's Description of His Instruments and Scientific Work as Given in "Astronomiae instauratae mechanica." (Wandesburgi, 1598)*. Copenhagen: I Kommission hos ejnar Munksgaard, 1946.

Rashed, Roshdi. "Optique géometrique et doctrine optique chez Ibn al Haytham." *Archive for History of Exact Sciences* 6 (1970): 271–298.

Redondi, Pietro. *Galileo Heretic*. Trans. Raymond Rosenthal. Princeton: Princeton University Press, 1987.

Reeds, Karen. "Renaissance Humanism and Botany." *Annals of Science* 33 (1976): 519–542.

Reif, Mary Richard. "Natural Philosophy in Some Early Seventeenth Century Scholastic Textbooks." Ph.D. diss., Saint Louis University, 1962.

Reif, Patricia. "The Textbook Tradition in Natural Philosophy, 1600–1650." *Journal of the History of Ideas* 30 (1969): 17–32.

Riccioli, Giovanni Battista. *Almagestum novum*. Bologna, 1651.

Ricoeur, Paul. *The Rule of Metaphor*. Toronto: University of Toronto Press, 1975.

Risnerus, Federicus, ed. *Opticae thesaurus: Alhazeni arabis libri septem, nunc primum editi*.... Basel, 1572.

Risse, Wilhelm. *Die Logik der Neuzeit, 1. Band 1500–1640*. Stuttgart–Bad Cannstatt: Friedrich Frommann, 1964.

———. *Die Logik der Neuzeit, 2. Band 1640–1780*. Stuttgart–Bad Cannstatt: Friedrich Frommann, 1970.

———. "Zabarellas Methodenlehre." In Luigi Olivieri (ed.), *Aristotelismo veneto e scienza moderna*, vol. 1, pp. 155–172. Padua: Editrice Antenore, 1983.

Roberts, Julian. "The Politics of Interpretation: Sacred and Secular Hermeneutics in the Work of Luther, J. S. Semler, and H. G. Gadamer." *Ideas and Production* 1 (1981): 15–32.

Rochemonteix, Camille de. *Un collège de Jésuites aux XVIIe & XVIIIe siècles: Le collège Henri IV de La Flèche*, 4 vols. Le Mans: Leguicheux, 1889.

Rochot, Bernard. "Comment Gassendi interprétait l'expérience du Puy de Dôme." In Pierre Costabel et al., *L'œuvre scientifique de Pascal*, pp. 278–301. Paris: Presses Universitaires de France, 1964.

Rose, Paul Lawrence. *The Italian Renaissance of Mathematics: Studies on Humanists and Mathematicians from Petrarch to Galileo*. Geneva: Droz, 1975.
Rose, Paul Lawrence, and Stillman Drake. "The Pseudo-Aristotelian *Questions of Mechanics* in Renaissance Culture." *Studies in the Renaissance* 18 (1971): 65–104.
Rosen, Edward. *Kepler's Conversation with Galileo's Sidereal Messenger*. New York and London: Johnson Reprint Co., 1965.
———. *The Naming of the Telescope*. New York: Henry Schuman, 1947.
———. "Renaissance Science as Seen by Burckhardt and His Successors." In Tinsley Helton (ed.), *The Renaissance: A Reconsideration of the Theories and Interpretations of the Age*, pp. 77–103. Madison: University of Wisconsin Press, 1961.
Rossi, Paolo. "The Aristotelians and the Moderns: Hypothesis and Nature." *Annali dell'Istituto e Museo di Storia della Scienza di Firenze* 7 (1982): fasc. 1:3–27.
———. *Francis Bacon: From Magic to Science*. Trans. Sacha Rabinovitch. Chicago: University of Chicago Press, 1968.
———. *Philosophy, Technology, and the Arts in the Early Modern Era*. Trans. Salvator Attanasio. New York: Harper and Row, 1970.
Ruby, Jane E. "The Origins of Scientific 'Law'." *Journal of the History of Ideas* 47 (1986): 341–359.
Rudwick, M. J. S. *The Great Devonian Controversy: The Shaping of Scientific Knowledge Among Gentlemanly Specialists*. Chicago: University of Chicago Press, 1985.
Russell, John L. "Catholic Astronomers and the Copernican System after the Condemnation of Galileo." *Annals of Science* 46 (1989): 365–386.
Sabra, A. I. "The Astronomical Origin of Ibn al-Haytham's Concept of Experiment." *Actes du XIIe Congrès International d'Histoire des Sciences, Paris 1968*, vol. 3 A, pp. 133–136. Paris: Albert Blanchard, 1971. Reprinted in Sabra, *Optics, Astronomy and Logic: Studies in Arabic Science and Philosophy*, chap. 6. Aldershot, U.K.: Variorum, 1994.
———, ed. and trans. *The Optics of Ibn al-Haytham, Books I-III: On Direct Vision*, 2 vols. London: Warburg Institute, 1989.
Salmone, M., ed. *Ratio studiorum: L'ordinamento scolastico dei collegi dei Gesuiti*. Milan, 1979.
Salomon-Bayet, Claire. *L'institution de la science et l'expérience du vivant: Méthode et expérience à l'Académie royale des sciences, 1666–1794*. Paris: Flammarion, 1978.
Sambursky, S. *The Physical World of Late Antiquity*. New York: Basic Books, 1962.
Sargent, Rose-Mary. "Scientific Experiment and Legal Expertise: The Way of Experience in Seventeenth-Century England." *Studies in History and Philosophy of Science* 20 (1989): 19–45.
Scarborough, John. *Roman Medicine*. Ithaca: Cornell University Press, 1969.
Schaffer, Simon. "Glass Works: Newton's Prisms and the Uses of Experiment." In Gooding, Pinch, and Schaffer, *The Uses of Experiment*, pp. 67–104.

———. "Making Certain," essay-review of B. Shapiro, *Probability and Certainty*. *Social Studies of Science* 14 (1984): 137–152.

Scheiner, Christophorus. *De maculis solaribus . . . accuratior disquisitio* (1612). Reprinted in Galileo, *Opere*, vol. 5, pp. 35–70.

———. *Disquisitiones mathematicae, de controversiis et novitatibus astronomicis . . . sub praesidio Christophori Scheiner . . . Nobilis et Doctissimis iuvenis, Ioannes Georgius Locher, Boius Monacensis, Artium et Philosophiae Baccalaureus, Magisterij Candidatus*. Ingolstadt, 1614.

———. *Oculus, hoc est, fundamentum opticum*. Innsbruck, 1619.

———. *Rosa ursina sive sol*. Bracciano, 1630.

———. *Tres epistolae de maculis solaribus* (1612). Reprinted in Galileo, *Opere*, vol. 5, pp. 21–33.

Schiebinger, Londa. *The Mind Has No Sex? Women in the Origins of Modern Science*. Cambridge, MA: Harvard University Press, 1989.

Schier, Donald S. *Louis Bertrand Castel, Anti-Newtonian Scientist*. Cedar Rapids: Torch Press, 1941.

Schmitt, Charles B. *Aristotle in the Renaissance*. Cambridge, MA: Harvard University Press, 1983.

———. *A Critical Survey and Bibliography of Studies on Renaissance Aristotelianism 1958–1969*. Padua: Editrice Antenore, 1971.

———. "Experience and Experiment: A Comparison of Zabarella's View with Galileo's in *De motu*." *Studies in the Renaissance* 16 (1969): 80–138.

———. "Galileo and the Seventeenth-Century Text-Book Tradition." In Paolo Galluzzi (ed.), *Novità celesti e crisi del sapere: Atti del Convegno Internazionale di Studi Galileiani*, pp. 217–228. Florence: Giunti Barbèra, 1984.

———. "The Rediscovery of Ancient Skepticism in Modern Times." In Myles Burnyeat (ed.), *The Skeptical Tradition*, pp. 225–251. Berkeley: University of California Press, 1983.

———. "Toward a Reassessment of Renaissance Aristotelianism." *History of Science* 11 (1973): 159–193.

———. "William Harvey and Renaissance Aristotelianism: A Consideration of the Praefatio to *De generatione animalium* (1651)." In Rudolf Schmitz and Gundolf Keil (eds.), *Humanismus und Medizin*, pp. 117–138. Weinheim: Acta Humaniora, 1984.

Schott, Gaspar. *Cursus mathematicus*. Würzburg, 1661.

———. [Aspasio Caramuello, pseud.]. *Ioco-Seriorum naturae et artis, sive magiae naturalis centuriae tres*. N.p., 1666.

———. *Magia universalis naturae et artis*, 4 vols. Würzburg, 1657–1659.

Schramm, Mathias. "Aristotelianism: Basis and Obstacle to Scientific Progress in the Middle Ages." *History of Science* 2 (1963): 91–113.

———. "Steps towards the Idea of Function: A Comparison between Eastern and Western Science of the Middle Ages." *History of Science* 4 (1965): 70–103.

Schüling, Hermann. *Die Geschichte der axiomatischen Methode im 16. und beginnenden 17. Jahrhundert*. Hildesheim: Georg Olms, 1969.

Schuster, John A. "Descartes and the Scientific Revolution, 1618–1634: An Interpretation." Ph.D. diss., Princeton University, 1977.

———. "Descartes' Mathesis universalis: 1619–1628." In Stephen Gaukroger (ed.), *Descartes: Philosophy, Mathematics and Physics*, pp. 41–96. Brighton, Sussex: Harvester Press, 1980.

———. "Methodologies as Mythic Structures: A Preface to the Future Historiography of Method." *Metascience: Annual Review of the Australasian Association for the History, Philosophy and Social Studies of Science* 1/2 (1984): 15–36.

———. "Whatever Should We Do with Cartesian Method?—Reclaiming Descartes for the History of Science." In Stephen Voss (ed.), *Essays on the Philosophy and Science of René Descartes*, pp. 195–223. New York: Oxford University Press, 1993.

Schuster, John A., and Graeme Watchirs. "Natural Philosophy, Experiment and Discourse in the 18th Century." In H. E. Le Grand (ed.), *Experimental Inquiries: Historical, Philosophical and Social Studies of Experimentation in Science*, pp. 1–47. Dordrecht: Kluwer, 1990.

Schuster, John A., and Richard R. Yeo, eds. *The Politics and Rhetoric of Scientific Method: Historical Studies*. Dordrecht: D. Reidel, 1986.

Segre, Michael. *In the Wake of Galileo*. New Brunswick, NJ: Rutgers University Press, 1991.

———. "Torricelli's Correspondence on Ballistics." *Annals of Science* 40 (1983): 489–499.

Selectae propositiones in tota sparsim mathematica pulcherrimae. Pont-à-Mousson, 1632. (Part of a volume of mathematical disputations and disquisitions in the British Library found under the catalogue listing "Albertus Linemannus, *Disputatio ordinaria.*")

Sempilius, Hugo. *De mathematicis disciplinis libri duodecim.* Antwerp, 1635.

Serene, Eileen. "Demonstrative Science." In Norman Kretzmann, Anthony Kenny, and Jan Pinborg (eds.), *The Cambridge History of Later Medieval Philosophy*, pp. 496–517. Cambridge: Cambridge University Press, 1982.

Shapin, Steven. "The Invisible Technician." *American Scientist* 77 (1989): 554–563.

———. "Pump and Circumstance: Robert Boyle's Literary Technology." *Social Studies of Science* 14 (1984): 481–520.

———. "Robert Boyle and Mathematics: Reality, Representation, and Experimental Practice." *Science in Context* 2 (1988): 23–58.

———. *A Social History of Truth: Civility and Science in Seventeenth-Century England.* Chicago: University of Chicago Press, 1994.

Shapin, Steven, and Simon Schaffer. *Leviathan and the Air-Pump: Hobbes, Boyle, and the Experimental Life.* Princeton: Princeton University Press, 1985.

Shapiro, Alan E. *Fits, Passions, and Paroxysms: Physics, Method, and Chemistry and Newton's Theories of Colored Bodies and Fits of Easy Reflection.* Cambridge: Cambridge University Press, 1993.

Shapiro, Barbara J. *John Wilkins 1614–1672: An Intellectual Biography.* Berkeley: University of California Press, 1969.

———. *Probability and Certainty in Seventeenth-Century England: A Study of the Relationships between Natural Science, Religion, History, Law, and Literature.* Princeton: Princeton University Press, 1983.

Shea, William R. "Author's Response" to Stephen Gaukroger's review of *The Magic of Numbers and Motion*. *Metascience*, n.s., no. 3 (1993): 27–32.

———. "Descartes as a Critic of Galileo." In Butts and Pitt, *New Perspectives on Galileo*, pp. 139–159.

———. "Galileo Galilei: An Astronomer at Work." In Trevor H. Levere and William R. Shea (eds.), *Nature, Experiment, and the Sciences: Essays on Galileo and the History of Science in Honour of Stillman Drake*, pp. 51–76. Dordrecht: Kluwer, 1990.

———. "Galileo, Scheiner, and the Interpretation of Sunspots." *Isis* 61 (1970): 498–519.

———. *Galileo's Intellectual Revolution*. London: Macmillan, 1972.

———. *The Magic of Numbers and Motion: The Scientific Career of René Descartes*. Canton, MA: Science History Publications U.S.A., 1991.

Shils, Edward. *Tradition*. Chicago: University of Chicago Press, 1981.

Shirley, John. *Thomas Harriot: A Biography*. Oxford: Clarendon Press, 1983.

Slawinski, Maurice. "Rhetoric and Science/Rhetoric of Science/Rhetoric as Science." In Pumfrey et al. (eds.), *Science, Culture and Popular Belief*, pp. 71–99.

Smith, A. Mark. "Alhazen's Debt to Ptolemy's Optics." In Trevor H. Levere and William R. Shea (eds.), *Nature, Experiment, and the Sciences: Essays on Galileo and the History of Science*, pp. 147–164. Dordrecht: Kluwer, 1990.

———. "Descartes's Theory of Light and Refraction: A Discourse on Method." *Transactions of the American Philosophical Society* 77 (1987), pt. 3.

———. "Ptolemy's Search for a Law of Refraction: A Case-Study in the Classical Methodology of 'Saving the Appearances' and its Limitations." *Archive for History of Exact Sciences* 26 (1982), 221–240.

Snyders, Georges. *La pédagogie en France aux XVIIe et XVIIIe siècles*. Paris: Presses Universitaires de France, 1965.

Solomon, Howard M. *Public Welfare, Science and Propaganda in Seventeenth Century France*. Princeton: Princeton University Press, 1972.

Sommervogel, Carlos, et al., eds. *Bibliothèque de la Compagnie de Jésus*, 11 vols. Brussels: Alphonse Picard, 1890–1932; reprinted Louvain: Éditions de la Bibliothèque S.J., Collège Philosophique et Théologique, 1960.

Sprat, Thomas. *History of the Royal Society*. London, 1667; facsimile reprint Saint Louis: Washington University Press, 1958.

Southern, R. W. "Commentary on Medieval Science." In Alistair C. Crombie (ed.), *Scientific Change*, pp. 301–306. New York: Basic Books, 1963.

Stough, Charlotte L. *Greek Scepticism: A Study in Epistemology*. Berkeley: University of California Press, 1969.

Strawson, Galen. *The Secret Connexion: Causation, Realism, and David Hume*. Oxford: Clarendon Press, 1989.

Sutton, Geoffey V. "A Science for a Polite Society." Ph.D. diss., Princeton University, 1981.

Taton, René. "L'annonce de l'expérience barométrique en France." *Revue d'histoire des sciences* 16 (1963): 77–83.

Thorndike, Lynn. *A History of Magic and Experimental Science*, vols. 7 & 8. New York: Columbia University Press, 1958.

Toulmin, Stephen. *The Philosophy of Science: An Introduction.* London: Hutchinson, 1953.
Vallerius, Haraldus. *Dissertatio physico-mathematica de camera obscura.* Uppsala, 1631.
Van Helden, Albert. "The Accademia del Cimento and Saturn's Ring." *Physis* 15 (1973): 237–259.
———. "'Annulo cingitur': The Solution to the Problem of Saturn." *Journal for the History of Astronomy* 5 (1974): 155–174.
———. "Saturn and His Anses." *Journal for the History of Astronomy* 5 (1974): 105–121.
———. "Telescopes and Authority from Galileo to Cassini." *Osiris*, n.s., 9 (1994): 9–29.
Van Leeuwen, Henry G. *The Problem of Certainty in English Thought 1630–1690.* The Hague: Martinus Nijhoff, 1963.
Vasoli, Cesare. "The Contribution of Humanism to the Birth of Modern Science." *Renaissance and Reformation,* n.s., 3 (1979): 1–15.
Wagner, David L., ed. *The Seven Liberal Arts in the Middle Ages.* Bloomington: Indiana University Press, 1983.
Walker, D. P. *Spiritual and Demonic Magic from Ficino to Campanella.* London: Warburg Institute, 1958.
Wallace, William A. "Albertus Magnus on Suppositional Necessity in the Natural Sciences." In James A. Weisheipl (ed.), *Albertus Magnus and the Sciences: Commemorative Essays 1980,* pp. 103–128. Toronto: Pontifical Institute of Medieval Studies, 1980. Reprinted in Wallace, *Galileo, the Jesuits, and the Medieval Aristotle,* chap. 9.
———. "Aristotle and Galileo: The Uses of *Hupothesis (Suppositio)* in Scientific Reasoning." In Dominic J. O'Meara (ed.), *Studies in Aristotle,* pp. 47–77. Washington, DC: Catholic University of America Press, 1981. Reprinted in Wallace, *Galileo, the Jesuits, and the Medieval Aristotle,* chap. 3.
———. "The Certitude of Science in Late Medieval and Renaissance Thought." *History of Philosophy Quarterly* 3 (1986): 281–291.
———. *Galileo and His Sources: The Heritage of the Collegio Romano in Galileo's Science.* Princeton: Princeton University Press, 1984.
———. "Galileo and Reasoning *ex suppositione.*" In Wallace, *Prelude to Galileo: Essays on Medieval and Sixteenth-Century Sources of Galileo's Thought,* pp. 129–159. Dordrecht: D. Reidel, 1981.
———. *Galileo, the Jesuits, and the Medieval Aristotle.* Aldershot, U.K.: Variorum, 1991.
———. *Galileo's Logic of Discovery and Proof: The Background, Content, and Use of His Appropriated Treatises on Aristotle's "Posterior Analytics".* Dordrecht: Kluwer, 1992.
———. "The Problem of Causality in Galileo's Science." *Review of Metaphysics* 36 (1983): 607–632. Reprinted in Wallace, *Galileo, the Jesuits, and the Medieval Aristotle,* chap. 2.
———. "Randall *Redivivus:* Galileo and the Paduan Aristotelians." *Journal of the History of Ideas* 48 (1988): 133–149. Reprinted in Wallace, *Galileo, the Jesuits, and the Medieval Aristotle,* chap. 5.

---. "Traditional Natural Philosophy." In Charles B. Schmitt, Quentin Skinner, Eckhard Kessler, and Jill Kraye (eds.), *The Cambridge History of Renaissance Philosophy*, pp. 201–235. Cambridge: Cambridge University Press, 1988.

Waller, Richard. *Essayes of Natural Experiments*. London, 1684; facs. reprint New York: Johnson Reprint Corp., 1964.

Watts, Isaac. *Logick: or, The Right Use of Reason*, 2d edition. London, 1726; facsimile reprint, New York and London: Garland, 1984.

Wear, Andrew. "William Harvey and the 'Way of the Anatomists'." *History of Science* 21 (1983): 223–249.

Webster, Charles. "The Discovery of Boyle's Law, and the Concept of the Elasticity of the Air in the Seventeenth Century." *Archive for History of Exact Sciences* 2 (1965): 441–502.

---. *From Paracelsus to Newton: Magic and the Making of Modern Science*. Cambridge: Cambridge University Press, 1982.

---. *The Great Instauration: Science, Medicine, and Reform, 1626–1660*. London: Duckworth, 1975.

Weisheipl, James A. "Aristotle's Concept of Nature: Avicenna and Aquinas." In Lawrence D. Roberts (ed.), *Approaches to Nature in the Middle Ages: Papers of the Tenth Annual Conference of the Center for Medieval and Early Renaissance Studies*, pp. 137–160. Binghamton, NY: Center for Medieval and Early Renaissance Studies, 1982.

---. "Classification of the Sciences in Medieval Thought." *Medieval Studies* 27 (1965): 54–90.

---. "The Nature, Scope and Classification of the Sciences." In David C. Lindberg (ed.), *Science in the Middle Ages*, pp. 461–482. Chicago: University of Chicago Press, 1978.

Westfall, Richard S. "The Development of Newton's Theory of Color." *Isis* 53 (1962): 339–358.

---. *Never at Rest: A Biography of Isaac Newton*. Cambridge: Cambridge University Press, 1980.

---. "Science and Patronage: Galileo and the Telescope." *Isis* 76 (1985): 11–30.

Westman, Robert S. "The Astronomer's Role in the Sixteenth Century: A Preliminary Study." *History of Science* 18 (1980): 105–147.

---. "Humanism and Scientific Roles in the Sixteenth Century." In Rudolf Schmitz and Fritz Krafft (eds.), *Humanismus und Naturwissenschaften*, pp. 83–99. Boppard: Harold Boldt, 1980.

---. "The Melanchthon Circle, Rheticus, and the Wittenberg Interpretation of the Copernican Theory." *Isis* 66 (1975): 165–193.

---. "Proof, Poetics, and Patronage: Copernicus's Preface to *De revolutionibus*." In Lindberg and Westman (eds.), *Reappraisals of the Scientific Revolution*, pp. 167–205.

Wilcox, Donald J. *The Measure of Times Past: Pre-Newtonian Chronologies and the Rhetoric of Relative Time*. Chicago: University of Chicago Press, 1987.

Winch, Peter. *The Idea of a Social Science and Its Relation to Philosophy*. London: Routledge and Kegan Paul, 1958.

Winkler, Mary G., and Albert Van Helden. "Representing the Heavens: Galileo and Visual Astronomy." *Isis* 83 (1992): 195–217.

Wisan, Winifred L. "Galileo and the Emergence of a New Scientific Style." In Jaako Hintikka, David Gruender, and Evandro Agazzi (eds.), *Theory Change, Ancient Axiomatics, and Galileo's Methodology* [= *Proceedings of the 1978 Conference on the History and Philosophy of Science*], vol. 1, pp. 311–339. Dordrecht: D. Reidel, 1981.

———. "Galileo's Scientific Method: A Reexamination." In Butts and Pitt (eds.), *New Perspectives on Galileo*, pp. 1–57.

———. "On Argument *ex suppositione falsa*." *Studies in History and Philosophy of Science* 15 (1984): 227–236.

Wood, P. B. "Methodology and Apologetics: Thomas Sprat's *History of the Royal Society*." *British Journal for the History of Science* 13 (1980): 1–26.

Woolgar, Steve. "Discovery: Logic and Sequence in a Scientific Text." In Karin Knorr-Cetina, Roger Krohn, and Richard Whitley (eds.), *The Social Process of Scientific Investigation*, pp. 239–268. Dordrecht: D. Reidel, Sociology of the Sciences Yearbook 4, 1980.

Yates, Frances A. *Giordano Bruno and the Hermetic Tradition*. Chicago: University of Chicago Press, 1964.

———. "The Hermetic Tradition in Renaissance Science." In Yates, *Collected Essays*, vol. 3, *Ideas and Ideals in the North European Renaissance*, pp. 227–246. London: Routledge and Kegan Paul, 1984.

Ziggelaar, August. *François de Aguilòn S.J. (1567–1617): Scientist and Architect* [= *Bibliotheca Instituti Historici S.I.*, vol. 44]. Rome: Institutum Historicum S.I., 1983.

———. *Le physicien Ignace Gaston Pardies S.J. (1636–1673)*. Copenhagen: Odense University Press, 1971.

Zilsel, Edgar. "The Genesis of the Concept of Physical Law." *The Philosophical Review* 51 (1942): 245–279.

———. "The Origins of William Gilbert's Scientific Method." *Journal of the History of Ideas* 2 (1941): 1–32.

INDEX

Accademia del Cimento, 22
Aguilonius, Franciscus: and contrived experience, 51, 53, 56n, 161–62; and disciplinary boundaries, 161–62, 166–67; and experience, 14n, 45, 147–48; and optics, 51, 53, 60–61, 168–69; and physicomathematics, 168–69; and repetition, 19, 44–45, 53, 66, 75, and science, 53, 55, 61
Air-pump, 228
Akagi, Shozo, 197n
Albertus Magnus, 24, 154
Alhazen (Ibn al-Haytham), 51–53, 55, 57, 61
Antiquity, 190–91, 196–97, 211–12. *See also* Humanism
Antonini, Daniello, 107
Apelles. *See* Scheiner
Aquinas, Thomas, 154
Archimedes, 43–44, 117, 119, 127n, 166
Aristotelianism, 3–4, 6, 246; and art/nature distinction, 153–58, 161–62; and Barrow, 224; and controversy over void, 188–91, 195; and demonstrative science, 22–23, 27, 36–46, 58–59, 63, 94–95, 153–55, 168; and disciplinary boundaries, 161–68; and experience, 4, 13n, 21–23, 42–46, 63; and experiment, 4, 6; and Galileo, 22, 44, 86–87, 92, 100–101, 124–27, 129, 138–43; and induction, 18–19, 26–30; and nature, 6, 18–21, 153–55; and Newton, 235–36; and Pascal, 182–82, 185, 197–98; and postulates, 219–21; and repetition, 66, 69–70; and scientific status of astronomy, 46, 48, 50, 61, 93; and scientific status of optics, 53–57, 61; and teleology, 154–55, 158, 175; and tradition, 97. *See also* Aristotle; Causes; Certainty; Common experience; Contrivance; Demonstration; Disciplinary boundaries; Evident experience; Experience; Jesuits; Mathematical sciences; Mixed mathematics; Natural philosophy; Repetition; Scholastic philosophy; Universals
Aristotle, 210; and art/nature distinction, 151, 154–56, 159; as authority, 23–25, 37–38, 54, 67, 69, 75, 80, 121, 144, 202, 226; on demonstration, 27, 36, 38, 40–42, 140n, 221, 224; on experience, 4, 14, 22–23, 42–43, 45, 54–56; and Galileo, 44, 86; and hypotheses, 48, 219, 221; and induction, 6, 26–27, 29–30, 59, 224–225n, 241; mathematics useful for understanding of, 35–36, 41; on mixed mathematics, 38–39, 143–44, 168–69; on science, 11, 36–42; and scientific status of mathematics, 36–40, 54
Arnauld, Antoine, 28–29
Arriaga, Roderigo de, 67–71, 75, 156n, 169; and controversy with Riccioli, 71, 73–75, 82, 84–85, 175, 248
Art/nature distinction, 6, 8, 151, 153–58, 161–62, 211

281

Artisanal knowledge, 144–45, 147
Assmus, Alexi, 158n
Astronomy: contrived experience in, 46–50, 61, 63; and disciplinary boundaries, 162, 164–65, 167; disciplinary status of, 38; as mixed mathematical science, 39, 48–49, 164–65, 168, 175; as physico-mathematics, 175; as private science, 50–51; scientific status of, 39, 46–50, 61, 93–96, 150, 175; and tradition, 93–96, 120. *See also* Comets; Jupiter; Saturn; Sunspots; Telescope
Authority: appeals to, 37–38; and expertise, 63, 66, 72, 75–76, 89, 91, 130, 246; and source of experience, 22–25, 75, 144, 149; and tradition, 115–22; weakening of, 7, 11
Aversa, Rafael, 156
Axioms, 182, 184–85, 216, 218–19, 222

Bacon, Francis, 18–20, 26, 58n, 76n; and antiquity, 117n, 190; and experimental philosophy, 22, 234n; and operational knowledge, 2, 9
Baldini, Ugo, 101n, 170n
Baliani, Giovanni Battista, 48–49n, 136, 141n, 143–44; and Riccioli, 78–80, 84
Barnes, Barry, 152n
Barometer. *See* Torricellian device
Baroncini, Gabriele, 21n
Barrow, Isaac: geometrical constructivism in, 222; on geometry, 29–30, 213; on induction, 28–29, 224, 241–42; on mathematical sciences, 40–41n, 178–79, 223–27, 242; on physico-mathematics, 178–79, 222–24, 242
Bechler, Zev, 237n
Beeckman, Isaac, 170–74
Bellarmine, Robert, 100–101n, 108
Bellutus, Bonaventura, 73
Bennett, J. A., 215
Bentley, Richard, 17
Black, Max, 152n
Blancanus, Josephus (Biancani), 60–61, 223; on causes in mathematics, 40, 47, 168; and disciplinary boundaries, 161–66; on "observations" and "phenomena," 47–51, 54–59, 61, 98, 115, 150, 161; and scientific status of mathematics, 37–38, 40–41, 49–50, 59, 61, 166, 168; and telescopic discoveries as "observations," 47, 97–99, 101–2, 161; and tradition, 98, 115, 120; and utility of mathematics, 36, 169
Blay, Michel, 234n
Bloor, David, 97
Boyle, Robert, 22, 153, 215n; compared with continental philosophers of nature, 4, 8, 206–9; and experimental philosophy, 8, 227–29, 242, 245; and expertise, 76n, 242; and mathematics, 2, 226–27; and matters of fact, 186n, 208; shortcomings of enterprise, 3, 210n, 231; and virtual witnessing, 60, 246
Burckhardt, Jakob, 118n
Buridan, Jean, 25

Cabeo, Niccolò: and astronomical tradition, 93–97, 99, 105, 120, 123; and controversy with Riccioli, 71–75, 82, 84–85, 175; and physico-mathematics, 170n; and use of mathematical form, 64–66, 68–71, 94–95, 127–29, 142–43, 173
Campanella, Tommaso, 7n
Campanus, Johannes, 217n
Cantor, Geoffrey, 152–53
Capaldi, Nicholas, 243n
Cassianus, Georgius, 79
Castel, Louis Bertrand, 18–19, 249
Catena, Pietro, 36n
Catholicism, 3, 7, 11, 32; role of tradition in, 116–17, 121
Cattenius, Otto, 101n, 218n
Causes: as criteria of scientificity, 27, 36, 101n, 150, 154, 174–75; and disciplinary boundaries, 101n, 140, 162–63, 166–67, 193–94, 237–38; in mathematics, 36, 40, 168; in physical explanation, 135–36, 138–44, 162–63, 166–67, 193–94, 201, 237–38; in pysico-mathematics, 170–72; and problem of induction, 17–19, 27–28, 243
Cave dwellers, 183
Certainty: of demonstrative premises, 23, 42; and evidentness, 42, 74–76; importance in seventeenth century, 11; of mathematics, 3, 31, 38, 65, 174–75; Newton on, 236–38, 242; of physico-mathematics, 175, 177–78; of physics, 3, 144, 177–78; of true science, 23, 27–29, 74–76
Cesarini, Virginio, 88–89

Chanut, Pierre, 187
Chauvin, Étienne, 57–58
Children's games, 147–49
Christina, queen of Sweden, 187n
Cicero, 26, 117–18, 152
Cigoli, Lodovico, 107
Clavius, Christopher, 86; and astronomy, 38, 46–47, 97–98, 108, 111, 120, 165–66; and certainty of mathematical demonstration, 38, 163–64, 174–75; and monsters, 20, 48, 97–98, 162; and place of mathematics in Jesuit curriculum, 33–36, 61, 166; and postulates, 217–21; and scientific status of mathematics, 35–42, 56, 61, 166, 175; and utility of mathematics, 34–36, 166, 169, 223
Cohen, I. Bernard, 240n
Coïmbra commentaries, 36
Collins, H. M., 6n, 97
Comets, 86–87, 164
Commandino, Federigo, 217n
Commentary genre, 24, 63, 65, 139
Common experience, 67–68, 144–49; and evidentness, 63, 75, 90, 221; and novelty, 64, 202. *See also* Evident experience; Experience
Community: in Aristotle, 23; and astronomical tradition, 93–97, 108–9, 120, 123; in Descartes, 123; in Mersenne, 130, 136; and metaphor, 152–53; telescopic, 108–9
Continuity thesis, 2, 15, 96
Contrivance: and art/nature distinction, 153–61; and authority, 66, 89, 91; and credibility, 25, 59, 89, 91; in Descartes, 137; and disciplinary boundaries, 161–63, 167; geometrical presentation of, 59–62, 87; methodological problems presented by, 32, 47–62, 65–66, 127; in Mydorge, 137–38; in Newton, 232, 242; in Pascal, 183, 185, 200, 203; problem of rendering evident, 47–48, 58–62, 65–66; and sunspot controversy, 87, 89, 91. *See also* Astronomy; Observations; Telescope
Copernicus, Nicolaus, 117–19
Corpuscularianism, 170–71, 173
Coyne, George V., 101n
Credibility: and expertise, 91, 246; in Royal Society, 228–30, 247–48; social dimensions of, 109, 133, 149–50, 248; sources of, 89, 109–11; and witnesses, 84–88, 104–5

Dainville, François de, 33n
Daston, Lorraine, 14n, 39n
De Pace, Anna, 38n
Definitions, 216
Della Porta, J. B., 146, 147n, 176n
Democritus, 191
Demonstration: in Aristotelian science, 6, 23, 27, 36–37, 39–42, 58, 94–95, 140n, 154, 221; and certainty of empirical principles, 23, 42, 55, 58, 139–43; certainty of, in mathematics, 3, 31, 38, 64–65, 163–66, 174, 212; and disciplinary boundaries, 163–66; and establishment of empirical principles, 42–46, 53, 55–58, 63, 94–95, 192; in Euclid, 41–42, 216; geometrical model for, 41–42, 45, 64–65, 140n; imperfection of, in natural philosophy, 23, 166; in mixed mathematics, 38–39; in Newton, 232, 234–35, 237, 239–40; in Pascal, 183–86, 192, 199–201; in physico-mathematics, 172–75
Demonstrative regress, 26–28, 44, 64–65
Descartes, René, 14, 75, 94, 152, 174, 183, 187n, 226; experience and physical causes in, 134–37; and geometry, 30, 34, 214n, 217, 220–21; and induction, 28–30; and mechanical explanation, 211–12; and method, 121–23, 136–37; and physico-mathematics, 170–72; and Torricellian device, 197, 200–201
Dijksterhuis, E. J., 197n
Diodorus, 191
Diophantus, 122
Disciplinary boundaries: maintenance of, 161–68, 248; transgression of, 8, 170–76, 211, 221–22, 246. *See also* Mathematical sciences; Natural philosophy; Physico-mathematics
Disciplines. *See* Astronomy; Mathematical sciences; Mechanics; Mixed mathematics; Natural philosophy; Optics
Drake, Stillman, 108
Duhem, Pierre, 20n, 118n

Eamon, William, 247
Epicurus, 191
Eschinardus, Francesco, 176
Essences, 3, 37, 40, 154, 174, 185, 243

Index

Euclid: on demonstration, 30, 41, 45, 216; and evidentness, 42–43; as humanist model, 117–18; and mixed mathematics, 43, 143–44; and optics, 43, 51, 53; postulates of, 48, 216–21

Eustace of Saint Paul, 156

Event experiment, 15, 25, 180, 242, 246–47. *See also* Experiment; Historical reports; Singular experience

Evident experience ("evidentness"): in astronomy, 46–47, 49, 59–62, 103; and expertise, 75–76, 88; in Newton, 238; in optics, 55, 59–62; in Pascal, 183–85, 187; place in construction of demonstrative science, 7, 42–46, 58–59, 63, 74–77; techniques for creation of, 59–62, 66–67, 75–76, 88. *See also* Common experience; Experience; Universals

Ex suppositione demonstration, 154, 201, 225n

Experience: Aristotelian conception of, 4, 6, 11–14, 20–25, 42–45, 50, 54–55, 125–26, 138–39; and Aristotelian view of nature, 153–55; claimed insufficiency for physical demonstration, 11–12, 71, 132, 134–36, 138, 140–43; and community, 23, 44, 94, 108–9, 136; in controversy over void, 190–96; in Descartes, 134–37; and evidentness, 42–43, 46, 58–62, 74–76, 94–95; and experiment, 4, 13–14, 21–22, 45, 56–57, 124–25; and expertise, 63, 75–76, 83–85; in Galileo, 87, 89–92, 108–9, 125–27, 129, 143–47; mathematical model for, 64–70, 79–80, 127–29, 198–200, 203–9; in Mersenne, 129–36; and novelty, 63–64, 75–76; in Pascal, 181–87, 197–201, 203, 205–6, 208–9; and postulates, 219–22, 226; and problem of astronomy, 46–47, 49–50, 57, 61, 93–95; and problem of optics, 51–57, 61; in Royal Society, 227–31; in scientific demonstration, 22–23, 25, 43–45, 55–57, 65–66, 94–95, 125–27, 199–201; and sense perception, 19, 22–23, 42–45, 54, 66, 134–35, 141, 143–44; sources of, 22–24, 144–49; tabular presentation of, 80–82, 84, 205–6. *See also* Common experience; Evident experience; Observations; Phenomena; Singular experience; Universals

Experiment: in Alhazen, 52–53, 55, 57; in Barrow, 224–25; versus "experience," 3–6, 12–14, 21–25, 45, 124–26, 245; as *experimentum crucis*, 22, 86, 155, 224–25; in Galileo, 5, 22, 124–26, 145–47; and geometrical construction, 59–61; as metaphor, 158–59; in modern sense, 2–6, 12–15, 21–22, 45, 62, 85–86; in Newton, 14–15, 180, 238–42; in Pascal, 180, 198, 200–201. *See also* Contrivance; Event experiment; Experimental philosophy; Historical reports; Singular experience

Experimental philosphy (English), 14, 149, 215, 248; compared with French practice, 206, 209; in Newton, 8–9, 180, 210, 235, 238–39, 242–43, 247–48; in Royal Society, 4, 8, 25, 227–31, 242, 245–48

Expertise, 76n; and artisanal knowledge, 149–50; and authority, 91, 246; and common experience, 63; and evidentness, 74–76, 83, 88; historical reports as tokens of, 7, 83–88, 91, 229, 231, 246; and priority, 188; role in controversy, 7, 72–74, 84–85, 91

Fabri, Honoré, 136, 138–43, 175, 178, 236
Falling bodies, 67–74, 76, 79–85, 130–39, 141–44. *See also* mechanics
Feldhay, Rivka, 162n
Feyerabend, Paul K., 152
Foucault, Michel, 5
Funkenstein, Amos, 153n, 214, 220n

Gabbey, Alan, 152–53n, 194n, 215n, 222, 223n
Galen, 24, 30n, 52, 117–21
Galileo Galilei, 38n, 41n, 173, 187, 230, 240; as anti-Aristotelian, 86–87, 100, 129; and Aristotelian model of science, 22, 44, 92, 101, 124–27, 129, 138–43; and Blancanus's meta-rules, 98–99, 101–4, 106–8, 115, 248; and contrived experience, 87, 100, 102, 111, 127; and controversy over comets, 86–91, 164; and controversy over sunspots, 7, 53, 100–104, 106–7, 248; and credibility, 108–11, 130; and Descartes, 134–36; and disciplinary status of mathematics, 101, 110; and experience, 87, 89–92, 108–9, 125–27, 129, 144–47; and experiment, 5, 22, 124–25, 145–47; and Fabri, 138–43; and falling

bodies, 67, 78–80, 82, 84, 124–36, 138–42; and historical reports, 89, 125–26, 206; and Mersenne, 129–39, 174; and novelty, 97–98, 103, 108–9; and patronage, 108–10; and relation of mathematics to natural philosophy, 140, 143, 172, 178, 226, 238, 247; and satellites of Jupiter, 98–99, 107–15, 123; and telescopic monsters, 20–21, 98, 115, 162; and tradition, 98–99, 108–9, 115, 123, 248; as university professor, 9, 86, 99
Galison, Peter, 158n
Garber, Daniel, 136
Garrison, James W., 214n, 215–16, 223n
Gassendi, Pierre, 118n, 119, 197n
Geminus, 219, 222n
Geometrical construction: and constructive presentation of experience, 229, 232–33, 235, 239; and constructivist view of knowledge, 8–9, 56, 211–17, 219–22; *problemata* as models for presenting experience, 59–63, 75, 173, 203; *problemata* and Riccioli, 77, 81, 83
Geometry: in Clavius, 38–39, 218–20; in Descartes, 220–21; in Newton, 211–13, 215–16, 237–38; as a science, 36–43; as a subordinating science, 38–39, 43, 49, 53–54, 169. *See also* Euclid
Ghetaldus, Marinus, 226
Ghisonus, Stephanus, 79
Giacobbe, Giulio Cesare, 36n, 42n
Gilbert, Neal, 121
Gilbert, William, 58n, 65n, 144–45, 147, 159–61
Glanvill, Joseph, 231
Gooding, David, 158n
Gordon, George, 248
Grant, Edward, 24n
Grassi, Orazio, 21, 86–91, 164
Gravity, 210–11
Grazia, Vincenzo di, 101n
Grienberger, Christopher, 104–5
Grimaldi, Francesco Maria, 78–80, 176–78, 235–36
Grimaldus, Vincentius Maria, 79
Grosseteste, Robert, 24
Guerlac, Henry, 224, 240n
Guidobaldo del Monte, 119, 127n, 178
Guiffart, Pierre, 149, 156–59, 161, 190–92, 196–97
Guldin, Paul, 105, 107, 172–73

Hacking, Ian, 16n, 23
Hall, A. Rupert, 236n
Hanson, Norwood Russell, 13
Harrington, Thomas More, 181n
Harriot, Thomas, 108n
Harris, Steven J., 33n
Harvey, William, 130
Henry, John, 151n, 230n
Hero of Alexandria, 147, 176n
Hesse, Mary, 151–52
Hintikka, Jaako, 241n
Hipparchus, 50, 52
Historical reports: and the barometer, 201–8; in controversy over void, 187–88, 192–93, 195–96; and credibility, 7, 78–79, 84–85, 87–89, 91, 228–29; in English experimental philosophy, 149, 206, 209, 227–31; and Galileo, 89, 124–26, 206; in Mersenne, 131–32; in Newton, 14–15, 209, 233–35, 238–39; place in Aristotelian science, 13–14, 67, 69–70, 88, 91, 125, 205–6, 208–9; and Puy-de-Dôme, 181, 196–201, 208–9; relation to experiment, 6, 13–14, 79, 83, 85–86, 124–25, 246; as tokens of expertise, 7, 77–88, 91, 229, 246. *See also* Singular experience; Event experiment; Experiment
History, 93–96
Hobbes, Thomas, 210n, 213, 220n, 231n, 245
Holy Spirit, 116, 121–22
Hooke, Robert, 22, 215, 224, 227, 236–37
Hooykaas, Rejner, 172n
Horky, Martin, 111
Horror vacui, 182, 189, 196–97
Hubin, 207
Humanism, 7, 26, 117–23
Hume, David, 15–19, 21, 25, 28n, 243
Huygens, Christiaan, 135, 210–11, 215, 237n
Hypotheses: in Aristotle, 48, 219–21; in Barrow, 222, 224; in Fabri, 139–43; in modern usage, 45; in Newton, 237–38; in Pascal, 184; in Royal Society, 230

Induction, 26–30, 59, 224, 240–43; Humean problem of, 6, 15–19, 30, 243

Jardine, Nicholas, 20n, 36n
Jesuits: and Aristotelianism, 6, 36, 41, 44, 58, 61; and Barrow, 223, 225; and educa-

Jesuits: and Aristotelianism (*continued*) tion, 7, 32–36; and evident experience, 7, 32, 42–47, 58–63, 91; and Galileo, 92; and humanism, 120; and King of China, 153; and Pascal, 180; and physico-mathematics, 170–78, 247; and place of mathematics in curriculum, 33–36, 61, 166; and scientific status of mathematics, 6–7, 31–32, 34–42, 58–63. *See also* Aguilonius; Aristotelianism; Arriaga; Blancanus; Cabeo; Clavius; Fabri; Grassi; Mathematical sciences; Mixed mathematics; Riccioli; Scheiner; Scholastic philosophy

Johnson, Mark, 159
Jung, Joachim, 14n, 28n, 221–22
Jupiter, satellites of, 98–99, 107–15, 123, 173

Kepler, Johannes, 13, 51, 53n, 161n; on satellites of Jupiter, 98, 105n, 109–11; on Galileo, 109–11
Kircher, Athanasius, 172–73, 178, 189
Koyré, Alexandre, 12, 76n, 124–25, 145–47
Krayer, Albert, 33n
Kristeller, Paul Oskar, 46n
Kuhn, Thomas S., 96, 152, 246

Lakatos, Imre, 119n
Lakoff, George, 159
Las Casas, Bartolomé de, 110n
Latour, Bruno, 158n, 189
Lattis, James M., 100n
Laws of nature, 157–58, 207–8, 225n
Lenoble, Robert, 194n
Leucippus, 191
Leurechon, Jean, 138n
Livy, 117
Lloyd, G. E. R., 42n
Loeck, Gisela, 152
Lucas, Anthony, 224
Lucretius, 191
Luther, Martin, 117n

Machamer, Peter, 44n
MacLachlan, James, 146–47
Magnenus, Iohannes Chrysostomus, 14n
Magnetism, 64–66, 144–45, 159–61, 172–73
Magni, Valeriano, 187–89, 192

Mali, Joseph, 96
Mariotte, Edme, 206–8
Mastrius, Bartholomaeus, 73
Mathematical natural philosophy. *See* Newton
Mathematical sciences: and art/nature distinction, 8, 151, 153, 158, 161–62; and Barrow, 178–79, 223–27; causes in, 36, 40, 47, 140–42, 150, 162–63, 168, 170; certainty of, 3, 31, 38, 163–65, 174–75, 177–78; as demonstrative, 6–7, 34–42, 46, 61, 140–44, 150–51, 166, 175, 193–95; disciplinary ambitions of, 8, 101, 143, 150, 168, 172, 175–76; and disciplinary boundaries, 161–68; disciplinary status of, 3, 8, 33–38, 61, 101, 150, 156n, 166, 170, 172, 247; evident experience in, 44, 46, 61–62; Jesuits as practitioners of, 31–34, 44, 46, 58, 61, 63, 91–92, 150, 247; as methodological model, 63–70, 127–29, 172–73, 198–200, 221–22; necessity in, 43–44; and Newton, 9, 180, 210–12, 235–40, 242, 247–48; in Pascal, 198–200, 205–6, 208–9; and physico-mathematics, 2, 8–9, 168–79, 246; and Royal Society, 227, 230–31; utility of, 34–36, 101, 166–67, 169–70, 223. *See also* Aristotelianism; Aristotle; Astronomy; Mechanics; Mixed mathematics; Optics
Matters of fact, 228n, 231, 239
Maurolico, Francesco, 136
Maxims, 182–87, 196, 243
Mazzoni, Jacopo, 38n, 75
McMullin, Ernan, 238n
Mechanical philosophy, 18, 151, 210–11, 215; and machine metaphor, 151–53
Mechanics, 169, 178, 211–16, 222, 226. *See also* Falling bodies
Medici, 108–10
Mersenne, Marin, 129–36, 141, 174–75; and physico-mathematics, 174–75, 178; and Torricellian device, 180, 187, 197
Metaphor, 151–53, 158–61
Method, 15, 121–23, 136, 145, 240
Middleton, W. E. Knowles, 189
Milton, J. R., 16n
Mimesis, 158–61
Miracles, 14, 116, 122, 154
Mixed mathematics: astronomy as, 48–49; Barrow on, 178–79, 223–24; in Cabeo,

64–65; and disciplinary boundaries, 162–65; in Galileo, 44, 125–26; optics as, 43, 51, 53–54, 56–57; in Pascal, 192, 199, 203, 205; and physico-mathematics, 168–70, 174, 178–79, 223–24, 235–37, 242; scientific status of, 39, 58, 61, 143–44, 162–65, 168–70. *See also* Mathematical sciences; Physico-mathematics

Molland, A. G., 221n

Monsters: singulars as, 14, 18–21, 225n, 249; telescopic discoveries as, 48, 97–98, 115, 162

Moss, Jean Dietz, 23n, 209, 235

Mydorge, Claude, 137–38, 146–47, 171–72, 174

Natural magic, 145–47, 173

Natural philosophy: as Aristotelian science, 2–4, 34–37, 43–44; and art/nature distinction, 8, 151, 153, 155, 157, 159–62; and disciplinary ambitions of mathematics, 8, 101, 143, 150, 168, 172, 175–76; and disciplinary boundaries, 150, 161–68; disciplinary status of, 3, 34–38, 101, 150, 156n, 161, 166, 170; and mathematical techniques in, 64–65, 127–29, 153, 172–73, 221–22; vs. mathematics, 36, 138–44, 150; and Newton's mathematical natural philosophy, 210–11, 235–40, 242, 247–49; and physico-mathematics, 8, 150, 168–79, 222–27; uncertainty of, 3, 144, 164–65, 174, 177–78, 227; utility of mathematics for, 34–36, 101, 166–67, 169–70, 223

Nature, 6, 18–21, 151, 153–55, 157–58, 225–26

Newton, Isaac: and Aristotelianism, 235–36, 243; and causes, 237–39, 243; and demonstration, 232, 234–35, 239–40, 242; and experiment, 8–9, 22, 180, 210, 224–25, 232–34, 238–42; and experimental philosophy, 8–9, 180, 215, 235, 238–39, 242–43, 247–48; and geometry, 29, 211–16, 226, 237–38; and historical reports, 14–15, 233–35, 238–39, 242; and hypotheses, 237–39; on induction, 240–43; instructional form in, 232–33, 235, 239; and mathematical natural philosophy, 179, 210–11, 235–38, 242, 247–49; and mathematical sciences,

210–12, 235–40, 242, 247; and mechanical philosophy, 210–11, 215; and mechanics, 211–16, 222, 226; and natural philosophy, 235–40, 247; on nature, 157; and optics, 209, 232–40; and postulates, 213, 215–16, 226; and Royal Society, 210, 227, 234–35, 239, 242, 247; and Rules of Philosophizing, 184, 241, 243; and singulars, 18, 224–25, 241–42

Nicole, Pierre, 28–29

Noël, Étienne, 184, 186, 193, 195–96, 200

Novelty: in astronomy, 92, 97–98, 102–3, 108–11; and falling bodies, 67, 76; as problem for demonstrative sciences, 7, 60–61, 63–64, 75, 97; and Torricellian device, 63, 190–91, 208

Observations (as term): in Blancanus, 49–50, 54, 56–57, 161; compared with Scheiner's "experience," 54, 56–57; in Ptolemy, 14n, 48–50; scientific status of, 59, 61–62, 67, 150; and telescopic discoveries, 98–99, 101–3, 161, 248

Occult properties, 17–19

Oldenburg, Henry, 209, 233, 236–38, 242

Omar, S. B., 51–52

Operational knowledge, 1–2, 155, 158–59

Optics: and contrivance, 51–57, 59, 63, 163; and disciplinary boundaries, 162–63; as mixed mathematics, 53–54, 168–69; Newton on, 232–35, 236–40; as physico-mathematics, 235–36; scientific status of, 40, 53–57, 59, 61, 63, 163

Ordinary course of nature, 6, 19–21, 47–48, 155

Oviedo y Valdés, Gonzalo Fernández de, 110n

Pallavicinus, Iacobus Maria, 79

Pappus, 122, 211, 214, 217–18n, 240

Pardies, Ignace Gaston, 238

Parker, Gary D., 98n

Pascal, Blaise: and Aristotelianism, 183, 185, 188–91, 195, 197; and barometer, 202–5; Boyle on, 208; and contrived experience, 183, 185, 203; and credibility, 185–86; and demonstration, 183–86, 192, 199–201; and evidentness, 183–85, 187; and experience, 181–87, 197–201, 203, 205–6, 208–9; and experiment, 198,

288   Index

Pascal, Blaise: and Aristotelianism (*continued*) 200–201; and historical reports, 181, 196–201, 205–6; and mathematical sciences, 192, 198–200, 203, 205–6, 208–9, 247; and maxims, 182–87, 196; and priority, 186–87, 197; and Puy-de-Dôme, 8, 180, 196–203; and scholasticism, 182–84, 201; Torricellian device, 180–81, 156, 159; universal truths, 183–84, 203; use of tables, 205–6; and the void, 182–84, 186–96
Patronage, 51, 108–10
Pecquet, Jean, 202–3
Pedersen, Olaf, 48n
Pemberton, Henry, 248–49
Pendulums, 14–15, 76–82
Pereira, Benito, 37, 54n, 223
Périer, Florin, 180, 196–201, 203, 205–6
Petit, Pierre, 180, 187–88, 192–93
Phenomena (as term), 48–50, 54, 98, 101–6, 108
Physico-mathematics, 2–3, 8, 150, 168–79; and Barrow, 223–24; and Jesuit mathematicians, 168–70, 172–78, 235–36, 247; and Newton, 210–11, 232, 235–38, 242, 247–48; and Royal Society, 227, 242, 247. *See also* Mixed mathematics
Physics. *See* Natural philosophy
Piccolomini, Alessandro, 36
Pierius, Jacques, 148–49, 186n, 189–93
Pinch, Trevor, 158n
Plato, 35–37, 41, 166
Polydore Vergil, 206
Port Royal Logic, 28, 30
Porter, Roy, 5n
Possevino, Antonio, 41
Postulates, 48, 213, 215–23, 226, 242
Priority, 186–88, 197
Private knowledge: place in science, 44, 50–51; as problem in astronomy, 47, 49; rendered public, 61–62, 66, 75, 185
Probabilism, 3, 23, 177–78, 201, 211–12
Proclus, 37, 43n, 218–19, 221, 222n
Projectile motion, 127–29
Protestantism, 3, 11, 32–33, 116–17, 121
Ptolemy, Claudius, 38, 144, 165; and contrived apparatus, 70; and evidentness, 43; as humanist model, 117–19; on optics, 43, 51–52, 53n, 81; on phenomena and observations, 14n, 48–50
Public knowledge, 44; and contrived experience, 59, 61–62, 66, 68, 185
Puy-de-Dôme, 8, 180, 189, 196–203, 205
Pythagorean ratios, 135

Quintilian, 26n, 118

Ramus, Petrus, 121
Reason, 134–36
Reformation, Protestant, 117
Reif, Mary Richard/Patricia, 21–22n, 156, 183, 222n
Remes, Unto, 241n
Renaudot, Théophraste, 190n
Renieri, Giovanni Battista, 129
Repetition, in Arriaga, 68–69, 85; in Cabeo, 66, 70, 128; creation of evidentness through, 44–45, 66–67, 75, 127–29; in Galileo, 125–26, 129; in Riccioli, 71, 78–80, 83–85; and unreliability of the senses, 19
Replication, 95, 110–11, 130–31, 205
Rey, Jean, 133
Riccioli, Giovanni Battista: and astronomical tradition, 120; and expertise, 71–85, 111n, 188, 229; and mathematical model, 104, 208, 248; and physico-mathematics, 175–77; and use of *experimentum*, 13–14n
Roberval, Gilles Personne de, 192–96, 201–2
Roderigus, Camilllus, 79
Rosen, Edward, 118n
Rossi, Paolo, 21n
Royal Society: and experimental philosophy, 4, 8, 25, 227–31, 242, 245–48; and Newton, 2, 210, 234–35, 239, 242, 247; and physico-mathematics, 2, 227

Sabra, A. I., 51–52
Sargent, Rose-Mary, 76n
Sarsi, Lothario. *See* Grassi
Saturn, 111, 113
Scepticism, academic, 19, 19n, 174
Schaffer, Simon, 158n, 228n; on Royal Society, 4, 8, 210n, 227, 245–46
Scheiner, Christophorus: and controversy

over sunspots, 7, 100–107, 109, 111–15, 248; and disciplinary boundaries, 161, 165, 167; and "experiment" vs. "experience," 14n, 55–57, 59, 66, 187n; on "phenomena" and "experiences," 54–59, 66–67, 102–4, 161; and scientific status of optics, 51, 53–57, 59–61; and tradition, 115, 123, 248; and utility of mathematics, 41, 167
Schmitt, Charles B., 21n, 126n
Scholastic philosophy: and disciplinary boundaries, 150, 168, 175, 177; experience in, 21–25, 32, 68, 74–76; and induction, 18–21, 26; and Pascal, 182–84, 201; and Scientific Revolution, 2, 15; and singular experience, 18, 20–21, 25, 46–47, 64; and the void, 191. See also Aristotelianism; Jesuits
Schott, Gaspar, 147n, 178
Schuster, John A., 4–5n, 137n
Sciences, subordinate. See Mixed mathematics
Scientific Revolution, 2, 12, 15, 22, 96, 248; compared with Reformation, 116–17
Sempilius, Hugo, 40, 73n, 120, 169–70
Sense perception: as basis of knowledge, 22–23, 42–45, 54, 134–35, 141, 143; unreliability of, 19, 24, 45, 134–35, 141, 143
Shapin, Steven, 60, 228n, 245; on Royal Society, 4, 8, 210n, 227, 245–46; on social criteria of truth, 133, 144n, 149–50, 248
Shea, William R., 137n
Simplicius, 202
Singular experience: in astronomy, 47, 93; and induction, 6, 16, 18, 30, 224–25, 241–42; in Newton, 14–15, 224–25, 241–42; in Royal Society, 25, 227–31; and universal knowledge claims, 6, 13–15, 25, 32, 44, 154–55, 206, 246. See also Contrivance; Event experiment; Experiment; Historical reports; Monsters
Smith, A. Mark, 53n
Southern, R. W., 23–24
Spinoza, Baruch, 8
Sprat, Thomas, 227–28
Stevin, Simon, 178, 226
Strawson, Galen, 17
Subordinate sciences. See Mixed mathematics

Sunspots: as phenomena, 100–107, 111–13, 115, 123; and physico-mathematics, 164, 167, 173

Tabular presentation, 59, 81–82, 84, 205–6
Tacitus, 117
Teich, Mikuláš, 5n
Teleology, 154–55, 155n, 157–58, 162, 175
Telescope: as contrivance, 98–100, 102–4, 161; and credibility, 109–11; and disciplinary boundaries, 167; and Newton, 227; as novelty, 97–98, 102–3, 111; and observational community, 108–11; and problem of evidentness, 47–48; as productive of monstrous appearances, 48, 98, 115, 162; and tradition, 98, 115
Terrella, 159–61
Thought experiments, 62, 124, 126
Tides, 230–31
Toricelli, Evangelista, 129, 148, 187, 202
Torricellian device: as event experiment, 8, 180, 187–88, 196–201, 206, 208–9; as instrument (barometer), 189, 202–3, 205, 207; as novelty, 63–64, 186–87, 190–91; priority disputes over, 186–88, 197; and the void, 180–96, 201–2; and the void in the void, 193–97; and the weight of the air (Puy-de-Dôme), 180, 189, 196–205, 207
Tradition, 7, 95–98, 108, 115–23, 248
Trust. See Credibility
Tycho Brahe, 13, 50–51
Tychonic system, 100–101n, 174

Universals: in Aristotelian science, 22–23, 25, 30–31, 44, 46, 94–95, 153–55; and art/nature distinction, 153–55; and authority, 22, 24–25; and community, 94; and contrived experience, 50, 54, 56, 66–67, 127; and controversy over void, 190–92, 195; in Fabri, 139–40; in Galileo, 127; and induction, 6, 13, 16, 18, 26, 28–30, 241–42; in Mersenne, 134; in Newton, 232, 234, 241–42; in Pascal, 181–84, 187, 197, 199, 201, 203, 206; in Scheiner, 105–6; and singular experience, 85–86, 153–55, 231, 241–42; techniques to create, 68, 70, 77–78, 83, 89–90

Urban VIII, 165

Vallius, Paulus (Valla), 48n
Van Helden, Albert, 98n
Vesalius, Andreas, 117–20
Vico, Giambattista, 96
Viète, François, 117–19, 121–22
Villalpando, Gaspar Cardillo de, 54n
Villalpandus. *See* Villalpando
Virtual witnessing, 60, 228n, 245–46
Viviani, Vincenzio, 129
Void, 181–82n, 182–84, 186, 188–96, 201; void in the void, 193–97

Wallace, William A., 48n, 154–55n
Wallis, John, 222, 227, 230–31
Watchirs, Graeme, 4–5n
Watts, Isaac, 249
Weisheipl, James A., 37n
Welser, Mark, 100–102, 104, 106, 111, 113–14

Westman, Robert S., 46n
Whewell, William, 184n
Wilkins, John, 2–3, 227, 238, 247
Wilson, C. T. R., 158n
Winch, Peter, 6n
Wisan, Winifred L., 155n, 230n
Witelo, 51, 81
Witnesses, 25, 66, 75, 129, 133; in Arriaga, 68–69, 85; in Grassi, 88–89, 91; in Périer, 197; and priority, 188; in Riccioli, 74, 78–79. *See also* Virtual witnessing
Wittgenstein, Ludwig, 6n, 97
Woolgar, Steve, 189
Wren, Christopher, 227

Yates, Frances, 117n

Zabarella, Jacopo, 21n, 26–28, 126n
Zahar, Elie, 119n
Zenus, Franciscus (Zeno), 78–79
Ziggelaar, August, 236n
Zoroaster, 120